Molecular Biology Biochemistry and Biophysics
19

AF845910

Martin B. Mathews

Connective Tissue

Macromolecular Structure and Evolution

With 31 Figures

Springer-Verlag Berlin Heidelberg GmbH 1975

Professor MARTIN B. MATHEWS, Ph. D.
The University of Chicago,
Department of Pediatrics,
5825 Maryland Avenue, Chicago, IL 60637/USA

ISBN 978-3-642-80906-4 ISBN 978-3-642-80904-0 (eBook)
DOI 10.1007/978-3-642-80904-0

Originally published by Springer-Verlag Berlin Heidelberg New York in 1975

Softcover reprint of the hardcover 1st edition 1975

Library of congress cataloging in publication data. Mathews, Martin B. 1912-Connective tissue. (Molecular biology, biochemistry, and biophysics; v. 19). Bibliography: p. Includes index. 1. Connective tissues. I. Title. II. Series. [DNLM: 1. Connective tissue. 2. Macromolecular systems. W1M0195t no. 19 / QS532.C7 M429c] QM563.M37. 596'.08'2. 74-28052.

Preface

This book deals primarily with the principal extracellular macromolecules of animal connective tissues. It attempts to answer some general questions about the biological organization of the tissues: What is the nature of this organization at various dimensional levels? What functions does the organization serve? How has it evolved?

I have given major emphasis to the structures and properties of the macromolecular components of extracellular matrices from a wide range of invertebrates and vertebrates. In doing so, however, I have treated cursorily many important aspects of connective tissue biology that appeared to be only indirectly relevant to the principal questions asked. On the other hand, I have not hesitated to search for broad biological principles outside the prevailing conceptual boundaries of animal connective tissues and the lower molecular dimensional levels. The numerous speculations presented will, I hope, stimulate the reader to further thought and investigation.

Acknowledgements

My previously unpublished data that have been included in this volume could not have been obtained without the generous gifts of specimens and other assistance by a number of individuals and institutions. These are Dr. G. BERENSON of Louisiana State University, Drs. W.P. BRAKER and D. ZUMWALD of the Shedd Aquarium, Dr. S.H. CHUANG of the University of Singapore, Dr. L.G. CLARK and Mr. U.M. VARELA-DIAZ of the University of Pennsylvania, the Government of Nicaragua, Dr. E. CLARK of Cape Haze Marine Laboratories, Mr. C.E. CUTRESS of the University of Puerto Rico, Dr. DE FIEBRE of Wilson and Co., Dr. D. DOCKINS of the Scripps Institute of Oceanography, Dr. J.H. HOWELL and Mr. T. DURKIN of the United States Fish and Wildlife Service, Dr. A.M. FISHER of Connaught Laboratories, Dr. R.B. GRINOLS of the University of Washington, Dr. R.E. HALL of Wisconsin State Department of Agriculture, Dr. W.D. HARTMAN of Yale University, Dr. E. HERALD of Steinhardt Aquarium, Dr. K. LIEM of the Field Museum, Profs. J. MILLOT and J. ANTHONY of the Museum National D'Histoire Naturelle (Paris), Dr. T. MOHSEN of the University of Dakar, Dr. P. PERSON of the Veterans Administration, Dr. T. THORSON of the University of Nebraska, and Dr. A. WILSON of Brandeis University. Financial support was received from the Chicago Heart Association and from the United States Public Health Service on grant nos. AM-05996 and HD-04583.

I am greatly indebted to many friends who read portions of the manuscript and freely offered advice and criticism. Drs. J.A. CIFONELLI, A. DORFMAN, P. PERSON, L. RODÉN, and A. VEIS detected my errors, contributed their own ideas and suffered my eccentricities of style. They have done wonders in protecting the reader against the limitations of my ignorance and in correcting mistakes but, of course, are not responsible for any that remain or for the views expressed. Numerous colleagues generously furnished copies of their manuscripts prior to publication. I am also grateful to Drs. S. SUZUKI and T. YAMAGATA of Nagoya University for invaluable analytical data. The quoted fragments by ANAXAGORAS are from translations by WHEELWRIGHT (1966).

Chicago, November 1974 MARTIN B. MATHEWS

Contents

Chapter 1

Introduction

"Die Intercellularsubstanz wird von den Zellen gebildet, steht unter deren dauernder Einwirkung und bildet mit ihnen zusammen ein lebendiges Ganzes." BAUER (1934).

"Most molecular biologists today focus their attention on those intracellular molecules which are involved in the storage of information and the control of cell processes. The assembly, organization, and specialization of cells in multicellular systems, however, occurs only when the intercellular matrix develops - protecting, connecting, binding and organizing the cells into one functional system of tissue, organ or organism. The intercellular matrix produced by the cells forms their external environment. This environment may play a primary role in the control of cell processes. Thus, the cellular function of the multicellular organism cannot be understood fully without complete knowledge of the structure and function of the matrix." BALAZS (1970).

1. General Considerations

The supporting tissues of higher animals had been recognized as a class by their characteristic fibrous appearance for a century before the name connective tissue, or *Bindegewebe*, was proposed by MÜLLER in 1830. Shortly after that cells were found to be present in these tissues and there ensued a prolonged debate over the precise relationship between the fibrous elements (primarily collagenous), the amorphous interfibrillar ground substance, and the cells. Today, it is generally agreed that the organic extracellular substances are secretory products of the connective tissue cells, yet the precise degree of molecular elaboration attained by these products prior to their release into the matrix is still not known. Additional historical details are given in excellent, brief accounts by KLEMPERER (1955), WASSERMAN (1956), FITTON JACKSON (1964), and DELAUNAY (1966).

For convenience, the mammals are the point of departure in our description of animal connective tissue, for it is this class of vertebrates that has been most intensively studied. In these higher animals the embryonic connective tissue, the mesenchyme, develops from the mesoderm immediately after formation of the germ layers, and gives rise to most of the collagenous tissues (notochord probably derives from the totipotent tissue of the blastopore). The cells primarily involved in the formation, growth and maintenance of connective tissues are the fibroblast (fibrocyte), the osteoblast (osteocyte), and the chondroblast (chondrocyte). These cells are probably modulations of the same cell type. Also present in connective tissues are: macrophages, leukocytes, lymphocytes and plasma, mast, pigment and fat cells. There is a constant exchange of cells between blood and connective tissues. It is worth noting that, in the earliest stages of development, endothelial cells of blood vessels and blood cells arise from mesenchymal elements.

The connective tissues are ubiquitous in the animal organism and extremely varied in morphology. However, all forms are characterized by an abundant extracellular matrix containing collagen and, frequently, elastic fibers imbedded in an apparently amorphous ground substance. Loose or areolar connective tissue fills the spaces between organs and penetrates with the blood vessels into the interior of organs. Besides serving a supportive function, this type of tissue is necessarily involved in the transport of metabolites between blood and lymph, and the parenchymatous cells of organs. Tissues with a high fiber content include tendons, ligaments, and the fibrous membranes surrounding various organs. Other tissues range in structure from gel-like vitreous humor and rubber-like cartilage to dense mineralized bone (MAXIMOW and BLOOM, 1948).

Although the central importance of the cellular elements of connective tissue was recognized shortly after the beginnings of microscopic morphology, interest in the extracellular matrix has not diminished with time. There are several reasons for this persisting emphasis. The centuries-old concept of the fiber as the basic unit of the living body continued to exert a persuasive influence on biological thinking through much of the nineteenth century, so that the characteristic extracellular matrices, comprising many types of fibers organized in a great variety of fashions, offered particularly attractive vistas for both microscopic and biochemical exploration. Interest was also greatly stimulated by Virchow's early demonstration of significant alterations of extracellular matrix in human disease. It appeared likely that the extracellular matrix (which in some instances, such as cartilage and bone, far exceeds in mass the cellular content of the tissue) was a major element in developing and shaping tissue functions.

Recent applications of biochemical and physical methods have revealed much of the chemical structure of the principal macromolecules that can be readily isolated from the matrices of connective tissues. Considerable advances have been made within the last two decades in elucidating the intracellular details of biosynthesis of these cell products. However, we have as yet little insight into either higher-order molecular organization or the physiological functions of connective tissues. This uneven progress stems from certain features of connective tissue research: 1. the newer knowledge, although rapid, revealed an unexpected variety and a bewildering complexity of macromolecular structures (MEYER, 1970); 2. relatively few attempts have been made to investigate in a comprehensive fashion the relationship of molecular organization at various levels to physiological functions (CHVAPIL, 1967; SCHUBERT and HAMERMAN, 1968; PEREZ-TAMAYO and ROJKIND, 1973).

Today a fresh viewpoint is needed - one that will help us to organize our knowledge in a more systematic biological fashion. SIMPSON (1962) pointed out that, in order to understand and explain biological organization, it is not enough to know what is organized, how it is organized, and what functions the organization serves; one should also learn how the organization came about. Thus, it is necessary to investigate the relationship of structure to function during development, not only of the individual organism, but also of the species and larger categories during evolution. This imperative is the central thread in the fabric of the present work.

2. Methodology

The success of twentieth-century chemistry and physics in interpreting biological structures and processes has led scientists to extend the molecular approach to problems of the origin and evolution of living organisms. Earlier expositions of comparative biochemistry and biochemical evolution by BALDWIN (1948) and by FLORKIN (1944) have since been considerably extended and, within the last decade, restated in terms of modern molecular genetics by ANFINSEN (1959); ALSTON and TURNER (1963), DIXON (1966), FLORKIN (1966), JUKES (1966), ALSTON (1967), WATTS (1968) and others (OPARIN, 1963; LEONE, 1964; BRYSON and VOGEL, 1965). DIXON (1966) writes: "The evidence grows increasingly that the sequence of bases in DNA (RNA in certain viruses), either nuclear or cytoplasmic, determines the ultimate properties of an organism in their entirety. If this hypothesis is accepted, a final and fundamental taxonomy and evolutionary phylogeny of organisms must be in terms of the similarities and differences of base sequences in nucleic acids." However, the relative inaccessibility of direct knowledge of nucleic acid structures dictates the necessity of an indirect approach to the evolutionary process.

In essence, current methodology aims to infer the history of the genotype from changes in evolution of molecular phenotypes. It is based primarily upon comparisons of chemical structures of homologous "informational molecules," or "semantides" (ZUCKERKANDL and PAULIN, 1965a, b), such as deoxyribonucleic acids, ribonucleic acids, and proteins. Since amino acid sequences reflect the structure of portions of the genetic nucleic acid according to a definite code, it follows that the primary structure of proteins can be directly translated into information of evolutionary significance.

We should at this point distinguish a hierarchy of molecular phenotypes. The proteins, which include the enzymes, bear a direct structural relationship to the genetic nucleic acid and occupy the first rank. Next are the products of the reactions catalyzed by enzymes. Higher orders may be represented by interactions of these products at various levels of molecular organization. Thus, the chemical structures of polysaccharides and other non-polypeptide molecules, the "episemantides" of ZUCKERKANDL and PAULING (1965a), are less directly related to the genome. According to FLORKIN (1966), a series of protein semantides that show a high degree of chemical similarity, or "isology," are assumed to be "directly homologous," or related genotypically. A series of chemically similar episementides are "indirectly homologous," provided the enzymes synthesizing them are chemical homologues.

It is essential to make the distinction between structures that are homologous and those of analogous function when considering gross morphological characters for taxonomic purposes. The application of the terms "homologous" and "analogous" to molecular phenotypes, however, poses special problems in interpretation (WATTS, 1968). As noted above, the term "homology" as applied to proteins refers to both similarity in amino acid sequence and genetic relationship.

NEURATH, WALSH, and WINTER (1967) distinguish two categories of homology of proteins. In "intraspecial homology" (proteins derived from individuals of the same species) it is possible to determine by genetic studies whether the proteins are derived from the same locus (allotypic β-chains of hemoglobins A and S) or from different loci (trypsin and chymotrypsin). The homology is "interspecial" when proteins from individuals of different species are involved. Of necessity, however, the determination of "interspecial homology" for a set of proteins depends upon functional analogy so that a genetic relationship cannot be rigorously established. These authors therefore take the term "homology" to connote the occurrence of a degree of structural similarity greater than might be produced by chance alone. Thus, genetic relationship is not a component of their definition.

As always, it is desirable to be as explicit as possible in terminology. We follow MARGOLIASH (1969) and NOLAN and MARGOLIASH (1968), who take the "classical biological viewpoint" (however, see BOYDEN, 1969). They offer the following definitions: "*Analogous* refers to structures which are similar but have derived from different evolutionary origins by functional convergence. *Homologous*, in accordance with its common biological usage, would imply that the genes coding for the polypeptide chains considered, in all the species carrying these proteins, had at one time a common ancestral gene and have evolved independently since their original divergence."

It may be noted that WINTER, WALSH, and NEURATH (1968) have rejected these definitions on the grounds that gene ancestries based upon polypeptide sequences are not susceptible to confirmation by independent evidence in the sense that morphological features can be traced to a common ancestry through fossil forms. However, if we assume that the genetic code has not changed and apply acceptable statistical methods in a proper manner, it is possible to demonstrate the ancestral homology of polypeptide sequences. The methodology used in determining homology of proteins is discussed in a comprehensive review by NOLAN and MARGOLIASH (1968). In essence, two conditions must be met by a set of proteins that are ancestrally related: 1. the observed similarities of amino acid sequence must not be the result of random occurrences, and 2. these similarities must be significantly greater in extent than any necessitated by the biological function of the protein. Where homology is not easily evident by direct comparison of polypeptide sequences, a rigorous statistical approach is applied to rule out chance. In order to distinguish between ancestral homology and evolutionarily convergent analogy, it is necessary to identify those sequences in a set of proteins that must have remained invariant in evolution because of critical biological requirements. The statistical tests are then applied to the remaining mutable sequences of the protein. Thus, in comparing two proteins, the number of positions occupied by the same amino acid residue in mutable sequences may be associated with the extent of ancestral relatedness of the proteins. This degree of relatedness is quantitatively expressed either in terms of variant residues or in terms of codon mutations. However, a valid conclusion with regard to phylogenetic relationships may be drawn only if the proteins are both homologous and "equivalent", in the sense that they are represented by a single gene in the ancestral form presumed common to both species being compared (FITCH, 1970).

Functional analogy is a basis for classification of proteins whose degree of homology in structure may then be investigated. Analogy can be established on the basis of many types of functional comparisons. For example, analogies may be recognized for enzymes by a similarity of catalytic functions or by a similarity in molecular mechanisms of catalysis. At a more complex level, analogies can be considered in relation to broad biological functions.

The degree of homology in structure within a group of proteins may indicate the extent of evolutionary relatedness between them. In the case of cytochrome c, where the analogous interspecial homology rests upon extensive functional and structural data on large numbers of proteins, it is possible to construct a phylogenetic tree of molecular evolution that is in accord with classical phylogeny (FITCH and MARGOLIASH, 1967). However, most proteins are known only in terms of approximately analogous functions and partial homologies. In such instances, classical taxonomy may point the way to molecular phylogeny.

Further interpretation of change in molecular phenotypes requires that the observed alterations in function and structure be related to the evolutionary mechanisms of modification and selection of the genome. At the level of the genetic nucleic acid, evolution of protein structure may occur by progressive modification of a single structural gene. Alternatively, novel structures and functions may arise by gene duplication, a process that appears to have played a major role in evolution, followed by parallel and divergent, or by divergent and convergent evolution of the daughter genes. It should also be noted that minor variations in structure may result from allelomorphism, which represents another mechanism for devergent evolution.

However, the fact that a gene is present does not guarantee its ready recognition. Expression of a particular gene, as indicated by the appearance of its specific product, is under genetic and other control and may vary in different organs and tissues of the same organism with embryonic development and with aging. Finally, we may comprehend the full biological significance of such products only if we see them as functional components in the integrated complex of elements that make up cell, tissue, organ, and complete organism. A broad view is indispensible when dealing with questions of phylogenetic change and of the adaptive significance of biological structures, because natural selection, which is essentially differential reproduction, operates as an evolutionary force primarily at the level of the population of individuals (SIMPSON, 1949).

The above methodology is directly applicable to analysis of the data on connective tissue biochemistry. However, some special considerations are worth noting. The relevant macromolecules primarily fulfill structural and other non-enzymatic functions, which depend in part upon the formation of covalent linkages as well as non-covalent interactions outside the cell. The natural variability and complexity of these extracellular events complicates analysis of functional analogy.

Further, the majority of connective tissue matrix macromolecules contain both polypeptide and polysaccharide or oligosaccharide moieties. This presents a mixed situation with respect to the possible structural homology of macromolecules from different tissues or of different species: while the polypepetide sequences may be directly homologous, structurally related polysaccharide moieties are indirectly homologous only if the enzymes synthesizing them are homologous (see above).

Although the peptide sequences of the polysaccharide-synthesizing enzymes are unknown, we shall be on relatively secure ground in inferring homology when the macromolecules studied derive from anatomically homologous organs and tissues of phylogenetically related species. However, rigid adherence to requirements for embryological homology may unnecessarily restrict the range of phylogenetic consideration. Macromolecules that are uniquely identified with animal connective tissue are synthesized by cells that are usually mesodermal, but may also be ectodermal in origin. Even within the same organism, hyaline cartilage may be produced by cells of either germ layer (DE BEER, 1958). CHAPMAN (1966a) regards the mesoglea of coelenterates, which contains collagen and is of epithelial origin, as the primitive equivalent of the mesoderm of tripoblasts. In general, both DE BEER (1958) and ANDREW (1959) consider current ideas of strict division of tissues based upon germ layers to be unduly restrictive. This it is likely that homology will be found independent of germ layer, and that comparative biochemistry will be instructive even in the absence of anatomic homology.

Finally, it should be emphasized that the above methodological considerations have been confined to the level of individual homologous macromolecules. In general, this restricted approach will suffice neither for understanding biological structure at higher levels of organization and function, nor for a complete interpretation of the evolutionary history of macromolecules (see Chap. 10).

Chapter 2

The Tissues

"Biology is largely the study of fibres." NEEDHAM (1968)

1. Principles

The comparative study of the tissues of living organisms is founded upon two related propositions that may be stated briefly as follows: 1. there is a "spatial hierarchy" (WOODGER, 1929; NEEDHAM, 1968) of organization from higher categories down to the molecular and atomic levels, and 2. there exist "biochemical universals" (GREEN and GOLDBERGER, 1967) such as cellular organization, chemical elements, specific types of molecules, etc., which are invariant features of all forms of life. These statements imply, for example, that the organization of tissue at the gross-anatomic and light-microscopic levels is directly and specifically traceable through successive, integrative levels to molecular structure. They suggest, too, that the basic properties and general principles of organization that characterize all living organisms will become evident when comparisons are made at equivalent spatial levels. The latter point is illustrated in striking fashion for some extracellular features of plant and animal forms in Fig. 2.1. (also see SIEGEL, 1962).

2. Molecular Unity and Diversity

All living organisms possess common structural features. FREY-WYSSLING (1957) observes that the basic element of the living substance consists of macromolecules capable of aggregating into reticulate, fibrillar, or lamellar structures. The release of some of these macromolecules from the cell produces the extracellular matrix. WILLMER (1965, 1970) suggested that the extrusion of proteins such as collagen and of acidic mucopolysaccharides arose phylogenetically as a primitive process of unicellular organisms for regulating ionic and non-ionic metabolite traffic across the cell membrane. By virtue of their special properties, these substances were utilized subsequently in evolution as protective agents, intercellular cements, and structural components. PICKEN (1960) views the major role of extracellular structures as the "spatial extension of cellular activities" (see Chap. 10).

The chief skeletal support of living organisms is associated with three different macromolecular components. These are the cellulosic system of plants, the collagenous system of animals, and the chitinous system. Chitin occurs as an alternate to cellulose in plants and as an alternate to collagen in animals (RUDALL, 1955, 1968; HUNT, 1970). The characteristic structural element is always associated with other macromolecules such as acidic polysaccharides, proteins, and glyco-

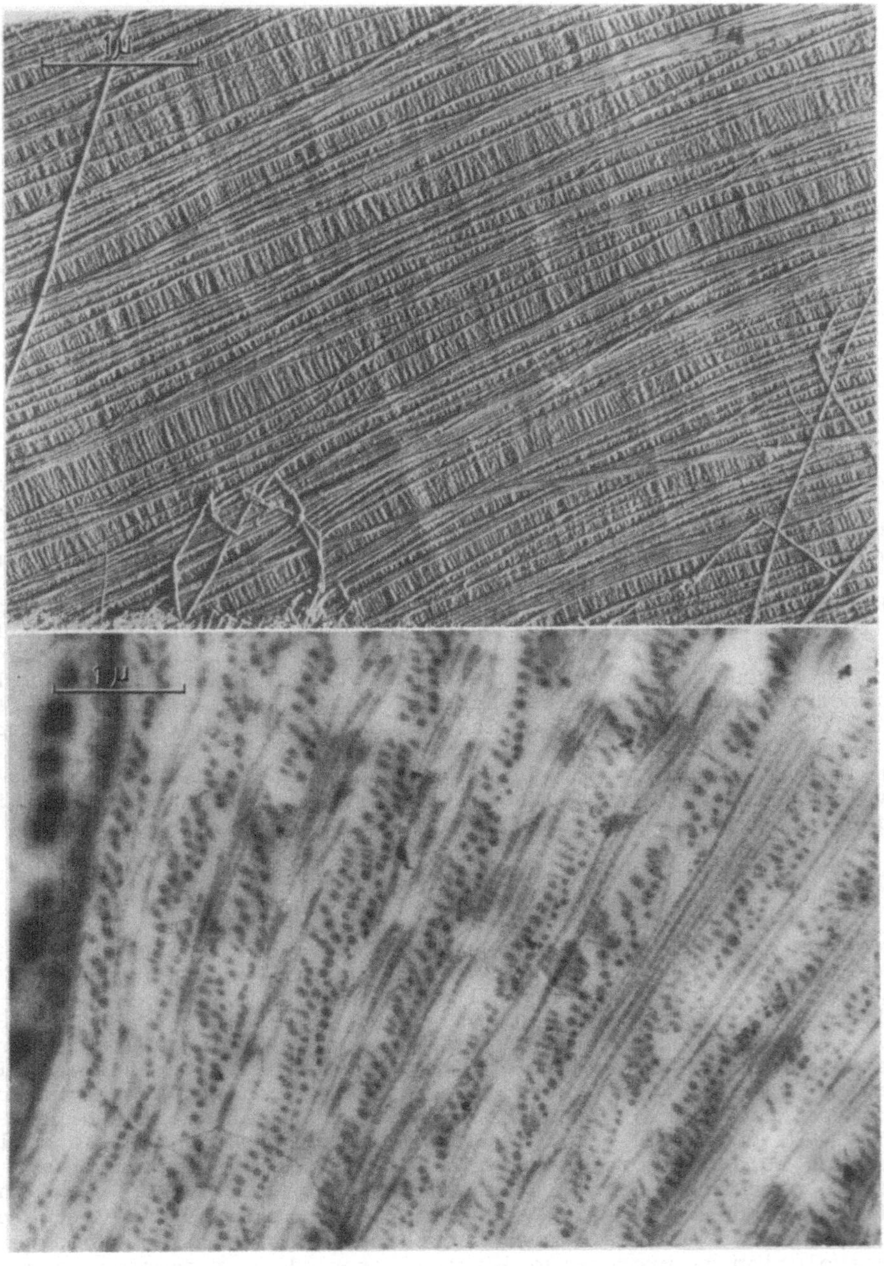

Fig. 2.1. Comparison of structural organization in plants and animals. Top, electron micrograph of cell wall of the alga *Valonia ventricosa* showing pattern of cellulose fibers (see PRESTON et al., 1948; photograph, courtesy of R.D. PRESTON). Bottom, electron micrograph of basement membrane of amphibian skin showing pattern of collagen fibers (see WEISS, 1956;, photograph, courtesy of P.A. WEISS). In both cases, the tissue is compounded of layers of fibers of similar diameter in parallel array, with the fiber direction of one layer approximately at right angles to the fiber direction of the adjacent layer

proteins, and frequently with minerals. Table 2.1 shows the general distribution of the principal types of structural macromolecules.

Table 2.1. Distribution of some secreted structural macromolecules

Macromolecule	Plants	Invertebrates	Vertebrates
Protein type			
collagen	-	+	+
elastin	-	?	+
basement membrane	-	?	+
fibroin	-	+	-
resilin	-	+	-
Polysaccharide type			
glycoprotein	+	+	+
polyanionic glycan	+	+	+
polyanionic proteoglycosaminoglycan	-	+	+
chitin	+	+	-
cellulose	+	+	-

The wide distribution of collagen in association with acidic polysaccharides in metazoan phyla indicates the existence of a broad plan of evolution involving combinations of these types of macromolecules. The plan is of ancient origin and great diversity of expression. Indeed, the presence of these characteristic components in intercellular tissues of Porifera and Coelenterata (CHAPMAN, 1966b), which develop from only two germ layers, suggests a primitive origin antedating the evolutionary appearance of the higher metazoa. The deposition of hydroxy apatite (bone salt) is typically associated with collagen in vertebrate tissues. However, this salt, as well as other crystalline forms of calcium phosphate, is also found in plants, Protozoa, Coelenterata, Arthropoda and Brachiopoda (PAUTARD, 1966; TRACEY, 1968).

It is interesting that among the Brachiopoda calcification of the shell is phosphatic in association with collagen in the class Inarticulata while the shell is calcitic and free of collagen in the class Articulata (JOPE, 1967a, b, 1969a, b; HARE and ABELSON, 1965). A primitive protist, *Hymenomonas*, utilizes an organic matrix consisting of a hydroxyproline-containing protein and a sulfated polysaccharide in its production of intracellular calcitic coccoliths (LAVINE and ISENBERG, 1964; ISENBERG, DOUGLAS et al., 1965; ISENBERG, LAVINE et al., 1965; WILBUR and WATABE, 1967). Such facts indicate that the ancestry of calcification processes associated with collagen must also be ancient. The evidence that other macromolecules may substitute for collagen has led PAUTARD (1966) to suggest the term "topographical matrix" to describe the skeletal, usually fibrous, framework laid down in connection with the salt phase. Besides mineralization, other functional aspects of matrix components are general biological phenomena in both plant and animal kingdoms (MOSS, 1964).

As in animals, supporting structures in plants generally consist of organized microfibrils embedded in a ground substance which may con-

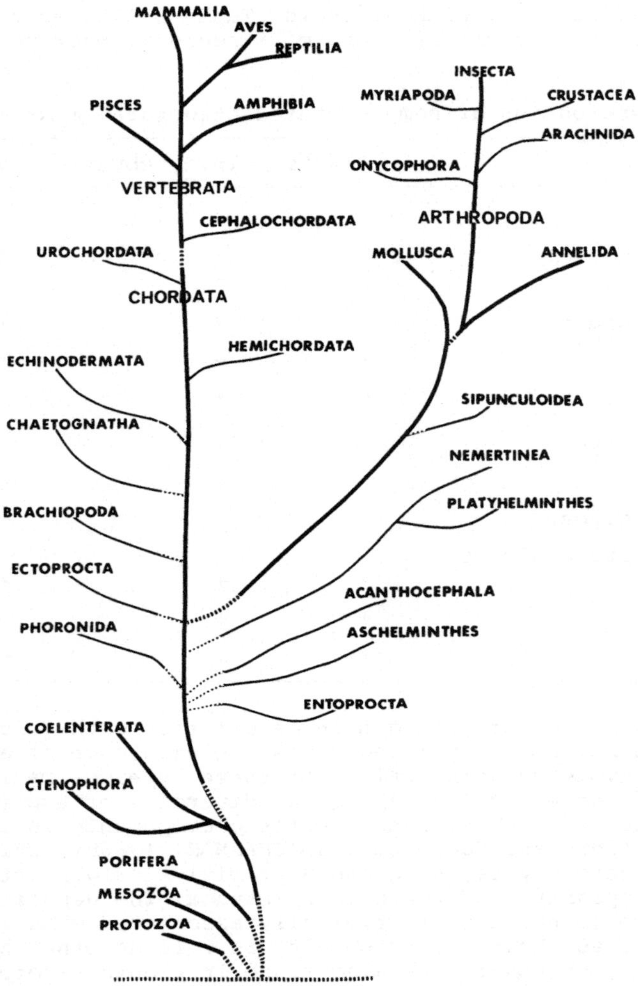

Fig. 2.2. Hypothetic relationships between major taxa of extant animals (adapted from STAFFORD, 1960)

tain acid polysaccharides and minerals (BONNER and VARNER, 1965; NORTHCOTE, 1963). The mechanical advantages of such an ordered biphasic system have been noted and are applied extensively in modern technology. In higher plants, the cell wall contains cellulosic fibrils in a ground substance consisting largely of uronic acid-containing pectic substances (WORTH, 1967) along with hemicelluloses, lignins, etc. The fibrils form highly ordered arrays (see Fig. 2.1) which are specific for particular plants or plant parts and largely determine the physical properties and biological functions of plants (MIA and SETTERFIELD, 1966). Although extensively studied, the process by which this ordering is achieved is not understood (COLVIN, 1972). However, a significant role for the pectic substances is suggested by the observations of GRANT et al. (1969) that long-range molecular order persists in seed mucilages and of BEN-HAYYIM and OHAD (1965) that the oriented aggregation of cellulose fibrils is facilitated by polyanionic polysaccharides (see also Chap. 10, Sect. 3c). The cell walls of many algae likewise contain cellulose and polyuronides; highly sulfated polysaccharides are present in some species (PERCIVAL and McDOWELL, 1967).

There are striking parallels in the chemical structure and properties of acidic polysaccharides of plants and those of animal tissues. For example, the alginic acids found in certain seaweeds are copolymers of D-mannuronic acid and the C-5 epimer of this monosaccharide, L-guluronic acid. Dermatan sulfate of vertebrate tissues is a repeating polymer of disaccharide units consisting of N-acetylgalactosamine 4-sulfate and a uronic acid, which is principally L-iduronic acid, but may also be D-glucuronic acid, the C-5 epimer of L-iduronic acid. For both plant and animal polysaccharides, binding of divalent cations increases with increase in the proportion of the L sugar in the polymer (TRIFFIT, 1968; see Chap. 10, Sect. 2c). However, sharp differences in sugar components of acidic polysaccharides also exist between plants and animals. Thus, although amino sugars are not components of acidic polysaccharides of plant cell walls, they are invariant features of acid mucopolysaccharides (glycosaminoglycans) of vertebrate connective tissues. Curiously, polyfucose sulfate containing α-L-(1 → 2) glycosidic linkages is restricted in distribution to echinoderms (KATZMAN and JEANLOZ, 1969) and to one division of plants (PERCIVAL and ROSS, 1950; CONCHIE and PERCIVAL, 1950).

Bacteria of widely differing species produce polyuronides (BARKER, 1963). Some species synthesize polymers of glucuronic acid and acetylglucosamine that are indistinguishable from hyaluronic acid of mammals (WARREN, 1950). A strain of *Bacillus subtilis* produces teichuronic acid, which has the same monosaccharide components as chondroitin of bovine cornea (JANCZURA et al., 1961). Such compounds might participate in ion-exchange processes and thus regulate the access of divalent cations in particular to the interior of the cell (HEPTINSTAL et al., 1970).

Proteins associated with cellulose in plant cell walls (PRESTON, 1961) have not been well characterized generally. Intensive investigations have been carried out recently upon a group of such proteins that were shown to contain hydroxyproline (KEEGSTRA et al., 1973); interest has been keen because this amino acid was previously known to occur only in collagen and elastin. The hydroxyproline arises from hydroxylation of previously incorporated proline, as in collagen. The enzyme in carrots that catalyzes the hydroxylation reaction has the same cofactor requirements as the hydroxylase of vertebrate tissues responsible for hydroxylation of prolyl residues of protocollagen (Chap. 3, Sect. 11b), and it readily hydroxylates prolyl residues in tropocollagen from chick embryos (SADAVA and CHRISPEELS, 1971a). Hydroxylation is followed by glycosylation, resulting in a glycoprotein consisting of 60% protein containing principally serine, hydroxyproline, lysine and glycine, and 40% carbohydrate containing principally arabinose (BRYSK and CHRISPEELS, 1972). The glycoprotein is then transported to the cell wall and covalently attached to the wall matrix (CHRISPEELS, 1970).

The linkages between polysaccharide and peptide have been extensively studied. Most occur as O-arabinosyl-hydroxyproline in green plants from algae to angiosperms (LAMPORT and MILLER, 1971) and less commonly as O-galactosyl-hydroxyproline linkages (MILLER et al., 1972) and O-galactosyl-serine (LAMPORT et al., 1973). Since the protein had been shown to be a stable moiety of cell walls, and because of its suggested role in cross-linking wall polysaccharides, it was given the name "extensin." A related protein is found in some algae in which a polymannan replaces cellulose (THOMPSON and PRESTON, 1967). Hydroxyproline peptides that are associated with polysaccharides and glycolipids have been isolated from leaves of *Vicia faba* (PUSZTAI and WATT, 1969). Potato lectin is a glycoprotein containing arabinose linked to hydroxyproline residues (ALLEN and NEUBERGER, 1973).

NAKAJIMA and VOLCANI (1969) reported finding a new amino acid, 3,4-dihydroxyproline, associated with the mature wall of the diatom, which also contains proteins and polysaccharides, some of which contain glucuronic acid and are sulfated (COOMBS and VOLCANI, 1968). A functional relation to mineralization was suggested. It is indeed curious that hydroxyproline occurs nearly exclusively in extracellular structural proteins of plants and animals. The observation has stirred speculations concerning the macromolecular similarities between plant cell walls and animal skeletons (AARONSON, 1970).

Another homopolysaccharide is a ubiquitous competitor of cellulose in supporting structures of plants and animals. Although some fungi contain cellulose, most fungal cell walls contain chitin, a linear polysaccharide similar to cellulose, but composed of N-acetyl-D-glucosamine in place of D-glucose in β-(1 $\rightarrow$ 4) linkage. Chitin is also the major component of the skeletal framework of arthropods in the virtual absence of collagen (KENT, 1964). Comparative studies by RUDALL (1955) showed that this structural polysaccharide of ectodermal origin is absent from vertebrates, echinoderms and sponges, groups which typically contain collagen, while both collagenous and chitinous structures are present simultaneously in individuals of other animal phyla. The typical chitinous system in animals - which is mainly, if exclusively, of ectodermal origin (JEUNIAUX, 1971) - broadly resembles the ground substance of collagenous connective tissue, being composed of polysaccharide covalently linked to protein and associated with other proteins (HUNT, 1970; RUDALL, 1969; ATTWOOD and ZOLA, 1967; KURUP et al., 1969). Because of their extreme insolubility, chitinous systems are resistant to investigation by biochemical techniques. They will not be discussed further.

Most, if not all, polysaccharides are associated with peptide. Like chitin, glucomannans of yeast cell wall (NICKERSON et al., 1961) and alginic acids (MICHEEL and MILLE, 1953) also complex with protein. In animals, cellulose occurs in association with protein in the tests of tunicates, possibly in certain worms, and in man (HUNT, 1970). HALL and SAXL (1961) consider reticulin, a collagen associated with carbohydrate in mammalian kidney basement membrane, related in structure to tunicin, the cellulose protein of tunicates. They suggest that the presence of cellulose in both chordate subphyla supports concepts of an evolutionary relationship between tunicates and vertebrates. However, it is not certain that mammals synthesize cellulose (JÜRASEK et al., 1967).

The general physical functions of supporting tissues in plants and in animals are served by specific types of macromolecules capable of forming ordered fibrous arrangements, frequently cross-linked covalently, but also involved noncovalently with additional components of matrices. Where there is a requirement for a high degree of elasticity, a highly branched molecule (for example, vertebrate elastin) may be found. However, the general requirements of relative inextensibility and high strength dictate the use of asymmetric macromolecules, many of which are capable of aggregating in crystalline order.

The major fibrous element in animals is an asymmetric protein molecule, collagen. It is closely associated in connective tissues of vertebrates (and probably also in invertebrates, although data are still meager) with linear polysaccharides that are liberally endowed with carboxyl and sulfate groups. Most of these polysaccharide molecules exist as very long side-chains convalently linked to a polypeptide. The glycosidic linkage in these acidic linear polysaccharides is invariably of the β-(1 → 4) and β-(1 → 3) types. Acidic glycoproteins may also be present in collagenous tissues.

In plant cell walls the major fibrous element is the linear polysaccharide, cellulose, which consists of D-glucose units linked β-(1 → 4). Chitin consists of N-acetyl-D-glucosamine units similarly linked in a linear fashion. The same linkage type is also found in more complex polysaccharides, mannan, xylan, and alginic acid, but not in pectic acid. It is unlikely that plant cell walls contain any simple linear polymer composed of single type of monosaccharide other than cellulose.

A noncellulosic polysaccharide may consist of a linear polysaccharide backbone, frequently of β-linked pentose units such as xylose, with oligosaccharide side-chains that may contain uronic acid units. Alternatively, such a polysaccharide may contain a linear polygalacturonide backbone with other sugars as side-chains. Overall, however, there appears to be an overwhelming preference for β linkages, in particular the 1 → 4 type, in the structural polysaccharides of higher plants and animals. This situation contrasts with the prevalence of α linkages in storage polysaccharides, for example, starch and glycogen, and suggests a particular contribution of β-linkages (the 1 → 4 type especially) to linear chain conformation and to aggregation of chains in fibrous structures.

A broad similarity in mechanisms for biosynthesis of polysaccharides in both plants and animals is indicated by the universal utilization of transferase enzymes in conjunction with nucleoside sugar diphosphates as sugar donors (STOOLMILLER and DORFMAN, 1969; HELMREICH, 1969; HASSID, 1969; SPIRO, 1969; COLVIN, 1972). By far the most common base is uracil; other bases such as guanine, adenine, and thymine may occur also, frequently in the same species. The common precursor for most of the polysaccharide constituents, that is, for D-galactose, D-glucuronic acid, D-galacturonic acid, D-xylose, and L-arabinose, is UDP-D-glucose. Although UDP glucose is the precursor substrate for biosynthesis of cellulose in microorganisms, it is not clear yet (COLVIN, 1972) whether biosynthesis of cellulose in higher plants requires a sugar nucleotide containing the guanine base as suggested by HASSID (1971). Secondary chemical alterations, such as methylation of carboxyl groups, O-methylation, and sulfation, generally take place subsequent to transfer of monosaccharide groups during polysaccharide synthesis.

3. Morphology

In applying what we know of the molecular structure and physiological role of macromolecules toward explaining connective tissue organization, we are heavily dependent upon information from higher categories of biological structure and function. Such information is derived from observations made at dimensional levels above the molecular level, including the electron-microscopic, the light-microscopic and the gross-anatomic, as well as all levels of the complex functioning of living organisms. Since microscopic morphology is outside the scope of this work, only the most pertinent generalizations drawn from this area of knowledge can be presented. Some views of the phylogeny of connective tissues based mainly upon morphology are summarized below.

The similarites in microscopic morphology between plant and animal supporting structures, observed by many early biologists, are discussed by PERSON and PHILPOTT (1967, 1969a, b) with particular reference to newer studies of cartilage. These workers note many correspondences at both light- and electron-microscopic levels in tissue organization and extracellular matrix structure between plant tissues and cartilages of the horseshoe crab *(Limulus)*, the squid *(Loligo)*, the feather-duster worm *(Eudistylia)*, and a gastropod genus (Busycon). Similarities in cell structure are also seen. For example, in cartilage of the horseshoe crab, cytoplasmic division and new cell-wall formation occur in a plane established by a structure which closely resembles the phragmasome observed in plant cell division. They conclude that common principles of organization unite plant and animal supporting tissues, and that 'true' cartilage as a tissue type may have originated with invertebrates or, alternatively, may have evolved independently in invertebrates and vertebrates.

The light microscopic morphology of animal connective tissues has been extensively described by early workers (LUBOSCH, 1938; SCHAFFER, 1930; MAXIMOW, 1927; NOWIKOFF, 1912). SCHAFFER (1930), in his comprehensive monograph, classified tissues of cartilage type in a hierarchic order of three types and nineteen subtypes, ranging from the most primitive "chordoid" type consisting of swollen cells, to the most advanced type of "ground substance-rich cartilage", or "true cartilage" of vertebrates. Both he and NOWIKOFF (1912) held that the distribution of the various types of connective tissue cannot be given a phylogenetic interpretation. They believed that the type of supporting tissue which appears at a given anatomical site and developmental stage of an animal organism is determined primarily by functional (i.e. mechanical) needs. Although offering little evidence in support of his thesis, LUBOSCH (1938) proposed a scheme of phylogenesis of cartilage types. A phylogeny of calcified tissues, in which the general trend is from cell enclosure to acellularity, was suggested by ØRVIG (1951, 1965). However, MOSS (1964, 1968a) found no discernible orthogenetic trends in the evolution of bone. PEDERSEN (1966) distinguished three different grades of connective tissue development in the class Turbellaria (phylum Platyhelminthes) ranging from the completely cellular, with intracellular polysaccharides, to the extracellular matrix-rich type. He questions whether these grades can be considered a phylogenetic line and concludes that the nature of the first primitive type of connective tissue in metazoa cannot be deduced from the data on comparative morphology and histochemistry.

Adaptive relations to the ecological environment may be paramount. This aspect is emphasized in the histochemical investigation by ELKAN (1968) who found that a matrix-rich connective tissue layer in the skin of Anura is nearly always present in terrestrial species, but virtually absent from aquatic groups. It should be noted at this point that histochemical methods, while of general value in preliminary characterizations of connective tissue components, cannot give wholly reliable quantitative or qualitative information (SZIRMAI, 1969; MEYER, 1969). For this reason, no attempt will be made here to evaluate literature based upon these procedures.

Chapter 3

Collagen

1. Identification

Collagens from various species differ in composition and possess complex structures with many features that are still obscure. Nevertheless, a reasonable working definition can be given, based upon certain common molecular characteristics of this class of protein. Details may be found in reviews by GROSS (1963), RAMACHANDRAN (1967), BAILEY (1968), TRAUB and PIEZ (1971), GALLOP et al. (1972), and E.J. MILLER (1973).

At the light-microscopic level, collagen is observed as extracellular fibers consisting of unbranched fibrils (with the single exception of spongin B in the Keratosa, or horny sponges) generally staining in a characteristic fashion. In the electron microscope most collagen fibrils have diameters in the range 100 to 2000 Å and appear banded with an axial repeating period of 650 to 700 Å. At high resolution, a fine structure is evident in the interband region. However, the most general and primary criterion for identification of fibrous collagen is a unique wide-angle X-ray diffraction pattern. Meridional spacings near 10 Å, 4 Å, and 3 Å, and an equatorial spacing of 10 to 15 Å are indicative of parallel-aligned, triple helical structures having central axes 10 to 15 Å apart, depending upon the degree of hydration of the fiber. A prime characteristic of collagen is that the glycine content is about one third that of the amino acid residues. Collagens typically contain a substantial amount of the amino acids, proline and hydroxyproline, usually near one fifth of the total amino acid residues.

The X-ray diffraction data for collagen indicate a compound helix, consisting of three individually helical polypeptide chains. Several structures that meet the stereochemical requirements have been proposed. As a minimum requirement, every third residue in each chain is glycine, since the triple helical conformation cannot accommodate an amino acid with a side-chain in this position in the interior of the compound helix. The conformation of the individual polypeptide chains is such that proline and hydroxyproline fit in naturally and stabilize the triple helix in sequences of the type: -Gly-Pro-Hypro-, -Gly-Pro-X-, and Gly-X-Pro, where X is some other amino acid.

Each chain is in the form of a left-handed, polyproline-II-type helix that advances approximately 3 Å per residue and requires 3.3 residues per turn. The chains are twisted into a supercoil (or coiled coil) so that each minor coil twists in a right-handed direction about the central axis at an angle of about 30°. RAMACHANDRAN (1967) had proposed that the structure contained two systematic hydrogen bonds of the type NH . . . O per three residues, but calculations by TRAUB et al. (1969), supported by data of CHAPMAN et al. (1970) and of KATZ (1970), strongly favor a structure maintained by only one such systematic hydrogen bond per three residues. The γ-hydroxyl group of a hydroxyproline residue is capable of forming an additional hydrogen bond via a water molecule in a modified 1-bonded structure (RAMACHANDRAN et al., 1973).

2. Chemical Structure

a) Primary and Direct-Coded Structure

The collagen fibril is an ordered assembly of a monomeric unit, the tropocollagen molecule. The constituent polypeptide chains of this molecule are synthesized at the ribosomal level by processes that result in main chain sequences that reflect the information coded in structural genes. Various alterations and additions to the molecular structure are apparently introduced by other processes both within and external to the cell. These so-called "secondary features" will be discussed separately from the "direct-coded" structure. The point of departure must be molecules from mammalian tissues, since they have furnished the vast bulk of structural data.

The tropocollagen molecule consists of three polypeptide chains. Each chain contains about 1000 residues extending the full length of the molecules, which has a molecular weight of slightly less than 300,000 and a length near 3000 Å. The earliest data indicated the presence of two types of chain, α1 and α2, in mammalian skin collagen. Three distinctly different α chains, α1, α2, α3, were first clearly identified in codfish-skin collagen by PIEZ (1965). Evidence for three different chains was obtained by PIKKARAINEN and KULONEN (1964) and by HEIDRICH and WYNSTON (1965) for calfskin collagen, and for chick-bone collagen by FRANCOIS and GLIMCHER (1967a, b). However, KANG et al. (1969a, b) noted that all other vertebrate collagens examined to date, with the exception of codfish-skin collagen, contain only two different α chains, and they questioned whether the apparent difference between α1 and α3 actually reflected differences in the primary structure of the main body of the polypeptide chains. MÜLLER and KÜHN (1968) and STARK and KÜHN (1968) suggested that the α1 and α3 chains of calfskin collagen differ in their peptide side-chains and not in the amino acid sequence of the main chains. Subsequently MARK et al. (1970) and RAUTERBERG et al. (1972) showed that the "α3" chain of soluble calfskin collagen was an artifact, derived by enzymatic removal of amino acid residues from the C-terminus of an α1 chain. However, insoluble collagen of skin (and dentine) contains a mixture of two collagen forms, one of which has three different α chains (VOLPIN and VEIS, 1973).

MILLER and MATUKAS (1969) detected in chick cartilage a second type of α1 chain, which clearly differed from the α1 chains of chick skin and bone collagens. The new chain was designated α1 type II, or α1 (II), the α1 chain of skin collagen being α1 (I). Accordingly, the chain composition of collagen in chick sternal (E.J. MILLER, 1971; TRELSTAD et al., 1970), human growth plate (E.J. MILLER et al., 1971), and bovine articular (STRAWICH and NIMNI, 1971) cartilages is $[\alpha 1\ (II)]_3$. Homologous collagens are present in sturgeon cartilage and notochord (E.J. MILLER and M.B. MATHEWS, unpubl.) and probably also in chick notochord (LINSENMAYER, TRELSTAD, and GROSS, 1973). A third type of α1 chain, α1 (III), occurs in chick embryo skin (TRELSTAD et al., 1971) and in the skin and aorta of human infants (E.J. MILLER et al., 1971; CHUNG and MILLER, 1974) in the form of molecules of composition $[\alpha 1\ (III)]_3$. A fourth type of α1 chain occurs in basement membranes as molecules of composition $\alpha 1\ (IV)_3$ (KEFALIDES, 1971).

Clearly, we are confronted with a multiplicity of structural genes for vertebrate collagens. For general descriptive purposes, it appears desirable to choose a simple structural prototype of the collagen molecule, i.e. one containing only two different main chains, say, two α1 chains and one α2 chain. Rat-skin collagen has been more thoroughly investigated than any other collagen and is the only molecule for which the order of peptide fragments (see Fig. 3.1) along the α chains is

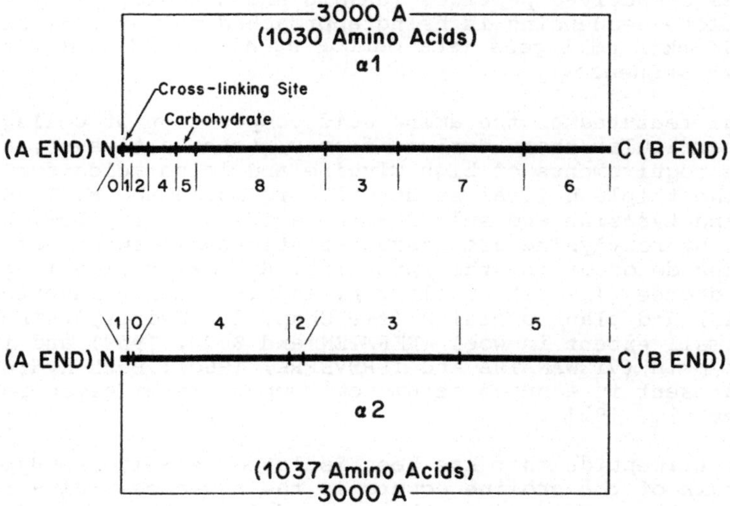

Fig. 3.1. Data on the α chains of rat collagen. Each chain is assumed to have a molecular weight of 95,000. Cyanogen bromide cleaves each chain at the carboxyl-side of methionyl residues. The resulting segments are designated α1-CBO (absent from acid-soluble rat skin collagen; see Fig. 3.7), α1-CB1, α-CB2... and α2-CBO, α2CB1, α2-CB2..., according to the chain of origin. The original numbering of the fragments is not related to their sequential order as shown here. The length of each segment is approximately proportional to the number of residues in the peptides. Animal collagenase cleaves the native molecule of tropocollagen one quarter of the way in from the carboxyl-terminus (C). Chymotrypsin acts on the native molecule to remove a small peptide containing the cross-linking site from the amino-terminus (N). Carbohydrate is linked to the N-terminal hydroxylysine of α1-CB5 (see Fig. 3.2). Adapted from PIEZ et al. (1968) with additional data from GALLOP et al. (1972). The sequence of α1-CB8 is (BALIAN et al., 1972)

known. In this collagen, the N-terminal (A) ends of the α1 and α2 chains (see Figs. 3.2 and 3.7) each contain a sequence of three to four peptide triplets; this sequence cannot exist in the triple-chain helical conformation of collagen, since glycine is not present as every third residue. In addition, such sequences differ in composition from the complete molecule in that they contain less glycine and amino acids, but more tyrosine and acidic amino acids; as a consequence, the N-terminal regions are much more susceptible to hydrolysis by proteolytic enzymes than are the interior triple helical regions. These sequences were named "telo-peptides" by RUBIN et al. (1965). Sequences incapable of participating in the triple helical conformation have also been found at the C-terminal (B) ends of α chains. The amino acid residues of α chains appear to be linked by regular peptide bonds only.

The determination of the complete primary structure of a collagen molecule has proved to be a formidable problem because of the large number of residues (about 1000) involved in each chain. The vast bulk of data on various collagens has been restricted to amino acid composition (EASTOE, 1967) and only occasionally extended to identification of sequences of derived peptides (HANNIG and NORDWIG, 1967; PIEZ, 1967). Complete sequencing is being approached only for α1 chains of rat and calf-skin collagens (see GALLOP et al., 1972, for a summary of all known sequences).

Some general features of the amino acid composition of collagens as a class, in particular those derived from vertebrate tissues, are worth noting. The requirements of high glycine and imino acids for stabilization of the triple helical structure were noted above. Tryptophan is absent and tyrosine and sulfur amino acids are very low. Hydroxyproline and hydroxylysine are characteristic constituents of collagen, although they do occur in other proteins. Hydroxyproline is present to a small degree (1 - 2%) in elastin, and in certain invertebrate (GROSS, 1963) and plant proteins (see Chap. 2). Hydroxylysine occurs to a very small extent in wool (KERSTEN and ZÜRN, 1959) and in trypsin and chymotrypsin (VISWANATHA and IRREVERRE, 1960). Both hydroxy amino acids are present in a human serum-complement system glycoprotein (YONEMASU et al., 1971).

A prominent tripeptide that has been isolated is -Gly-Pro-Hypro-. However, the bulk of the proline occurs in the sequences -Gly-Pro-X-, where X is an amino acid other than one of these three. These findings are consistent with the proposed structural models of collagen. Glycine is restricted to position 1 of the triplet but proline may be in position 2 (preferentially) or 3. Proline in position 3 (but only rarely in position 2) is generally hydroxylated. Hydroxylysine appears to occur only in position 3. Some other amino acids may also have preferred distributions between positions 2 and 3 (BALIAN et al., 1972; FIETZEK et al., 1972a, b).

b) Secondary and Non-Coded Structure

i) Cross-Links

The cross-links of collagen, recently reviewed by TANZER (1973) and by GALLOP et al. (1972), may be divided into two types: intramolecular and intermolecular.

The intramolecular cross-links are produced in the nonhelical portion of the collagen molecule and join two α chains to produce dimers designated β_{11} and β_{12} and a trimer designated γ_{112}, as illustrated in Fig. 3.3. The first step is the formation of residues of α-aminoadipic acid -δ-semialdehyde (allysine) by the enzymatic oxidative deamination of the ε-amino group of specific lysine residues, i.e. those at position 9 in α chains (see Fig. 3.2). Aldol condensation of two residues of allysine on adjacent chains then produces a cross-link, which is subsequently dehydrated to yield an unsaturated aldehyde structure. The condensation step occurs spontaneously *in vitro*.

The existence of intermolecular cross-links in collagen was inferred from observations that components such as β_{22} and triple-chain γ_{111},

```
[PCA-Met-Ser-Tyr-]-Gly-Tyr-Asp-Glu-Lys-Ser-Ala-Gly-Val-Ser-Val-Pro-Gly-Pro-Met-
                   5              10                  15                     ↑
                   Gly-Pro-Ser-Gly-Pro-Arg-Gly-Leu-Hyp-Gly-Pro-Hyp-Gly-Ala-Hyp-
                   20                                  30
                   Gly-Pro-Gln-Gly-Phe-Gln-Gly-Pro-Hyp-Gly-Glu-Hyp-Gly-Glu-Hyp-
                                           40
                   Gly-Ala-Ser-Gly-Pro-Met-Gly-Pro-Arg-Gly-Pro-Hyp-Gly-Pro-Hyp-Gly-
                   50                  ↑                   60
                   Lys-Asn-Gly-Asp-Asp-Gly-Glu-Ala-Gly-Lys-Pro-Gly-Arg-Hyp-Gly-
                                   70                                      80
                   Gln-Arg-Gly-Pro-Hyp-Gly-Pro-Gln-Gly-Ala-Arg-Gly-Leu-Hyp-Gly-
                                                               90
                   Thr-Ala-Gly-Leu-Hyp-Gly-Met-Hyl-Gly-His-Arg-Gly-Phe-Ser-Gly-
                               100          ↑  O                          110
                                         Gal
                                     Glc
                   Leu-Asp-Gly-Ala-Lys-Gly-Asn-Thr-Gly-Pro-Ala-Gly-Pro-Lys-Gly-
                                                               120
                   Glu-Hyp-Gly-Ser-Hyp-Gly-Glx-Asx-(Gly,Ala,Hyp)-Gly-Gln-Met-
                               130                                        ↑
```

Fig. 3.2. Amino acid sequences at the N-terminal end of the α1 chain of rat-skin collagen. The tetrapeptide in brackets is present in the α1 chain of rat-tail tendon. Its absence from the α1 chains of skin collagen is due to an artifact of isolation. PCA is pyrrolidone-5-carboxylic acid, which may have derived by cyclization from either an N-terminal glutamine or a glutamic acid residue. Cyanogen bromide cleaves the peptide chain at the carboxyl end of methionyl residues (indicated by arrows). According to the modified numbering system used here, residues 5-19, 20-55, 56-102, 103-139 yield, respectively, the fragments α1-CB1, α1-CB2, α1-CB4, and α1-CB5 (PIEZ et al., 1968).

The sequence for α1-CB5 shows schematically the attachment of a glycosyl-galactose moiety to the hydroxyl group of the N-terminal hydroxylysine. This group derives from a lysine residue which is nearly 100% hydroxylated and glycosylated, in contrast to lysine residues elsewhere in the α1 chain that undergo little chemical modification. Additional sequence data up to residue 238 (portion of α1-CB8) are cited by GALLOP et al. (1972). The composition of α1-CB3 is given by BUTLER (1972)

γ_{222}, and γ_{122} components may be obtained upon denaturation of insoluble collagen. Tetramers and higher polymers are also found. Such bonds might arise from interaction of hydroxylysine- and lysine-derived aldehydes and amino groups of lysine and of hydroxylysine to form Schiff bases. With aging, the unsaturated bond so formed becomes reduced by an unknown mechanism. Intermolecular cross-links may also arise by aldol condensation between the aldehydes of lysine and hydroxylysine residues in bone and dentine collagen, but the major cross-links are derived from Schiff-base precursors in this collagen (FOWLER and BAILEY, 1972). A new trifunctional crosslink containing histidine was recently identified in bovine-skin insoluble collagen (TANZER et al., 1973).

The extent to which the two types of hydroxylysine- and lysine-derived cross-linking occur in collagen is species-variable and also varies in different tissues in the same animal (the primary dependence on aldehyde formation will be discussed shortly). Whereas intramolecular cross-links occur mainly at the A end of the molecules, intermolecular cross-links may occur at both the A and B ends and along the length of molecules of collagen. Compared to chick-skin collagen,

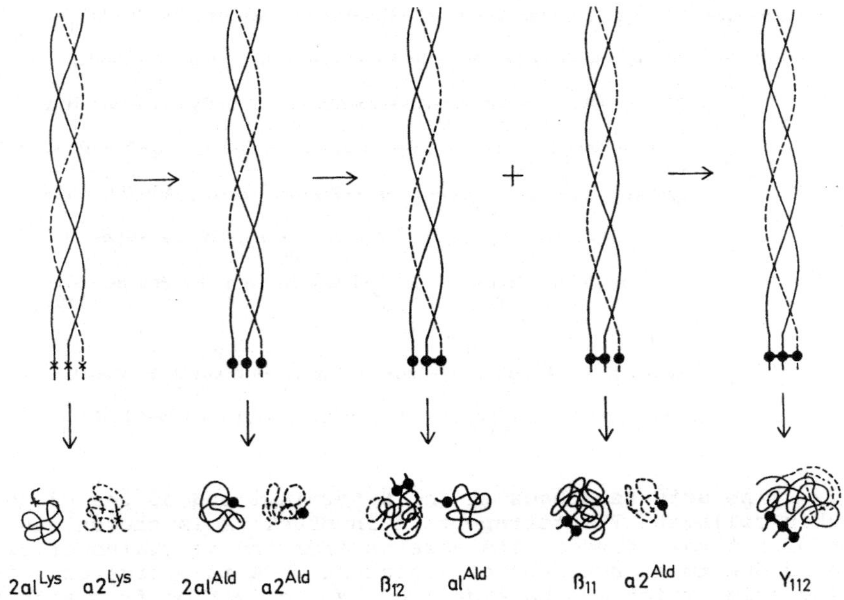

Fig. 3.3. Representation of intramolecular cross-linking of collagen (PIEZ, 1969; with permission). The lysyl residue at a specific position near the N-terminal end of each chain (Fig. 3.7) is converted to an aldehyde (●, allysine), two of which react to form β_{12} or β_{11}. The cross-link in γ_{112} has not been characterized. Denaturation of the soluble collagen yields the random coil structures at the bottom. The solid lines represent the two α1 chains, the dashed line the α2 chain

chick-bone collagen contains fewer intramolecular bonds and mainly intermolecular bonds. In an attempt to define a relationship between intra- and intermolecular cross-links, VEIS and ANESEY (1965) speculated that the two processes of intramolecular and intermolecular cross-linking are competitive and under extrinsic control. The intramolecular cross-link is probably not a separate entity but only an intermediate step in the formation of intermolecular cross-links in native fibrillar collagen.

The type of intermolecular cross-link that is formed spontaneously depends upon the spatial relationship between the two molecules of collagen involved. For example, the cross-linking residues are located in the nonhelical regions of the N-terminus of one α1 chain and the C-terminus (RAUTENBERG et al., 1972) of a second α1 chain (see Fig. 3.1). Reference to Figs. 3.3. and 3.4. shows that these two regions of different α1 chains are juxtaposed in the overlap zone. The spatial relationships between molecules suggest that intermolecular cross-linking may also involve interior sites in the helical regions of collagen. The chemistry of the cross-links appears to be similar in invertebrates and vertebrates (BAILEY, 1970, 1971).

During maturation and aging, the collagen fiber steadily increases in stability to external influences, e.g. mechanical forces, thermal denaturation, swelling, solubility, enzymes, etc. (SINEX, 1968; VIIDIK, 1969). Such changes may be ascribed principally to an increase in number and stability of intermolecular cross-links. PIEZ (1969) notes, however, that lysine-derived cross-links form early as part of a maturation process and may not be involved in late changes. Other kinds of cross-links are known to occur.

ii) Aldehydes

The main aldehydic components involved in cross-linking (discussed above) are present in normal collagen to an extent of 2 or 3 residues per collagen molecule. They occur at the B end as well as the A end of α chains (STARK, RAUTENBER, and KÜHN, 1971). Additional aldehydes can be present (DESMUKH and NIMNI, 1972). The role of these aldehydes in cross-linking is further clarified by observations that collagen of lathyritic animals has a very low aldehyde content and is deficient in cross-links.

iii) Hydroxyl Groups

Hydroxyproline and hydroxylysine derive from proline and lysine, respectively, by selective oxidative processes after or during biosynthesis of α chains (GOULD, 1968a; GRANT and PROCKOP, 1972). Thus, the occurrence of these amino acids in collagen is subject to greater variability than is expected for other amino acids, and this leads to microheterogeneity of collagen primary sequences.

Incomplete hydroxylation of proline was found in rat tissues, where tendon collagen was less hydroxylated than skin collagen (BORNSTEIN, 1967). MILLER et al. (1969) found chick-bone "telopeptide" lysine to be hydroxylated to the extent of about 50% whereas skin "telopeptide" lysine remains unhydroxylated. Collagens from bone and skin of the rat also differ in degree of hydroxylation of lysine (BARNES et al., 1971; STOLTZ, FURTHMAYER and TIMPL, 1973). Collagens of human articular and costal cartilage are hydroxylated to a greater degree than dermal collagen. Basement-membrane collagens are characteristically very high in hydroxylysine.

Hydroxylation is determined by many factors, including primary structure. For example, hydroxyproline and hydroxylysine appear to be largely confined to position 3 of tripeptide sequences beginning with glycine (BUTLER, 1968; NORDWIG and PFAB, 1968; FIETZEK et al., 1973) in vertebrates, but not in invertebrates (see Annelida, Sect. 6d).

iv) Carbohydrate Groups

The principal carbohydrates in vetebrate collagens are D-glucose and D-galactose, which are present as galactosyl residues and glucosyl-galactosyl residues linked glycosidically to the hydroxyl group of hydroxylysine. The structure of the isolated fragment was determined by SPIRO (1967) to be -O-α-D-glucopyranosyl-(1 → 2) -O-β-D-galactopyranosylhydroxylysine. SPIRO (1969a) distinguishes between vertebrate fibrillar collagens, which contain less than 2% carbohydrate, and basement membranes, which represent a specialized form of collagen with approximately 10% carbohydrate. Basement membranes are discussed later. The hexose content of collagen from various sources is given in Table 3.1. The degree of glycosylation of hydroxylysine and the relative proportions of monosaccharide and disaccharide are obviously highly variable (SPIRO, 1972).

Table 3.1a. Hydroxylysine-linked carbohydrate of vertebrate collagens

Source	Total no. of carbohydrate units linked to hydroxylysine	Total hydroxylysine	Hydroxylysine involved in linkage of carbohydrate units	Unsubstituted hydroxylysine	Distribution of carbohydrate between disaccharide and monosaccharide units	
					Glc-Gal-Hyl	Gal-Hyl
VERTEBRATE	units/1000 amino acids	residues/1000 amino acid residues	%	residues/1000 amino acid residues	%	%
Calf skin, citrate-soluble	2.3	8.7	26	5.5	54	46
Rat skin						
Citrate-soluble	1.3	7.6	17	5.6	56	44
Citrate-insoluble	2.4	7.2	33	5.4	71	29
Bovine Achilles tendon	1.8	12.6	14	11.2	64	36
Rat-tail tendon						
Citrate-soluble	0.40	7.5	5	6.6	40	60
Citrate-insoluble	1.0	8.6	12	6.9	74	26
Carp swim bladder						
Citrate-soluble, (ichthyocol)	1.4	11.2	13	9.5	83	17
Citrate-insoluble	1.0	8.0	13	7.0	89	11
Carp skin	0.9	5.7	16	4.8	89	11
Tuna skin	1.0	6.3	16	5.3	61	39
Rabbit sclera						
Citrate-soluble	1.0	6.4	16	5.4	56	44
Citrate-insoluble	1.0	7.3	14	6.6	63	37
Rabbit cornea, citrate-insoluble	5.8	10.4	56	4.5	60	40
Calf cornea						
Citrate-soluble	4.2	8.7	48	4.0	70	30
Citrate-insoluble	5.0	9.6	52	4.2	74	26
Bovine glomerular basement membrane	17.3	22.3	78	5.2	97	3
Calf anterior lens capsule	31.6	36.7	86	4.1	94	6

Table 3.1b. Hydroxylysine-linked carbohydrate of invertebrate collagens

Source	Total no. of carbohydrate units linked to hydroxylysine	Total hydroxylysine	Hydroxylysine involved in linkage of carbohydrate units	Unsubstituted hydroxylysine	Distribution of carbohydrate between disaccharide and monosaccharide units	
					Glc-Gal-Hyl	Gal-Hyl
INVERTEBRATE	units/1000 amino acid residues	residues/1000 amino acid residues	%	residues/1000 amino acid residues	%	%
Octopus (body wall)	5.6	16.5	34	10.9	95	5
Squid (body wall)	7.2	15.8	46	8.6	93	7
Abalone (foot)	6.6	14.0	47	7.4	92	8
Clamworm (Intestine)	8.3	19.6	42	11.3	92	8
Clamworm (cuticle)	0	0	0	0	0	0
Lobster (subcuticular tissue)	19.2	25.8	74	6.6	94	6
Crab (subcuticular tissue)	24.7	34.9	71	10.2	95	5
Ascidian (muscle layer)	20.3	24.2	84	3.9	92	8

Data on carp and tuna skin and invertebrate collagens from KIMURA (1972a, b); remaining data from SPIRO (1969). Glucose and galactose in collagens from sponge, coelenterate and echinoderm appear mainly as disaccharide units linked to hydroxylysine (KATZMAN et al., 1972). Note: annelid cuticular collagens contain carbohydrate linked to threonine and serine (see Sect. 6).

A substantial portion of the disaccharide side-chain of soluble guinea-pig-skin collagen is linked to hydroxylysine in the sequence Gly-Met-Hyl(Gal-Glc-Gly-His-Arg (BUTLER and CUNNINGHAM, 1966). An identical glycopeptide region was shown by BUTLER (1970) to be present in rat-skin collagen (see Fig. 3.2) and by MORGAN et al. (1970) to be present in collagens of carp and man. MORGAN and coworkers also found an additional major glycopeptide in carp and human collagens with the structure Gly-Ile-Hyl(Gal-Glc)-Gly-His-Arg and Gly-Phe-Hyl(Gal-Glc)-Gly-Ile-Arg for carp swim bladder collagen and human skin collagen, respectively. The occurrence of hexapeptides of similar general structure, Gly-X-Hyl(O-Gly)-Gly-Y-Arg, in invertebrate collagens (ISEMURA et al., 1973) emphasizes a narrow substrate specificity for the glycosylation process. It is not surprising, therefore, that in rat-skin collagen there are only four major sites of glycosylation in the molecule (AGUILAR et al., 1973).

The precise role of carbohydrates in collagen is not clear. It is evident that the hydroxylysyl glycoside groups are involved in cross-linking (EYRE and GLIMCHER, 1973). Also, it is probable that bound carbohydrates have an important influence upon the three-dimensional packing of collagen molecules. Signigicantly, the hydroxylysine-linked carbohydrate of the $\alpha 1$ chain occurs in a "hole" region of the quarter-stagger array (see Fig. 3.4) and would not be expected to impair fibril formation (BUTLER, 1970a; MORGAN et al., 1970). However, selective glycosylation may inhibit fibril formation, determine fibril diameter, or direct the type of organization of collagen. For example, unusual morphological organization is associated with large amounts of hydroxylysine-linked hexose in cartilage collagen and in basement membranes such as lens capsule and glomerular membrane.

In addition to glucose and galactose, a number of other sugars such as hexosamine, mannose, fucose and sialic acids are present in some collagen preparations. The content of these carbohydrates is usually very low in acid-soluble collagens but frequently much higher in insoluble collagens. Mannose and glucose are linked in bovine-skin insoluble collagen as $N\varepsilon$-glycosylamine derivatives of lysyl and hydroxylysyl residues (ROBINS and BAILEY, 1972; TANZER et al., 1972). The hexoses may possibly be terminal residues of larger carbohydrate chains. Heteropolysaccharides may be covalently linked to collagen in sponges and in coelenterates (Sect. 6). Cuticle collagen of annelids contains D-galactose as di- and trisaccharides linked to serine and threonine residues in *Lumbricus*, and a glucuronosylmannose disaccharide linked to threonine in *Nereis* (Sect. 6).

v) Other Functional Groups

The presence of phosphate groups in collagens has been extensively discussed by GLIMCHER and KRANE (1968). About two atoms of phosphorus per collagen molecule have been found in collagens from soft tissues and from bone. Covalent binding to serine and to carbohydrate has been suggested. Dentine collagen (Sect. 9f) contains a much larger amount of phosphorus bound to a phosphoprotein, which is apparently linked to a hydroxylysine residue (CARMICHAEL and DODD, 1973).

Disulfide links are definitely known to exist in invertebrate collagens. In vertebrates, cysteine appears to be confined to basement-membrane (Sect. 10) and biosynthetic-precursor (Sect. 11) types of collagen.

3. Aggregate Forms

Collagen is found in tissues mainly as fibrils of 50 to 2000 Å diameter. In most cases, it has been possible to demonstrate at the electron microscope level a marked major axial period of 640 Å (690 Å in the wet state) and, frequently, a highly characteristic series of minor bands after staining. Recent studies have revealed the principal features of the arrangement of collagen molecules in fibrils and the relationship of minor bands to chemical structure. The brief account that follows is based largely upon the reviews by HODGE (1967) and by BAILEY (1968).

A portion of the collagen in tissues can be extracted by aqueous salt and acidic solvents at low temperatures. After extensive fractionation procedures, highly purified collagen can be obtained in solutions in a mainly monomeric state, but accompanied by some dimer and trimer. Under various conditions, it is possible to precipitate the collagen molecules in three main states of aggregation that are interconvertible: 1. native-type fibrils, indistinguishable in the electron microscope from fibrils in the native tissue, 2. segment long-spacing (SLS) aggregates, and 3. fibrous long-spacing (FLS) aggregates (see also ARMITAGE and CHAPMAN, 1971).

The packing arrangement of molecules in the native fibril was elucidated when the band pattern of SLS aggregates was correlated with that of the native fibril and hence with the main periodicity of 640 Å. The main scheme proposed is illustrated in Fig. 3.4. Such a two-dimensional arrangement, however, cannot be extended regularly to three dimensions. Various extensions of the original scheme for molecular ordering have been made by COX et al. (1967), J.W. SMITH (1968), MILLER and PARRY (1973), VEIS (1967), and VEIS et al. (1967) who proposed that oligomeric aggregates act as units in fibril self-assembly processes. The unit microfibrils contain 5 or 6 collagen molecules in quarter-stagger arrangement. The amino terminal regions of collagens are of major importance in determining the packing arrangements in aggregates (YUAN and VEIS, 1973; VEIS et al., 1973), which differ in mineralized and in non-mineralized tissues (KATZ and LI, 1972), between skin and cartilage collagens (BRUNS et al., 1973), and between acid-soluble and insoluble collagen (VOLPIN and VEIS, 1973).

4. Stability

Various experimental parameters associated with the stability of the collagen molecule are readily determined and furnish insights regarding general features of molecular structure and fibrillar organization for different collagen molecules. The following summary relies mainly upon the detailed reviews of this area by VON HIPPEL (1967) and by BAILEY (1968).

The most widely used indices of stability are the shrinkage temperature (T_s) of macroscopic bundles of collagen fibers and the denaturation temperature (T_d) of collagen molecules in solution. The values obtained depend not only upon the particular specimen used but also, to a minor degree, upon the solvent, the rate of heating, and the particular physical parameter used to measure structural alteration. The thermal shrinkage of the fiber depends upon the energy of interaction of molecules as well as upon the forces stabilizing the colla-

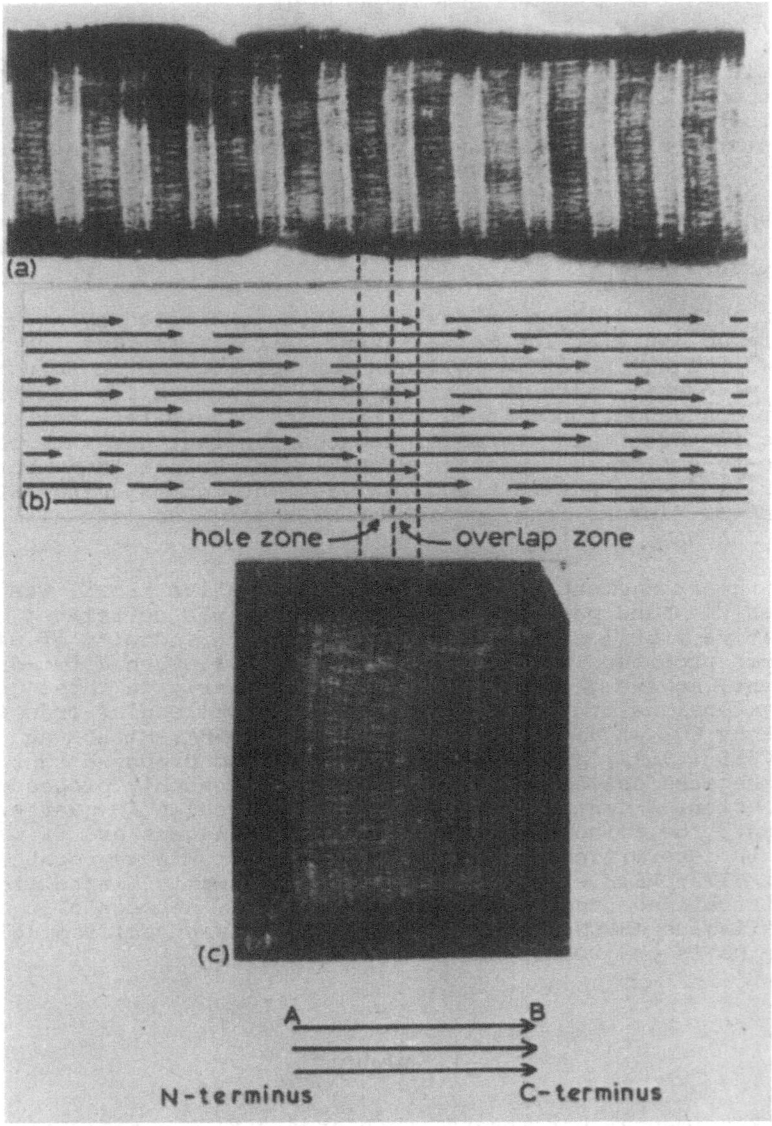

Fig. 3.4. Organization of collagen in fibril (from HODGE and PETRUSKA, 1963; reproduced with permission). (a) Native collagen fibril negatively stained with phosphotungstic acid; (b) schematic of organization of molecules in native fibril; (c) SLS aggregate with molecules aligned in parallel with ends in register

gen molecule. Since T_s-T_d is essentially constant ($27^o \pm 3^o$) for most collagens, it is evident that the variation in T_s of different collagens is due mainly to differences in the thermal stability of individual molecules. Thus, measurements of T_d, which are not complicated by superimposed intermolecular interactions, are more suitable than measurements of T_s for investigating the intramolecular interactions responsible for maintaining the structure of the collagen molecule.

The heat-induced collagen-to-gelatin transition in solution can be observed by various physical chemical methods to yield values of T_d. These methods, which may be applied in either equilibrium or kinetic modes, include determination of changes in optical rotation, viscosity, light scattering, sedimentation in the ultracentrifuge, susceptibility to proteolytic enzymes, etc. A method devised by RIGBY (1967a) permits the preliminary determination of T_d on tissue specimens without the need to extract soluble collagen (except for *Ascaris* cuticle collagen). The essential molecular changes observed upon denaturation are essentially those of a triple helix → random coil transition. The triple helical structure is stabilized largely by interchain hydrogen bonds and stereochemical restrictions to the rotation of pyrrolidine-containing residues in the random polypeptide chains. Some stabilization is also possible by hydrophobic and other interaction. Owing to the restricted rotation of the pyrrolidine ring, the conformational entropy change per pyrrolidine-containing residue for the transition is near zero. Consequently, an increase in pyrrolidine content decreases the total entropy change of collagen. Since at equilibrium, $T_d = \Delta H/\Delta S$, where ΔH is the enthalpy change per mole of collagen, a decrease in ΔS will result in an elevation of T_d, providing ΔH remains constant, or decreases only to a minor degree due to loss of H-bonding (BAILEY, 1968; VON HIPPEL, 1967).

This interpretation of the contribution of the imino acids to the stabilization of collagen is supported by the observation made for various collagens that T_d is correlated positively with the sum of proline and hydroxyproline residues per molecule. The finding also implies that certain interchain hydrogen bonds are less significant than pyrrolidine groups, since an increase in the number of pyrrolidine groups reduces the number of such bonds that can be formed (COOPER, 1971). However, these broad considerations ignore the many complexities arising from the presence of a large variety of tripeptide sequences in collagen (which may contain no, one, or two imino acid residues) and from binding of water molecules (BAILEY, 1968; SUSI et al., 1971). An essential role for water structure was suggested by PRIVALOV and TIKTOPULO (1970).

More recently, evidence was obtained that hydroxyproline makes a specific contribution to the thermal stability of synthetic polypeptides, apparently via hydrogen bonding involving the hydroxyl group (SAKAKIBARA et al., 1973), as does collagen (BERG and PROCKOP, 1973; ROSENBLOOM et al., 1973; see Sect. 1).

It is of major biological significance that denaturation is reversible, at least in part. The phenomena, as studied in the laboratory, are extremely complex and not amenable to ready interpretation. However, it is likely that the collagen molecule is formed initially by self-assembly of pro- α chains followed by secretion into the intercellular matrix and proteolytic removal of most of the peptide of nonhelical regions (Sect. 11). Continuing aggregation leads to fibril production and finally to the creation of higher-order structures. Both intracellular structures and other components of the extracellular matrix may contribute significantly to these processes.

5. Comparative Biochemistry

X-ray diffraction techniques were employed by BEAR (1952) and by RUDALL (1955) to study the distribution of collagen in animal groups (see Fig. 2.1). Collagen was absent only in the Protozoa. Failure to detect collagen or even collagen-like proteins in the cellular slime mold *Dictyostelium discoideum* and in *Tetrahymena pyriformis*, organisms that produce other fibrous proteins, led NORDWIG et al. (1969) to suggest that collagen occurs only in multicellular organisms. However, such generalizations on negative evidence may be premature. The protein produced by the protist *Hymenomonas* (ISENBERG et al., 1965) contains hydroxyproline and requires further investigation. Also, the existence of fibrils with a 670 Å period in the foraminifer *Haliphysema* suggests the possible presence of a collagen-type molecule in this protozoan (HEDLEY and WAKEFIELD, 1967).

Collagen is present in the simplest metazoans and is widespread in the animal kingdom. The mesoglea of medusoid forms of the Coelenterata contains abundant quantities of this protein, which is absent from hydrazoan polyps whose mesoglea is poorly developed. In general, chitin is produced where collagen is scarce or absent. Collagen is common in the Porifera and chitin apparently absent; Nematoda and Annelida possess complex cuticles containing collagen. Supportive tissues in the Mollusca contain either collagen or chitin, but rarely both in the same structure. Collagen comprises the bulk of the mesodermal connective tissue of Echinodermata and is also found in the Onychophora (dermis of *Peripatus*) and in association with chitin in the Brachiopoda (stalk of *Lingula*). It is present in the Arthropoda, where chitin is the principal skeletal component, including the Insecta whose connective tissue is sparse. The structural protein is especially predominant in the Chordata, including tunicates, hemichordates, cephalochordates and vertebrates, which contain no chitin.

In most instances, the identification of collagen is based upon its unique wide-angle X-ray diffraction pattern and confirmed by electron microscopy, amino acid analyses, and physicochemical studies on highly purified soluble preparations. Occasionally, reports of the presence of collagen have been founded upon observations of periodic fibril banding or upon amino acid analyses alone; in these instances, the evidence can be considered only presumptive. However, this general skepticism should be tempered by the realization that the essential form and properties of the collagen molecule may often lie concealed within an extremely complex biological structure. Identification and characterization of collagen become particularly difficult when secondary covalent features, such as carbohydrate content and intermolecular cross-links, are prominent. Also, strong associations with other components of tissues, as in the case of basement membrane and ectodermal collagens, hinder solubilization and purification of native structures. In consequence, data on a particular collagen is often fragmentary and sometimes limited to denatured and partially degraded structures.

In spite of considerable differences in molecular structure, collagens from various species appear reasonably homologous. Their varied composition and properties suggest the general outlines of a plan of molecular evolution. GROSS (1963) noted that glycine is a nearly constant component with a variability of less than 4%. Thus, the stereochemical requirement for a triple helical structure, i.e. glycine at every third position, is adhered to rather closely. However, the content of the imino acids, although of special structural significance, has varied greatly. A similar observation applies to hydroxylysine. The compositional variation in these amino acids and in carbohydrate content is shown for a range of collagens in Tables 3.2 and 3.3. It appears that the degree of variability of total imino acids is greatest in invertebrates, intermediate in fish, and lowest in mammals, in accordance with the range of the taxonomic group (GROSS, 1963). Nevertheless, collagen from different tissues of the same organism may vary in imino acid content. In general, however, the content of imino acids increases in the order invertebrates < fish < land vertebrates.

The following sections discuss with the composition and properties of collagens from many sources. Most of the preparations to be described have been extensively purified and are reasonably free from contaminants, but in most cases the protein has probably been altered from its "native state". Collagens solubilized by treatment of tissues with added proteolytic enzymes are known to have lost much, if not all, of the non-helical regions, both N- and C-terminal. Even extractions employing dilute acetic acid or neutral salt solutions at low temperature may result in loss of some terminal sequences.

6. Invertebrate Collagen

The early data on invertebrate collagens have been reviewed extensively by GROSS (1963) and by EASTOE (1967). However, much of this older information on amino acid and carbohydrate composition was obtained on preparations that had not been subjected to extensive fractionation and do not meet currently accepted criteria for homogeneity. Thus, analytical values for component amino acids are uncertain. Considerable doubt also attaches to carbohydrate contents, particularly to sugar residues such as hexosamines, mannose, arabinose, fucose, and others not found in purified preparations from vertebrates. The Protista contain collagen-like proteins but are considered elsewhere (Chap. 9, Sect. 1). Phylogenetic relationships among the major invertebrate groups are shown in Fig. 2.2.

a) Porifera

Two types of collagen have been identified in the body of sponges (see GROSS, 1963). Spongin A is present as a mesh of very fine fibrils while spongin B consists of large, branched fibers. Only spongin A shows the typical collagen banding in electron micrographs and low-angle X-ray diffraction pattern; both spongins give typical wide-angle X-ray diffraction patterns. The composition of each type of insoluble fiber was determined by PIEZ and GROSS (1959). Spongin B (Table 3.2) had a moderate content of carbohydrate consisting mainly of glucose and galactose with very little hexosamine. KATZMAN and JEANLOZ (1970a) found that highly purified spongin B from *H. gossypina* contained 3.0% of galactose, 2.3% of glucose and 0.3% of mannose. Glucosylgalactosylhydroxylysine (KATZMAN et al., 1972) was isolated in good yield. It was identical in structure with the compound isolated from mammalian collagen. Spongin A, however, contains much more carbohydrate, including hexosamine (2%), mannose, arabinose, and fucose in addition to glucose and galactose. Cystine contents are high. Although the preparations contain some contaminants, their amino acid composition is typical of collagen (see further, Chap. 9, Sect. 1).

Table 3.2. Composition of some invertebrate collagens (res./1000)

	Porifera	Coelenterata		Platyhelminthes	Annelida
	Sponge	Sea anemone	Sea anemone	Liver fluke	Earthworm cuticle
	PIEZ and GROSS (1959)	PIKKARAINEN et al. (1968)	NORDWIG and HAYDUK (1969)	NORDWIG and HAYDUK (1969)	JOSSE and HARRINGTON (1964)
Glycine	323	325	309	305	334
Proline	73	67	75	111	7.7
3-hydroxyproline		6	15	0	a
4-hydroxyproline	94	64	77	95	165
Hydroxylysine	24	16.5	28	26	(0)
Lysine	24	23	19	17	16
Histidine	3.2	4.5	3	5	1.0
Arginine	43	61.5	62	48	22
Aspartic acid	97	71.5	79	64	62
Glutamic acid	86	90	88	91	85
Serine	24	46.5	38	32	83
Threonine	27	39	38	25	49
Cystine (1/2)	12	5.5	0	0	0
Tryptophane					
Methionine	3.1	7	5	18	1.0
Alanine	94	71.5	66	57	100
Valine	24	27.5	31	35	20
Leucine	24	33	34	31	30
Isoleucine	17	23.5	23	26	16
Phenylalanine	10	8.5	8	7	5.7
Tyrosine	4.0	7.5	4	5	2.3
Amide N	(90)				(94)
Hexose (%)	5.0		14.1	12.8	8.8

[a] included with the value for 4-hydroxyproline

[b] see text

b) Coelenterata

Recent studies of collagen from the body wall of *Metridium senile L.* by PIKKARAINEN et al. (1968), from *M. dianthus* by KATZMAN and KANG (1972) and from the mesoglea of *Actinia equina* by NORDWIG et al. (1973) and NORDWIG and HAYDUK (1969) have yielded amino acid analyses (Table 3.2) in good agreement with earlier data (EASTOE, 1967). Cross-linking of α chains was indicated by presence of β and γ components after denaturation of the collagens. All groups of workers found that only

Aschelminthes		Arthropoda	Mollusca		Brachiopoda	Echinodermata
Roundworm cuticle JOSSE and HARRINGTON (1964)	Roundworm muscle[b] FUJIMOTO (1968)	crab KIMURA et al. (1969)	mussel byssus PIKKARAINEN et al. (1968)	oyster mantle KIMURA et al. (1969)	Lingula mantle JOPE (1969a)	sea cucumber KATZMAN et al. (1969)
261	326	314	318.5	325	259	348
291	103	101	82	97	123	101
a	a	1.0	0	0.5	a	a
19	122	77.5	65	83.3	101	37
(0)	40	34.9	2	15.7	12	15
49	19	15.5	33	10.0	33	3.7
10	3	3.9	12.5	1.3	4	1.6
32	39	58.2	36	53.9	49	76
64	57	65.8	78.5	62.5	67	45
72	80	96.8	79.5	112	107	94
17	16	50.4	51.5	39.4	43	45
16	13	23.2	34	22.2	25	34
16	3	3.9	<0.5	2.6	2	0
3.8	6	5.2	8	13.5	12	0
78	63	46.5	88	65.4	77	138
1	13	19.4	34	23.1	18	26
20	54	46.5	27.5	36.8	34	17
10	24	23.2	21	20.0	12	12
10	14	11.6	18	11.5	14	6.3
4.0	6	3.9	10	6.3	7	0.9
(5.3)		(65.8)		(70.8)		
0.4	12.6	6.3	2.4	3.1		0.2

one type of α chain was present. Although derived from specimens belonging to the same order, the collagens show appreciable differences in composition, particularly with respect to cystine.

The collagen (actinicol) from *Actinia* possesses all the characteristic features of vertebrate collagens. A molecular weight of 310,000 was determined. The pattern of sub-banding in long-spacing segments was identical for actinol and vertebrate collagen.

A protein with a high content of hydroxyproline, presumed to be collagen, is present in nematocysts of coelenterates. It is stabilized by disulfide bonds and is accompanied by a hexosamine-containing polysaccharide (BLANQUET and LENHOFF, 1966). Protein resembling collagen

Table 3.3. Compositon of some chordate collagens (res./1000)

	Cyclostomata		Chondrichthyes		
	Hagfish skin	Lamprey skin	Dogfish skin α1 chain	α2 chain	Elastoidin, coll. fr.
	PIKKARAINEN (1968)		LEWIS and PIEZ (1964)		KIMURA and KUBOTA (1966)
Glycine	381	356	343	345	329.0
Proline	92	107	108	81	109.0
3-hydroxyproline	a	a	1.2	0	a
4-hydroxyproline	62	66	60	48	76.0
Hydroxylysine	3	4	5.7	6.8	7.7
Lysine	14	23	27.8	20.3	24.4
Histidine	7	6	9.0	18.5	4.8
Arginine	54	56	51	56	45.6
Aspartic acid	50	49	45	41	41.7
Glutamic acid	59	67	69	67	65.3
Serine	71	58	51	78	30.7
Threonine	27	23	22.5	26.1	20.1
Cystine (1/2)					0.0
Tryptophane					0.0
Methionine	11	10	18.0	16.5	7.2
Alanine	116	121	117	93	115.0
Valine	17	17	25.7	31.6	17.0
Leucine	16	19	18.8	36.9	17.8
Isoleucine	10	8	14.7	19.3	21.1
Phenylalanine	8	9	11.6	12.5	10.9
Tyrosine	2	1	2.2	2.6	5.0
Amide N			(39)		(46)
Hexose (%)					

[a] Included with 4-hydroxyproline

[b] See text

in composition is present within spicules of *Briareum asbestinum* (SILBERBERG et al., 1972).

Collagen from *Metridium dianthus* contain a glycoprotein side-chain, consisting primarily of fucose, mannose and glucosamine (KATZMAN and JEANLOZ, 1970b; KATZMAN and KANG, 1972). However, the major oligosaccharide present was isolated as glucosylgalactosylhydroxylysine of the same structure as the compound obtainable from sponge and mammalian collagens (KATZMAN et al., 1972).

Osteichthyes			Amphibia	Reptilia	Aves	
Codfish skin			Toad skin	Crocodile skin	chicken skin	
α1 chain	α2 chain	α3 chain			α1 chain	α2 chain
	PIEZ (1965)		EASTOE (1967)	EASTOE (1967)	KANG, PIEZ and GROSS (1969a)	
339	348	347	301	324	328	328
98	97	96	109.7	127.9	120	117
0	0	0	a	a	0.9	1.0
55	52	58	77.5	92.8	107	97
5.5	9.5	5.3	4.3	4.9	5.2	10
31.3	20.6	30.3	29.1	25.3	30	21
5.2	11.5	7.0	6.5	4.7	1.9	7.6
51	54	51	49.2	49.5	50	50
50	54	51	54.9	45.5	42	50
76	62	78	77.9	72.8	77	67
70	73	73	66.3	42.1	27	31
23.4	26.9	24.5	26.4	22.0	19	20
0	0	0				
0.5	0.2	1.0				
16.7	18.3	16.8	8.7	6.5	8.4	4.7
119	107	101	98.0	114.0	129	104
15.5	19.5	20.1	21.9	15.4	14	28
18.8	24.4	17.5	28.8	20.1	20	32
10.8	9.2	8.9	14.0	11.4	6.4	16
13.2	9.1	11.0	19.3	17.7	12	13
1.8	4.7	2.6	6.1	3.3	1.7	1.8
(43)	(46)	(54)	(53.9)	(25.5)		
b	b	b				

c) Platyhelminthes

Soluble collagen was extracted from whole liver flukes *(Fasciola hepatica)* by NORDWIG and HAYDUK (1969). The molecular weight of 500,000 for fasciocol is difficult to explain in the absence of S-S bridges (Table 3.2). Nevertheless, the identity of sub-banding of fibrils and segments for fasciocol and vertebrate collagens indicates a normal triple helical structure and α-polypeptide chain length.

d) Annelida

The cuticle in the Oligochaeta is a fibrous structure of ectodermal origin lying above the epidermis. Analyses by WATSON (1958) of earthworm *(Lumbricus sp.)* cuticle showed that the dry cuticle consisted of

Table 3.3. Continued

	Aves	Mammalia			
	Chicken cartilage α1 (II) chain	Rat skin α1 chain	α2 chain	Human skin α1 chain	α2 chain
	MILLER (1973a,b)	PIEZ, EIGNER, and LEWIS (1963)		BORNSTEIN and PIEZ (1964)	
Glycine	329	330	336	333	337
Proline	115	129	113	135	120
3-hydroxyproline	2.2	1.0	0.0	0.8	0.9
4-hydroxyproline	103	96	86	91	82
Hydroxylysine	23	4.3	8.0	4.4	7.6
Lysine	13	30.4	22.4	30.0	21.6
Histidine	2.0	1.9	8.5	3.0	9.7
Arginine	50	49	51	50	51
Aspartic acid	42	46	44	43	47
Glutamic acid	87	74	66	77	68
Serine	26	42	43	36.8	35.1
Threonine	26	19.9	19.8	16.5	19.5
Cystine (1/2)					
Tryptophane					
Methionine	12	8.0	6.1	4.9	5.2
Alanine	104	112	102	115	105
Valine	16	19.6	32.0	20.5	33.3
Leucine	26	18.1	32.4	19.5	30.1
Isoleucine	7.8	6.4	16.1	6.6	14.8
Phenylalanine	15	11.6	10.1	12.3	11.7
Tyrosine	2.2	2.1	2.4	2.1	4.6
Amide N	(50)	(42)	(43)	(37.9)	(45)
Hexose (%)	3				

80% protein of high glycine and hydroxyproline content and a remainder that yielded mainly galactose upon hydrolysis. Although the fibrils observed in the electron microscope showed no periodic structure, the chemical analyses and the characteristic wide-angle X-ray diffraction pattern indicated the presence of the collagen structure. Chitin was absent. Subsequent work has revealed many of the complex features of this unusual collagen (RUDALL, 1968).

The amino acid composition of a highly purified preparation from *Lumbricus terrestris* (JOSSE and HARRINGTON, 1964) is given in Table 3.2. In comparison with collagens of other species, the hydroxyproline content is very high, as is also that of serine and threonine, whereas the proline content is unusually low, and hydroxylysine is absent. The collagen contained mainly D-galactose residues, which were re-

Human glomerulus	Dog Vitrosin	Cephalochordata Amphioxus
KEFALIDES (1971)	KEFALIDES (priv. comm.)	MATHEWS (unpubl.)
310.0	310	323
61.0	69	116
11.0	a	a
130.0	140	75
44.6	38	9.5
10	10	25.4
10.4	10	1.2
33.0	30	47.5
51.0	52	60.4
84.0	79	78.4
37.0	40	43.7
23.0	22	25.9
8.0	8	0
10.0	7	7.2
33.0	34	101
29.0	28	26.6
54.0	58	25.3
30.0	27	14.6
27.0	29	16.8
6.0	5	3.2
12.0	8.8	1.9

leased as reduced di- and trisaccharides of galactose by treatment with alkaline borohydride. The simultaneous destruction of threonine and some serine residues suggested a glycosidic linkage to the hydroxyl groups of these amino acids (MUIR and LEE, 1970). The structures of the oligosaccharides were shown by MUIR and LEE (1969) to be O-α-D-galactopyranosyl-(1 → 2)-D-galactose and O-α-D-galactopyranosyl-(1 → 2)-O-α-D-galactopyranosyl-(1 → 2)-D-galactose. This collagen is also unusual, and possibly unique, in that an appreciable amount (about 10%) of hydroxyproline residues occur in position 2 of tripeptide sequences beginning with glycine (GOLDSTEIN and ADAMS, 1970).

The molecular properties and parameters of acetic acid-extracted earthworm cuticle collagen were extensively studied by MASER and RICE (1962, 1963a, b). The protein had a molecular weight of about 640,000, molecular dimensions of about 5400 · 15 Å, and an intrinsic viscosity of 28 - 35 dl/g, values about twice those of monomeric vertebrate tropocollagen. Heat-denatured preparations showed three components in the ultracentrifuge, analogous to α, β, and γ components of vertebrate collagen. When solutions were sheared, it was possible to precipitate SLS particles that were similar to SLS from vertebrate tropocollagen in dimensions and in major banding features. Apparently, earthworm cuticle collagen is a dimer with properties similar to vertebrate tropocollagen. The absence of periodicity in the native fibril might be accounted for by the inability of the dimer to assemble in the same manner as monomeric collagen (see Fig. 3.4), possibly due to blocking of molecular assembly by carbohydrate.

Somewhat different molecular parameters were found for earthworm cuticle collagen obtained by salt extraction of cuticles (JOSSE and HARRINGTON, 1964). The intrinsic viscosity was 62 dl/g at an average shear gradient of 300 sec^{-1}. The calculated particle length of 9400 Å was considered to be a minimum value. From the above value of the intrinsic viscosity and a $S^{o}_{20,w}$ of 7.0, the molecular weight was calculated as $1.9 \cdot 10^{6}$. However, the denatured collagen gave only a single peak in the ultracentrifuge with $S^{o}_{20,w} = 5.1$ and a molecular weight about 600,000. Further studies are needed to clarify the complex structure of this unusual collagen.

In contrast to cuticle collagen, the collagen fibrils made by cells of mesodermal origin in the abundant connective tissue of oligochaetes readily reveal periodic banding in the electron microscope. A 560 Å spacing was observed in fibrils of the subepidermal layer of *Lumbricus* (RUSKA and RUSKA, 1961; RUDALL, 1968). A similar period was shown for fibrils of basement membranes, sheaths and supporting structures of *Octolasium complanatum* (BACCETTI, 1967). The same nine bands typical of mammalian collagen were resolved within a period. In these, as well as other respects, the fibrils resemble those in insects. FUJIMOTO and ADAMS (1964) reported that collagen extracted from the cuticle-free bodies of *Lumbricus* differed in amino acid composition from the cuticle collagen, especially in the hydroxyproline-to-proline ratio, which nevertheless remained unusually high.

The giant earthworm *Rhinodrilus fafneri* has a cuticular collagen quite different in composition from that of *Lumbricus* (MASER and RICE, 1962). The glycine and total imino acid contents are, respectively, only 255 and 120 per 1000 residues; cystine is present. Like *Lumbricus* collagen, this collagen has an unusually low proline-to-hydroxyproline ratio of 0.21. The content of 30 residues hexosamine per 1000 residues indicates a possible association of the collagen with glycoprotein or heteropolysaccharide. Similar results were obtained by RIGBY (1968a) for cuticle collagen of another giant earthworm, *Digaster longmani*. Its body-wall collagen differed in amino acid composition from its cuticle collagen, as was the case with a small earthworm, *Allobophora* sp. For all earthworm collagens examined, a very high proportion (65 to 95%) of the total imino acids are hydroxylated.

In the subepithelial zones of the proventriculus of *Syllis*, collagen fibers form a highly organized network embedded in an acid polysaccharide-containing ground substance. They show a periodicity of 560 Å (BOILLY, 1970), a value similar to that of body collagen in the Oligochaeta. Collagen is also associated with polyanionic polysaccharides (see Chap. 7), forming cartilage-like tissues in the tentacular complexes of *Eudistylia* and of *Sabellaria*.

KIMURA (1971a, b) studied in detail the collagens from several polychaeta and compared cuticular and intestinal collagens from *Neanthes* (Nereis) *diversicolor*. The cuticular collagens of polychaetes resemble oligochaete *(Lumbricus)* cuticular collagen in containing very low amounts of lysine, histidine, and aspartic acid, no hydroxylysine, and very large amounts of 4-hydroxyproline (about five times more hydroxyproline than proline) and serine and threonine. They differ in containing significant amounts of 3-hydroxyproline and very little carbohydrate. The carbohydrate consisted mainly of galactose mono- and oligosaccharides apparently linked to threonyl and seryl residues (KIMURA, 1973). Molecular characteristics of *Neanthes* cuticular collagen were most unusual but similar to those of *Lumbricus* cuticular collagen.

Intestinal collagen was markedly different from cuticular collagen in *Neanthes* with respect to both chemical composition and molecular characteristics. The ratio of 4-hydroxyproline to proline was 1.6 for intestinal collagen (4.8 for cuticular collagen) and hydroxylysine was present in appreciable amount (absent from cuticular collagen); the hexose content was approximately double. Molecular weight and dimensions were comparable to those of vertebrate collagens.

e) Aschelminthes

Like the Oligochaeta, the Nematoda possess a solvent-resistant cuticle secreted by epidermal cells. Analyses by WATSON and SILVESTER (1958) of the cuticle of roundworms *(Ascaris lumbricoides)* indicated that about one half of the tissue consisted of a protein with high glycine, very high proline, very low hydroxyproline, and no hydroxylysine content. The protein was identified as a collagen on the basis of its wide-angle X-ray diffraction pattern, characteristic high glycine content, and infrared absorption band at 3330 cm^{-1}.

JOSSE and HARRINGTON (1964) studied a purified preparation of collagen from *Ascaris lumbricoides* of unusual amino acid composition (Table 3.2). A molecular weight of 900,000 and a length of 4100 Å were calculated from an $S^{o}_{20,w}$ of 7.0 S and an intrinsic viscosity of 12.3 dl/g. Denaturation in solution occurred at 52° (a very high value for T_d) without a corresponding drop in molecular weight.

A reasonable physical model of the *Ascaris* particle emerged as a result of further studies by McBRIDE and HARRINGTON (1967a, b). In the presence of a disulfide-reducing agent, the intrinsic viscosity dropped greatly. The resulting solution contained molecules relatively homogeneous with respect to size (molecular weight, 62,000) and retained over 75% of the collagen-fold conformation. The reduced alkylated product had a T_d of 33° (19° below that of the native protein). Presumably, the subunit of *Ascaris* collagen is a single polypeptide chain of 900,000 molecular weight that folds back upon itself to form a stable, collagen-type, single-stranded triple helix.

Collagen isolated by pepsin action from the muscle layer of *Ascaris* has an amino acid composition (Table 3.2) quite different from that of cuticle collagen (FUJIMOTO, 1968). A subsequent study of the preparation by FUJIMOTO et al. (1969) corrected the content of half-cystine to 8 residues per 1000 and the carbohydrate content to about 5%. Treatment of the preparation with β-mercaptoethanol reduced the viscosity of solutions without affecting a change in optical rotation or in T_d. The collagen component appeared to be closely associated with another protein *in situ* and was stabilized by disulfide bonds. The noncollagenous protein is rich in acidic amino acids and is similar in composition to those (Chap. 5, Sect. 3c) of vertebrate tissues (FUJIMOTO and IIZUKA, 1972).

The acanthocephalans bear some structural resemblance to the nematodes but are placed in a separate phylum. Collagen was readily isolated (CAIN, 1970) from the subsurface layer of oriented fibers of specimens of *Macrocanthoryhnchus hiradinaceus* infesting the pig intestine. The amino acid composition was comparable to that of vertebrate collagens with a low value (140 residues/1000) of total imino acid residues. Molecular characteristics were comparable to those of vertebrate skin collagen except for unusually high values of T_d, 46° and T_s, 68°.

f) Arthropoda

Electron microscopy of the acellular stroma of various insect organs reveals a fibrous component of collagen imbedded in a relatively structureless matrix. However, biochemical investigations have been hampered by the paucity of connective tissue and the insolubility of collagen of insects (ASHHURST, 1968). This problem was explored by HARPER et al. (1967) in a study of the cockroach *Leucophaea moderae*, whose tissues were shown to contain a protein having the compositional characteristics of collagen. The crossband periods of 570 to 610 Å fall below the value of 640 Å that is characteristic of vertebrate collagens.

A collagen-type diffraction pattern is given by fibers drawn from the silk glands of mature larvae of the common gooseberry sawfly *Nematus ribesii*. The amino acid composition is of the collagen-type; hydroxylysine is present but hydroxyproline is absent. However, only a portion of the structure is likely to be in the triple helical conformation (LUCAS and RUDALL, 1968).

The insoluble tissue residue of a crustacean, the white shrimp *Penaus setiferous*, gave a wide-angle X-ray diffraction pattern typical of collagen (THOMPSON and THOMPSON, 1968, 1970). This residual "collagen" contained only 150 residues of glycine per 1000. While hexose was virtually absent, the tryptophan content of 119 residues per 1000 was extraordinarily high. It is apparent from the very low glycine contents that only a portion of the preparations of these authors consisted of collagen.

Soluble collagen preparations were readily obtained in good purity by pepsin treatment of collagenous tissues of other crustaceans such as the spiny lobster *Panulirus japonicus* and the blue crab *Portunus trituberculatus* (Table 3.2) by KIMURA et al. (1969) and KIMURA (1972b). These collagens were similar to vertebrate collagens in their physical properties but differed in containing cystine and more hydroxylysine, amide nitrogen and carbohydrate (Table 3.1).

g) Mollusca

Compositional data and molecular properties of collagens obtained by pepsin solubilization of the mantle of the pearl oyster *Pinctada morterisii*, of the body wall of the squids *Todarodes pacificus* and *Loligo pealii*, and of the octopus *Octopus vulgaris*, resemble those of the crustacean collagens mentioned above (KIMURA et al., 1969; HUNT et al., 1970). T_d is in the range 27 - 34° and T_s 49 - 58°. Collagen of the abalone, a gastropod, also resembles the crustacean collagens in many respects (KIMURA and KUBOTA, 1968c). The amino acid composition of four gastropod collagens was compared by RIGBY and MASON (1967). The marine species differed from the terrestrial snails mainly in that their collagens had more lysine, hydroxylysine, leucine and methionine, but less hydroxyproline and serine. Collagen of the garden snail *Helix aspersa*, contains 12% hexose (WILLIAMS, 1960).

Wide-angle X-ray diffraction patterns and amino acid analyses have indicated that the secreted byssus threads that serve to attach bivalve mollusks to rocks contain collagen (RUDALL, 1955). However, an unusual organization of monomeric molecules is indicated by the absence of a small-angle X-ray pattern and of striations in the electron microscope. Also, the byssus of the sea mussel *Mytilus edulis* L. exhibits the high shrinkage temperature of 89 - 92° compared to 56 - 59° for skeletal collagen in the same organism. Unfortunately, an amino acid analysis was not done on this skeletal collagen. However, the gelátin from byssus threads (Table 3.2) differs greatly in composition from the mantle collagen (Table 3.2) of another pelecypod, the oyster, which has a T_s of 58° (PIKKARAINEN et al., 1968). It is reasonable, therefore, to infer that other components besides collagen contribute in an important way to the thermal stability of byssus. The highly calcified (aragonite) byssus of *Anomia ephippium* L., however, does not contain collagen. The protein component is free from hydroxyproline and hydroxylysine residues and contains only a modest complement of glycine residues. Surprisingly, about one fourth of the organic fraction is carbohydrate, principally glucosamine (PUJOL et al., 1970). The proteins of *Mytulis* shells are similarly free from hydroxyproline and hydroxylysine but may have a high glycine content, between one quarter and one half of the total amino acid residues (HARE and ABELSON, 1963). Molluscan shell proteins and carbohydrates appear to have been little studied beyond the collection of data on composition (WILBUR and SIMKISS, 1968; see Chap. 9, Sect. 2f). These proteins commonly have very high glycine contents (approaching 50 mole %) and, in a few species, small contents of hydroxyproline and hydroxylysine (DEGENS and SPENCER, 1967).

h) Brachiopoda

The collagen of the Brachiopoda is of particular interest because of the association of this protein with phosphatic calcification among members of the phylum. The excellent preservation of fossil representatives back to the Cambrian period (RUDWICK, 1970) permits study of their chemical-paleontological history over an extended period of geologic time (JOPE, 1967b, 1969b). There are two main classes of Brachiopoda, the Inarticulata, commonly with chitinous phosphatic shells, and the Articulata, with nonchitinous carbonate shells. The Craniacea, a small group, is included taxonomically among the Inarticulata although they possess a carbonate shell. While the shell proteins of members of both classes have a high glycine content (ranging from 130 to 400 residues per 1000) only those from the Inarticulata with phosphatic calcification contain both hydroxyproline and hydroxylysine (JOPE, 1971). For example, these amino acids are absent from the carbonate-shelled Craniacea (order Aerotritida) but are found in a related genus (Discina) of the same order that has phosphatic shell (JOPE, 1967a).

The structural proteins of the mantle of Brachiopoda were also investigated by JOPE (1969a). Although no evidence for fibrillar periodic structure was found in the epithelial layers of the mantle, fibril banding was seen in electron micrographs of the connective tissue that lies between these layers. The amino acid composition of the protein, or proteins, of the mantle of *Lingula anatina* Lamarck, a typical member of the Inarticulata, is given in Table 3.2. Surprisingly, the amino acid composition of mantle proteins from various articulates is similar except that the contents of hydroxyproline and hydroxylysine are somewhat reduced. Thus, the outstanding difference between inarticulates (Table 3.8) in shell protein, i.e. the fact that these two amino acids are present only in inarticulates, does not extend to mantle protein.

The composition of mantle protein of *Lingula* is characteristic of collagen having a somewhat lowered glycine content. Only a single component was observed during electrophoresis in polyacrylamide gels (4 - 25% polymer) in buffer, pH 8.9, containing 7 *M* urea. Under these conditions, the protein showed a mobility identical with that of shell protein and of ox-bone collagen. Disulfide crosslinks and ester-type bonds appeared to be absent. However, JOPE's conclusion that the shell protein and mantle protein preparations consist of homogeneous polypeptide chains of the order of 3000 Å length and 100,000 molecular weight requires substantiation.

The cuticle of the peduncle or stalk of *Lingula*, when freed of epithelia and subepithelial tissues and stretched, yields well-oriented diffraction patterns indicative of α-chitin and in addition a well-defined meridional spacing of 2.86 Å (RUDALL, 1955, 1968). On a dry-weight basis, the cuticle contains about 20% of chitin, about 5% of sulfated polysaccharide (to be described later), and 0.3% of hydroxyproline, suggestive of a small collagen content (MATHEWS, unpubl.). The microscopic morphology of the stalk suggests that all of these cuticle components are secreted by epithelial cells.

i) Echinodermata

The secreted filaments of the sea cucumber were shown to contain collagen by ASTBURY (1938) who based his conclusion upon wide-angle X-ray evidence. WATSON and SILVESTER (1958) found that the fibers that can be teased out from Cuverian tubules of *Holothuria forskali* also gave evidence of large-scale periodicity with an axial repeat of about 650 Å in both electron microscopy and low-angle X-ray diffraction. The amino acid composition and the moderate hexose content of the fibers were characteristic for collagen.

Extensive studies have been carried out on mesodermal collagen of the body wall of echinoderms. However, the purity and homogeneity of the protein (Table 3.2) cannot be assured unless extensive fractionation is done on the gelatin products. The composition of such a homogeneous gelatin fraction from *Thyone* is given in Table 3.2. Glucose and galactose are linked as disaccharide units to hydroxylysine residues (KATZMAN et al., 1972), and are contained in cross-links derived from such residues (EYRE and GLIMCHER, 1973). Another fraction that was free of hydroxyproline and hydroxylysine contained all of the 'extraneous' sugars. The molecular relationships of these fractions to each other and to the native structures in the tissue are unknown. That these relationships may be complex is suggested by the observations of FOUCART (1966) on the sea urchin *Paracentratus lividus*, whose test is composed of sets of plates sutured together by ligaments. Although fibrils showing characteristic collagen-type banding are evident at the electron-microscopic level in the ligaments, none with periodic structure are detected in the fibrillar organic matrix of the plates. Fibrils with a periodicity of 620 to 650 Å have been observed in various mineralized and unmineralized tissues of a number of other echinoderms (TRAVIS et al., 1969; GADEYNE and FRANCOIS, 1970). The collagen components are insoluble in a number of solvents except at high temperature. They have been isolated from the organic matrix of spicules of sea urchin embryos (PUCCI-MINAFRA, CASANO, and LA ROSA, 1972).

7. Deuterostomia

There are animal groups which occupy a borderline position with respect to the traditional division of the animal kingdom into invertebrates and vertebrates. Recognition of common relationships of these groups with vertebrates has led to the creation of a new assemblage, the Deuterostomia. Most of the remainder of the animal kingdom falls into an assemblage called the Protostomia. The Deuterostomia have been extensively discussed by BARRINGTON (1965), who subdivides this grouping as follows:

Deuterostomia

- Phylum Echinodermata
- Phylum Pogonophora
- Phylum Hemichordata
  - Class Pterobranchia
  - Class Enteropneusta
- Phylum Chordata
  - Subphylum Urochordata (Tunicata)
    - Class Ascidiacea
    - Class Thaliacea
    - Class Larvacea
  - Subphylum Cephalochordata (Acrania)
  - Subphylum Vertebrata (Craniata)

In the present work the Echinodermata have been included among the invertebrates in accordance with traditional procedure. The absence of biochemical data on connective tissues of Pogonophora and Hemichordata relieves us for the moment of the need to consider these groups further. However, special consideration must be accorded the Urochordata and the Cephalochordata. These groups, although lacking a vertebral column, possess at some stage, or throughout life, the following important vertebrate features: an axial rodlike notochord for support of the body, a single dorsal and tubular nerve chord, and paired gill slits between the pharynx and exterior. Such strong points of resemblance compel acceptance of the new grouping, the Chordata.

Finally, it should be noted that the classification Deuterostomia implies that the constituent groups have shared a common ancestry and that study of these groups may illuminate questions of the origin and development of vertebrate organization. In particular, BARRINGTON (1965) suggests a close affinity between Vertebrata and Cephalochordata, which may have shared a common origin from a stock separate from the urochordate line. In view of the great significance but mainly unsettled character of problems of vertebrate origins, one may hope that the near relatives of the vertebrates will receive more attention from biochemists in the future.

8. Chordata Acrania Collagen

a) Urochordata (Tunicata)

The tunicates, which are frequently included in the grouping Acrania, are unusual in containing cellulose as a structural support. Collagen (detected as hydroxyproline) is apparently present in very small amounts only in the body wall of ascidians (PIKKARAINEN et al., 1968). A preparation of collagen (Table 3.1B) from the muscle layer of the ascidian *Halocythia roretzi* gave a high content of glycosylated hydroxylysines (KIMURA et al., 1972).

b) Cephalochordata

The animals of this subphylum, commonly known as amphioxus, are divided into only two genera. The amino acid composition (Table 3.3) of collagen from the *Branchiostoma* is like that of other chordates except for the unusually high content of total acidic amino acids (MATHEWS, unpubl.); similar high contents are common among invertebrate collagens, however. The solubilized collagen appears to be highly cross-linked. A T_d of 31° (uncorrected) is consistent with a total pyrrolidine content of 191 residues per 1000. Hexose consists of 0.9% each of galactose and glucose and 0.1% of mannose, corresponding to a presumed glycosylation of about 5 residues of hydroxylysine per 1000 (see also Chap. 7, Sect. 3).

9. Chordata Craniata (Vertebrate) Collagen

The vertebrates have been classified into eight classes, one of which, the Placodermi, has become extinct. They have also been divided in a less formal fashion into Pisces ("fish"), including all primitive aquatic groups, and Tetrapoda ("four-footed"), the terrestrial classes:

Pisces	Class Agnatha Class Placodermi Class Chondrichthyes (Elasmobranchii) Class Osteichthyes
Tetrapoda	Class Amphibia Class Reptilia Class Aves Class Mammalia

The general relationships between the major vertebrate taxonomic groups are outlined in Fig. 3.5. Collagen has been isolated from many representative species of all extant classes and clearly identified in some younger fossil specimens. For convenience, the comparative data on collagens to be presented have been assembled by vertebrate classes, except for fossil collagens and for some unusual collagens such as basement membranes and collagens of nonmesodermal origin.

The physical parameters, such as molecular weight, intrinsic viscosity, optical constants, etc., of collagen solutions are not readily determined with accuracy. In addition to doubtful purity, there is reason

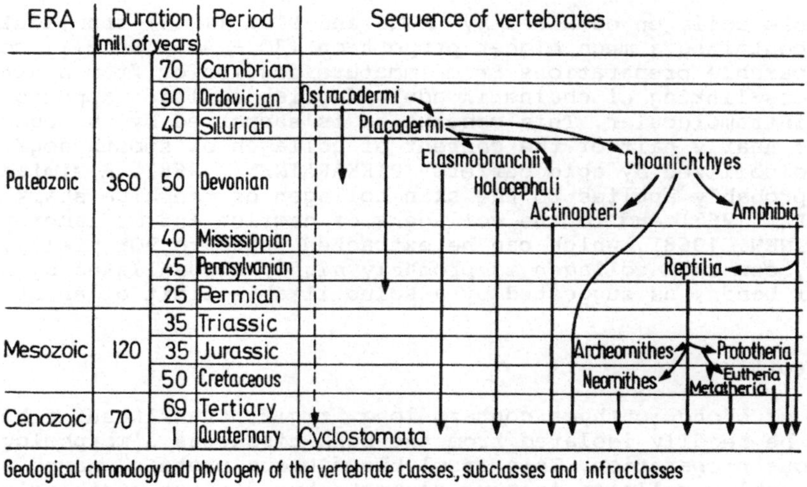

Fig. 3.5. Vertebrate evolution (from SMITH, 1960; redrawn with permission)

to question the degree of molecular homogeneity of the solutions. Thus, there appears little justification for extensive citation of molecular-weight data at this time. We may assume that all vertebrate collagens have approximately similar molecular weights (in the range 260,000 to 290,000). Since optical rotatory constants for vertebrate collagens are very similar, it is likely that their molecular conformations are also very similar. The reader is referred to the tabulation of such data by PIKKARAINEN (1968). Finally, it should be noted that the conversion of procollagen to collagen (Sect. 11a) may occur to a variable and incomplete extent in different species and tissues (VEIS et al., 1972).

a) Cyclostomata

The hagfish *Myxine glutinosa* L. and the lamprey *Petromyzon fluviatilis* L. are representative of the two orders of cyclostomes. Their skin collagens (Table 3.3) were reported by PIKKARAINEN (1968) to contain only a single type of α chain.

b) Chondrichthyes (Cartilage Fish)

i) Dermal Collagen

Collagen from spiny dogfish *(Squalus acanthias)* skin (LEWIS and PIEZ, 1964) has an unusually low T_d (Table 3.5). Chromatography applied to the gelatin revealed a mixture of chains which were identified as α1, α2, β_{11}, β_{12}, and the triple chain γ_{112} component (Table 3.3). PIKKARAINEN (1968), however, was unable to detect more than one α component in skin collagen of the smooth dogfish *Mustelus mustelus* L. by gel electrophoresis. This discordance may be due to some unsuspected biological variation, or simply to insufficient resolving power of the method (note also that this collagen differs from spiny dogfish collagen in composition and in T_d).

The soluble collagen obtained by LEWIS and PIEZ (1964) from adult spiny dogfish contained a much higher proportion (30 - 40%) of γ_{112} component than comparable preparations from immature dogfish or from adult mammals. Cross-linking of chains in adult dogfish collagen appears to be largely intramolecular. This conclusion is supported by the observation that nearly half of the content of collagen of smooth dogfish skin can be solubilized by acid buffers (PIKKARAINEN, 1968). A similar conclusion probably applies to the skin collagen of the blue shark (KUBOTA and KIMURA, 1967) and to the collagens of hagfish and of lamprey skin (PIKKARAINEN, 1968), which can be extracted in 80 to 90% yields. By contrast, cow-skin collagen is probably highly cross-linked by intermolecular bonds, as suggested by a solubilizable limit of about 3%.

ii) Elastoidin

The fins of chondrichthyes contain long, rigid, translucent fibers that can be readily isolated from the bulk tissue as a morphologically homogeneous preparation. This insoluble fibrous protein, named elastoidin, resembles collagen in some respects but differs strikingly in others, and has been the subject of extensive study since 1885 (see review by BEAR, 1952). Sections of a fiber show the characteristic main banding of collagen with a regular 600 to 700 Å periodicity with, however, an unusual symmetric interbanding.

In addition, the wide-angle X-ray pattern is characteristic of collagen except for the absence of an equatorial maximum at 10 to 16 Å. This suggests that, in contrast to tendon collagen, there is a lack of order perpendicular to the fiber axis, which may be due to a unique arrangement of tropocollagen molecules (McGAVIN and PYPER, 1964; MAHANARADHAKRISHNAN and RAMANATHAN, 1966). It has been suggested also that protofibrils may be arranged in concentric sheets (McGAVIN, 1965) and that fibers and fibrils may be surrounded by sheaths (SAHA et al., 1964).

In the first of a series of studies on elastoidin, KIMURA and KUBOTA (1966) reported the isolation and characterization of the tropocollagen component obtained by pepsin treatment of fibers from the great blue shark *Prionace glauca*. The molecular parameters and amino acid composition (Table 3.3) of this protein were typical of vertebrate collagens. In particular, neither cystine nor tryptophan were present; tyrosine content was very low. Electron micrographs of the reconstituted fibrils and of the fragmented original elastoidin fiber after pepsin treatment showed a typical 650 Å period. Subperiod banding also appeared characteristic of native-type collagen.

The collagen component comprised about three fourths of the protein of elastoidin; the remaining quarter consisted mainly of noncollagenous protein (nearly free of tryptophan) of which only a few per cent could be extracted by dilute acetic acid at low temperature. This extractable fraction contained 2.74% of hydroxyproline and 19% of tyrosine and appeared to be identical to the protein residue that GROSS and DUMSHA (1958) obtained after autoclaving the elastoidin. It consisted of two components: a soluble collagen (20%) and a thermostable collagen-noncollagenous protein complex with a ratio of collagen to noncollagenous protein of about 1 : 2 (KIMURA and KUBOTA, 1968a).

The elastoidin fiber shrinks at a temperature of 61°, e.g. within the range of T_s values of mammalian skin; this is surprising since the collagen component of the fiber has a much lower imino acid content than mammalian skin collagens. The pepsin-solubilized collagen has a

T_d of 28°, a value which correlates well with its imino acid content whereas reconstituted fibrils have a reasonable T_s of 49°. Accordingly, KIMURA and KUBOTA (1967) suggested that special crosslinkages may contribute toward stabilization of the fiber. Such crosslinkages could include disulfide bonds but seem to involve mainly tyrosine residues (cf. Chap. 4, Sect. 3a). This conclusion received support from the observation that the T_s values of elastoidins from eight species of sharks, which varied from 57.3° to 67.2°, showed a positive correlation with the sum of three amino acid components: proline, hydroxyproline, and tyrosine (KIMURA and KUBOTA, 1968b). The difference in T_s between elastoidin and skin collagen of the same individual correlated positively with the tyrosine content of the elastoidin.

iii) Ovokeratin

The term ovokeratin is applied to the fibrous layers of the skate egg capsule. A typical wide-angle X-ray pattern, albeit faint, can be shown, yet neither small-angle X-ray patterns nor electron microscopy reveal axial periodicity. A low content of hydroxyproline indicates the presence of a high proportion of noncollagenous proteins (GROSS, 1963).

iv) Cornea

Collagen of corneal stroma of the shark *Scyliorhinus canicula* L. contains a much smaller proportion of hexose than does collagen from mammalian corneas (Chap. 10, Sect. 4b).

c) Osteichthyes (Bony Fish)

Investigations have been carried out on collagens and gelatins from several different species of bony fish, including cod, carp, sturgeon, pike, hake, halibut, lungfish, flounder, turbot, eel, and herring. The significant data have been tabulated by EASTOE (1967) and PIKKARAINEN (1968). The composition of α1 and α2 components of hake *(Merluccius hubsii)* swim-bladder collagen have been determined (PAZ et al., 1967). Of particular interest is a group of fish of the order Perciformes, found exclusively in the Antarctic and adjacent waters at temperatures of -1° to 3°. The skin collagen of various species had values of T_s in the range of 23° to 29° and values of T_d in the range of 5.5° to 6° (RIGBY, 1968b). The composition of collagens from deep-sea species is cited by FUJIMOTO (1972).

The collagen of codfish (unidentified species) skin is of particular interest since it is composed of three different α chains (Table 3.3) each of about 95,000 molecular weight (PIEZ, 1965). The hexose contents of α1, α2 and α3 chains were found by FRANCOIS and GLIMCHER (1966) to be 1.9, 0.9, and 1.7 residues per 1000 amino acids, respectively.

The collagens from cartilage and from notochord of the sturgeon *S. platorhyncus* are identical molecules having the composition $[\alpha 1\ (II)]_3$ and a neutral carbohydrate content of about 5% (E.J. MILLER and M.B. MATHEWS, unpubl.).

d) Reptilia and Amphibia

Amino acid analyses are available for skin gelatin of crocodile, python and toad (Table 3.3).

e) Aves

Collagen from skin, bone and tendon of the chicken has been extensively investigated. The amino acid composition of purified α chains from the skin of lathyritic white Leghorn chicks is given in Table 3.3 (KANG, PIEZ, and GROSS, 1969a). The cyanogen bromide peptides of the α1 chain (KANG, PIEZ, and GROSS, 1969b) and of the α2 chain (KANG, IGARASHI, and GROSS, 1969; IGARASHI et al., 1970) are clearly homologous to the corresponding peptides of the α1 and α2 chains, respectively, of rat-skin collagen and are similarly ordered (Figs. 3.1 and 3.6).

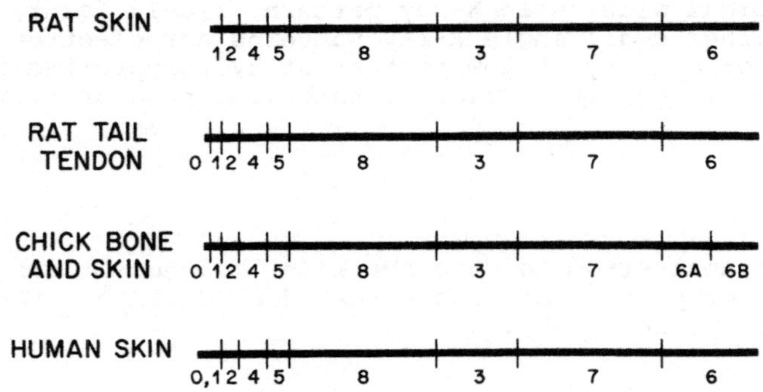

Fig. 3.6. Ordering of peptide fragments in the α1 (I) chains of five collagens. The numbers identify the peptides produced by CNBr cleavage at methionyl residues (vertical marks). The length of each segment is approximately proportional to the number of residues in the peptides. Additional details on the above, as well as α1(I) and α2 chains of collagens from other species, are given by GALLOP et al. (1972)

Differences in sequence (Fig. 3.2) are much greater between α2 chains than between α1 chains (KANG and GROSS, 1970a). HIGHBERGER et al. (1971) determined the primary structure of the 30-residue α1-CB2 sequence and showed that only three substitutions have occurred within the homologous segments between chick and rat-skin collagens. This contrasts with the six substitutions between the two species in the 15-residue α2-CB1 sequence from the NH_2-terminal region (KANG and GROSS, 1970a). The skin of chick embryos contains a newly detected type of α1 chain, α1(III), occurring as $[\alpha1(III)]_3$ collagen, which is replaced during development by the $\alpha1(1)_2\alpha2$ collagen characteristic of the adult (TRELSTAD, KANG, and GROSS, 1971).

E.J. MILLER (1973) noted that the low solubility of normal chicken-bone collagen was due to extensive intermolecular cross-linking (see also MECHANIC, GALLOP, and TANZER, 1971). When such cross-linking was inhibited by administration of a lathyrogenic agent, relatively large amounts of undenatured collagen could be extracted. Only two different α chains were found, respectively nearly identical in amino acid composition to the α1 and α2 chains of skin collagen (Table 3.3).

$\alpha 1$ CHAIN

Rat Skin	H-Gly-Tyr-Asp-Glu-Lys-Ser-Ala-Gly-Val-Ser-Val-Pro-Gly-Pro-Met-
Rat Tail Tendon	PCA-Met-Ser-Tyr(Gly, Tyr, Asp, Glu)Lys(Ser, Ala, Gly, Val, Ser, Val, Pro, Gly, Pro)Met-
Chick Skin	PCA-Met-Ser-Tyr-Gly-Tyr-Asp-Glu-Lys-Ser-Ala-Gly-Val-Ala-Val-Pro-Gly-Pro-Met-
Chick Bone	PCA-Met(Ser, Tyr, Gly, Tyr, Asp, Glu)Lys(Ser, Ala, Gly, Val, Ala, Val, Pro, Gly, Pro)Met-
Human Skin	(PCA, Leu, Ser, Tyr, Gly, Tyr, Asp, Glu)Lys-Ser-Thr-Gly-Gly (Ser, Val, Ile, Pro, Gly, Pro) Met-
	1 2 3 4 5 6 7 8 9 10 11 12 13 14 15 16 17 18 19 20

$\alpha 2$ CHAIN

Rat Skin	PCA-Tyr—— Ser-Asp-Lys-Gly-Val-Ser-Ala-Gly-Pro-Gly-Pro-Met-
Human Skin	PCA(TYR)——(Gly, Asp, Lys, Gly, Val, Gly, Leu, Gly, Pro, Gly, Pro)Met-
Rabbit Skin	PCA(Phe)——(Gly, Asp, Lys, Gly)——(Gly)——(Gly, Pro, Gly, Pro)Met-
Chicken Skin	PCA-Tyr-Asp-Pro-Ser-Lys-Ala-Ala-Asp-Phe-Gly-Pro(Gly, Pro) Met-
	1 2 3 4 5 6 7 8 9 10 11 12 13 14 15

Fig. 3.7. Amino acid sequences at the N-terminal ends of $\alpha 1$ and $\alpha 2$ chains of collagen. Sequences within parentheses have not been determined but a probable sequence in the order shown was made on the basis of homology (GALLOP et al., 1972).

Although not shown here, the remainder of the sequence of the $\alpha 1$ chain of chicken-skin collagen up to residue 55 has been determined. For the first 55 residues, the only differences from the sequence of the $\alpha 1$ chain of rat-skin collagen, including the polypeptide shown in brackets in Fig. 3.2, is in the substitution of alanine for serine at positions 14 and 22. This close similarity between $\alpha 1$ chains of the two species is in sharp contrast with the wide divergence in the sequence of the first 15 residues in the $\alpha 2$ chains. Note, however, possibility of variation in degree of hydroxylation of proline and lysine; see Chap. 3, Sect. 2b. The N-terminal tetrapeptide of the $\alpha 1$ chain that is present in rat-tail tendon collagen may be absent from rat-skin collagen due to a more extensive proteolysis of the precursor procollagen (see Chap. 3, Sect. 11). This peptide, which is identical with the N-terminal tetrapeptide of rat-tail tendon collagen $\alpha 1$ chain, is present in urea-extracted skin collagen $\alpha 1$ chains (STOLTZ, TIMPL, and KÜHN, 1972)

The isolation of ten unique polypeptides in equimolar amounts after cyanogen bromide cleavage of the α1 chain supports the conclusion that bone collagen contains only two types of α chains. A variation was noted in the degree of hydroxylation of lysine in position 9 (Fig. 3.7). Otherwise, the peptides were identical with the cyanogen bromide peptides from chicken-skin collagen. Complete identitiy of primary structures of collagen from both tissues was established by the finding that the six cyanogen bromide peptides from the α2 chain of bone collagen were identical with the peptides of the α2 chain of skin collagen (see E.J. MILLER, 1973). Chicken bone and skin collagens were both denatured at 41° (T_d) (OREKHOVITCH et al., 1960). The hydroxylysine content of newly synthesized collagen decreased with age while the sum of lysine and hydroxylysine remained constant.

MILLER and MATUKAS (1969) found an unusually high proportion of α1 to α2 chains in collagen of lathyritic chick epiphyseal cartilage. They showed the presence of two homologous forms of α1 chains and proposed that the α1 chain that is common to bone, skin, and cartilage be designated Type I and written α1 (I), and that the chain that is unique in cartilage be designated Type II and written α1 (II). The differences (albeit small) in amino acid composition between the two α1 chains were such that the synthesis of the α1 (II) chain must be due to the expression of an additional structural gene. Thus cartilage contains a mixture of two different collagen molecules, one with the chain structure $[\alpha 1(I)]_2\alpha 2$ and the other composed of three identical chains $[\alpha 1(II)]_3$.

The above findings were confirmed and extended by TRELSTAD et al. (1970) for xiphoid cartilage of lathyritic chicks. The two types of collagen, although nearly identical in molecular parameters (IGARISHI et al., 1973) were separable by salt fractionation (TRELSTAD et al., 1972). Galactose and glucose (total 5.5% by weight), the only carbohydrate residues found, were present in a molar ratio of 2 : 1. Definitive studies (see review by E.J. MILLER, 1973) showed that xiphoid-cartilage collagen (Table 3.3) consisted nearly entirely of 3 chains of α1(II), each similar in length and molecular weight to α1(I). About two thirds of the lysyl residues are hydroxylated; of these, about 40% are glycosylated. Of the 13 peptides obtained by cyanogen-bromide cleavage, only 5 appear to be homologous to those from α1(I) chain of chick bone and skin collagens.

When their peptide compositions are compared, cartilage collagens from chick and man differ less from each other than either collagen does from skin collagen of the same species. Comparison of long-spacing segment banding patterns shows that bone, skin, and cartilage collagens have a very similar linear distribution of charged residues, although the cartilage-collagen molecule is about 10 nm longer at the C-terminal end. However, fibrils formed from cartilage collagen are narrow and only faintly stained, probably because of the large number of carbohydrate groups present (STARK, MILLER, and KÜHN, 1972). These differences in chemical composition and fibril structure are probably significant for the calcification process since the cartilage region and the ossifying region of chick long bone contain the $[\alpha 1(II)]_3$ and $[\alpha 1(I)]_2\alpha 2$ types of collagen, respectively (TOOLE et al., 1972).

Chick notochordal tissue in culture produces a collagen composed of three identical α chains resembling those of chick-cartilage collagen (LINSENMAYER et al., 1973). This similarity is very likely an identity, in view of the identity of collagens from cartilage and notochord of the sturgeon (E.J. MILLER and M.B. MATHEWS, unpubl.). A collagen with three identical α chains is synthesized also by embryonic spinal chord epithelium (TRELSTAD et al., 1973).

The cross-links of chicken (and bovine) cartilage collagen appear to be localized primarily in the nonhelical terminal regions of the molecule. This is suggested by the enzymic solubilization of most of the insoluble collagen as monomeric collagen molecules denaturable into single α chains. In contrast, pepsin solubilizes acid-insoluble chicken-skin collagen by removal of nonhelical terminal regions but produces significant proportions of γ as well as β subunits (BANNISTER and BURNS, 1972). Interior cross-linking may be blocked by carbohydrate residues.

f) Mammalia

Among the mammals, the rat is the experimental tool most commonly used for exploring the structure of collagen. In this species, the α1 and α2 chains of tail-tendon collagen differ somewhat in amino acid composition from the corresponding α1 and α2 chains of skin collagen (Table 3.3). However, the differences become minor if comparisons are based upon total imino acids and the sum of lysine and hydroxylysine. Thus, the same structural genes are probably utilized for synthesis of collagen in both tissues. Data on the peptide sequences are given in Figs. 3.1 and 3.2.

Guinea-pig-skin collagen is composed of two α1 and one α2 chains (NOLD et al., 1970). This collagen is clearly homologous to other tetrapod collagens, except for the number and distribution of methionyl residues in the α1 and α2 chains (CLARK and BORNSTEIN, 1972).

The ready availability of tissues and a continuing economic interest have led to a huge literature on bovine collagens. Investigation of molecular structure has progressed rapidly. The α1 and α2 chains of acid-soluble calf-skin collagen were shown by VOLPIN and VEIS (1971a) to contain CNBr fragments homologous to those isolated from rat skin, human skin, chicken skin, and bone collagens. Also, the order of the CNBr-peptides is the same in α1 chains from bovine and rat-skin collagens. However, the N-terminal end of the α1 chain is variable in that it may lack a N-terminal tetrapeptide (Glu, Leu, Ser, Tyr); the dipeptide α1-CBO, present in chicken and in rat tendon but absent in human-skin collagen, is not present in bovine-skin collagen. The C-terminal sequence in α1 chains extracted in 8 *M* urea is -Ala-His-Asp-Gly-Gly-Arg-Tyr-Tyr (RAUTERBERG et al., 1972). The complete sequences of α1-CB3 and α1-CB6 are known (WENDT et al., 1972; FIETZEK et al., 1972a, b).

Use of crude bacterial amylase permitted isolation of insoluble or "polymeric" collagen of high purity (GRANT et al., 1969). The hexose content generally varied inversely with collagen fibril diameter in different tissues of the same organism. Insoluble collagen may contain α chains having a telopeptide sequence longer than soluble collagen α chains (VOLPIN and VEIS, 1973).

BOSMANN and JACKSON (1968) reported that polymeric collagen of bovine cornea contained 5.8% of carbohydrate residues, consisting mainly of galactosamine but also including small amounts of glucosamine, galactose, glucose, xylose, mannose, fucose, and sialic acid. However, more extensive purification of a similar preparation by FREEMAN et al. (1968) led to nearly complete removal of all sugars except galactose and glucose (20 and 9 residues, respectively, per 3000 amino acid residues). The amino acid composition was close to that of bovine-tendon collagen.

Collagen isolated from bovine articular cartilage by STRAWICH and NIMNI (1971) consisted of only a single type of α chain, distinctly different in composition from the α1 chain of bovine-skin collagen. It was designated $[\alpha 1(II)]_3$, in accordance with the new nomenclature for cartilage collagen. The most striking compositional features of the α1(II) chain relative to the bovine α1(I) chain were high contents of hydroxylysine (28 residues per 1000) and glycosidically bound carbohydrate (8 - 9 glucose and 13 galactose residues per α chain). Twelve CNBr peptides were characterized (MILLER and LUNDE, 1973; DESMUKH and NIMNI, 1973).

Particular attention has been directed to bovine dentinal collagen because of its extremely high content of organic phosphate groups: about 55 moles per mole of collagen are present compared to 1 - 2 moles per mole of soft tissue or cortical bone collagen (GLIMCHER and KRANE, 1968). CARMICHAEL and DODD (1973) showed that the organic phosphorus is contained in a polypeptide rich in serine, phosphoserine, and aspartic acid that is apparently covalently linked via an oligosaccharide sequence to the side-chain of a hydroxylysine residue in an α chain of tropocollagen. The phosphoprotein chain may have an important role in regulating such matrix properties as swelling and solubility and in providing sites for the nucleation of the epitactic mineralization of the dentin.

The amino acid composition of dentin collagen varies little among mammals, including cow, dog, rabbit, guinea pig, rat, and man, and is very similar to that of collagen from other tissues which have comparatively lower contents of hydroxylysine. The CNBr peptides of rat dentin and skin collagens are identical (BUTLER, 1973).

Another highly mineralized tissue, tooth enamel, also contains significant amounts of protein-bound phosphate groups that occur as phosphoserine. However, in the case of this tissue, which is of ectodermal origin, the structural protein is not collagen. Investigations of the organic matrices of enamel of a wide range of mammalian species indicate that they contain a separate class of structural proteins with a distinctive amino acid composition (LEVINE et al., 1967; SCHIFFMAN et al., 1970). The proteins of shark enamel (LEVINE et al., 1966; MOSS, 1968a) are somewhat related to the mammalian enamel proteins in amino acid composition.

Normal human-skin collagen was studied by BORNSTEIN and PIEZ (1964) and CLICK and BORNSTEIN (1970) with regard to amino acid composition (Table 3.4), partial amino acid sequence (see Fig. 3.7), and molecular properties; the results were very similar to those from other mammalial collagens. EPSTEIN et al. (1971) compared the CNBr peptides of human and baboon soluble skin collagens with those from rat skin collagen. The collagens of the two primate species differed very little, if at all. All the peptides isolated are homologous to individual peptides of rat collagen except that α1-CB(0,1) is homologous to α1-CB0 plus α1-CB1. (Fig. 3.7) This result is due to the substitution of a leucyl residue in primate collagen for a methionyl residue near the NH_2-terminus of the α1 chain of rat collagen. In the peptides α1-CB2, α1-CB4, α1-CB4, and α1-CB5 derived from the helical portion of the chain, only one substitution (in α1-CB5) apparent between primate (Ala) and rat (Thr) collagens. However, between primate and chicken collagens three substitutions are apparent for these homologous peptides. Several deletions and substitutions are evident at the NH_2- terminal end of the α2 chain between human or baboon collagen and rat, chicken, and rabbit skin collagens (Fig. 3.7).

Table 3.4. Classification of basement membranes by cell-type association (KEFALIDES, 1973)

Cell type	Examples of tissue
Epithelial	Anterior lens capsule; Bowman's capsule; respiratory, intestinal, and glandular epithelium, kidney tubules
Mesenchymal	Capillaries of skin, muscle, and choroid plexus; Descemet's membrane of cornea
Epithelial and mesenchymal	Capillaries of glomerulus and alveolar septum

Cyanogen-bromide cleavage of the insoluble collagen of skin and cartilage of newborn infants permitted good purification of peptides showing the presence of new α1 chains, viz. α1(II) in cartilage and α1(III) in skin (MILLER et al., 1971) and aorta (CHUNG and MILLER, 1974). The collagen of human newborn skin comprised 30% of $[\alpha1(III)]_3$ and 70% of $[\alpha1(I)]_2\alpha2$. Essentially all the collagen of cartilage consisted of $[\alpha1(II)]_3$. The CNBr peptides from this collagen are closely homologous to those derived from chick and bovine cartilage α1(II) (MILLER, 1971a, b; MILLER and LUNDE, 1973) and differ from those derived from α1(I) chains of both species. Cartilage collagens appear to represent a series of closely homologous collagens specifically adapted to function in cartilagenous structures.

The whales are a set of populations of aquatic mammals which diverged from the eutherian (placental) stock at a very remote date. They have been of particular interest because their specializations for aquatic life involve divergences from the typical mammalian plan greater than those found in any other order (YOUNG, 1962). Analyses by EASTOE (1967) indicate that the amino acid composition of whale skin gelatin corresponds typically to that of mammals. However, SUMMERS and CLAYCOMB (1965) reported an unusually low hydroxyproline content (recalculated as 77 residues per 1000) for the gelatin from gingival tissue of the sperm whale *Physter catadon*. The total imino acid content was within the mammalian range. Collagen solubilized by pepsin treatment from hide of the dolphin *Tursiops truncatus* contained 212 total imino acid residues per 1000. Except for the virtual absence of methionine, it resembled whale gelatin in composition (EASTOE, 1967) and had a T_d and T_s of 31° and 59°, respectively (KUBOTA et al., 1967).

Ununsual preparations were obtained from the cartilage of whales. Neutral salt-soluble collagen from nasal cartilage of the sperm whale showed a single sharp peak in the ultracentrifuge and consisted of 70% of collagen, 25% of noncollagenous protein and 5% of a mucopolysaccharide containing hexuronic acid and hexosamine (KUBOTA et al., 1969). Pepsin treatment removed the noncollagenous protein, but failed to separate either the acid mucopolysaccharide or the 3% of hexose. The resulting collagen was similar in amino acid composition to whale-skin collagen (EASTOE, 1967) in most respects but differed remarkably in a few amino acids. The unusual chromatographic pattern of subunits was ascribed to associated noncollagenous material (KUBOTA et al., 1970a). In addition, the very high proportion of α chains as compared to β and γ chains suggests a situation similar to that found in chicken cartilage. By contrast, collagens from other types of connective tissue, including skin, were similar to each other but different from the cartilage collagen of the same animal (KUBOTA et al., 1970b). Thus, it is possible that the unusual differences in composition between collagens of whale skin and cartilage are due to the presence of Type-II collagen (see Sect. 9e) in the cartilage. This cartilage-derived collagen contains 223 residues per 1000 of total imino acids, but has a T_d of only 30°.

10. Basement Membranes and Vitrosin

The presence of a collagen component in an unusual relationship to noncollagenous elements suggests that basement membranes have to be regarded as specialized forms of connective tissue matrices. The collection of these tissues of different embryological origin and anatomical location (Table 3.4) into a single group is based upon certain shared features, viz. morphology, chemistry, and general functional roles in transport.

Major advances in the molecular characterization of basement membranes have been achieved recently (see reviews by KEFALIDES, 1971, 1973; SPIRO, 1972). The data are, however, limited to mammalian species. Generally, the isolated basement membranes are difficult to solubilize and completely intact and undenatured macromolecular components have never been isolated. Although far from complete, the chemical studies support some plausible inferences regarding molecular structure and organization.

The various basement membranes possess common features with respect to amino acid and carbohydrate contents. The collagen content, estimated from the proportion of glycine, makes up between two thirds and three fourths of the dry mass. The total carbohydrate content is high, varying from 6 to 14%, consisting largely of equimolar amounts of D-glucose and D-galactose. The considerable content of sugars not usually associated with collagen, such as mannose, fucose, hexosamine and sialic acids, indicates the presence of a sialoprotein. Half-cystine, absent from the ordinary collagens, is present as 1 to 2% of the total amino acid residues. Both the 3-hydroxyproline and total hydroxyproline contents (which may exceed the proline content) are very high; hydroxylysine contents of basement membrane collagens approach four times the proportion of this residue in tendon collagen. The high carbohydrate contents are associated with a high degree of glycosylation of hydroxylysine residues (Table 3.1). Such striking differences from ordinary mammalian collagens are less marked for alveolar and choroid plexus basement membranes than for the others. It is very likely that additional structural genes are involved in the biosynthesis of the main polypeptide chains of the collagen component of basement membranes.

The collagens solubilized after pepsin treatment from human glomerulus, sheep anterior lens capsule, and Descemet's membrane appear to consist of three identical α chains of molecular weight of 108,000 (KEFALIDES, 1973). The difference in molecular weight against that of interstitial collagens was attributed to hexose contents near 12%. Amino acid compositions were characteristic of basement membrane collagens (Table 3.3). Lens-capsule collagen of cow, isolated in the same way by PIKKARAINEN and KULONEN (1971), had a T_d of 34.6° and an amino acid composition close to that of lens-capsule collagen of dog (Table 3.3). However, two different α bands were noted in gel electrophoresis. Segment long-spacing crystallites had the same length as those from skin collagen and there were many other similarities but also some conspicuous differences in charge profile. Cyanogen-bromide peptides included one which contained 2 residues of mannose, 1 of glucosamine, 4 of half-cystine, 11 of hydroxylysine and 7 of glucosyl-galactose (KEFALIDES, 1973).

Ultrastructural studies of basement membranes reveal a network of fibrils 30 to 40 Å in diameter embedded in an amorphous matrix. Although the molecular fragments isolated from membranes retain the major features of the triple helical structure of collagen, they are not aggregated into fibrils having the characteristic 600 to 700 Å periodicity. Furthermore, the various types of basement membranes are solubilized to different degrees by dissociative reagents and physical conditions. The structural organizations of these matrices thus seem to depend upon unique interactions and are probably related to specific functional roles (KEFALIDES, 1973; OLSEN et al., 1973), but the precise nature of the interactions remain unknown. It has been suggested that the high hexose content of the collagen component interferes with the formation of the classic type of collagen fibril. The sialoprotein component may be involved through disulfide bonds (KEFALIDES, 1973) or other covalent linkages (SPIRO, 1972) as well as noncovalent interactions.

The vitreous body, which probably originates from ectodermal elements during embryological development, contains structural elements laid down as a meshwork of very thin fibrils embedded in viscous fluid. The structure and biochemistry of the vitreous body have been reviewed by BERMAN and VOADEN (1970). MATOLTSY (1952) developed procedures for the isolation and purification from the vitreous of the insoluble native fibrous protein, vitrosin. Partial amino acid analyses of the preparations revealed a strong resemblance to collagen. A shrinkage temperature of 60° was found with 7 - 9% of carbohydrate residues. However, no axial repeating period could be seen in the electron microscope.

Further study of vitrosin by GROSS et al. (1955) demonstrated a wide-angle X-ray pattern typical of collagen. Small-angle diffraction patterns were poor but suggested an average period of 625 Å. OLSEN (1965) solubilized vitrosin by treating the insoluble protein with trypsin. Sebsequent addition of ATP at low pH produced a "dimeric" type of SLS, which had an interband structure identical to that produced from calf-skin collagen treated in the same manner. OLSEN concluded that the morphology of the macromolecular building unit was the same as that of tropocollagen, and that the lack of the characteristic 700-Å axial periodicity was due to an unusual packing arrangement.

Published chemical studies of vitrosin are far from adequate for explaining fibril structure. Recently, KEFALIDES (priv. comm.) has obtained analytical data on purified products obtained after pepsin treatment of dog vitrosin. The amino acid composition and carbohydrate content of this preparation (Table 3.3) are characteristic of basement-membrane collagen; the data are nearly identical to the analytical data for the collagen components of dog anterior lens capsule. Quite different analytical data were obtained by PIEZ (unpubl.; quoted by GROSS, 1963) for bovine vitrosin. Recent data obtained by SWANN et al. (1972) showed a strong resemblance in amino acid and carbohydrate composition between insoluble vitreous and soluble cartilage collagens of cow and it was suggested that vitreous collagen has the composition $[\alpha 1(II)]_3$.

11. Biosynthesis

Collagen polypeptides are synthesized by cells and ultimately play a physiological role as components of fibrillar elements in an extracellular matrix. The complex process of formation of extracellular matrices may be conveniently divided into two phases: intracellular and extracellular. This separation is not absolute; extensive interactions occur between phases (FITTON JACKSON, 1968). The interrelationships between phases and between noncollagenous components and collagen within the extracellular phase are discussed in Chap. 10; present considerations are confined primarily to intracellular events.

The precise biosynthetic boundary between phases is not yet delineated. It is clear that primary polypeptide chain synthesis and the major covalent transformations of secondary nature occur within the cell. Secretion of products might occur either at the level of single α chains, as triple helical tropocollagen molecules, or even as oligomeric assemblies of tropocollagen (M. WEINSTOCK, 1972). All of these entities are capable of automatic assembly into highly-ordered native fibrils and are therefore plausible secretory products. Recent results indicate biosynthesis and secretion of a precursor collagen molecule, procollagen.

Most, but not all, of the reactions involved in the synthesis of the covalent structure of collagen fibrils occur within the cell. These intracellular steps include a) synthesis of the main polypeptide chains, b) hydroxylation of selected proline and lysine residues, c) glycosylation of selected hydroxylysine residues, and d) aldehyde formation. The following is a brief summary of the main findings which are developed fully in recent reviews (CHVAPIL and HURYCH, 1968; ROSENBLOOM and PROCKOP, 1969; GALLOP et al., 1972). Various controversial and unsettled questions are discussed by PROCKOP (1970) and GRANT and PROCKOP (1972).

a) Polypeptide Synthesis

The differentiated cells that are actively engaged in synthesis and secretion of particular groups of macromolecules are associated with the presence of well-developed subcellular systems, in particular the rough endoplasmic reticulum and Golgi apparatus (ROSS, 1968). Like other proteins, the large polypeptide precursors of collagen are synthesized on polyribosomal complexes of the endoplasmic reticulum (KERWAR et al., 1972). The especially large number of residues per chain appears to require a greater number of ribosomes in each complex than is usual for proteins. The messenger RNA for α chains appears to be monocistronic. Upon completion of polypeptide synthesis, chains are released into the cytoplasm to undergo further chemical modification of the side-chains prior to secretion from the cell. The entire process of polypeptide chain synthesis, which includes the sequential addition of amino acids beginning at the amino-terminal end and the hydroxylation of certain prolyl and lysyl residues, requires about 4.8 minutes; for the larger precursor (procollagen) molecule 6.0 minutes is required (VUUST and PIEZ, 1972).

Consideration of the rapid rate at which α chains are correctly assembled by cells compared to the very slow rate with which separate α chains form the native triple helical structure *in vitro* led SPEAKMAN (1971) to propose a novel mechanism for assembly of α chains: synthesis of α chains with telopeptides covalently attached at one end of the chain. The additional polypeptides would serve to bring the appropriate α chains into exact register for rapid triple-helix formation, after which they were moved enzymatically to result in a tropocollagen molecule. In support of this hypothesis were older observations suggesting the existence of components of molecular weight greater than α chains, substantiated by many newer results.

The procollagen α1 and α2 chains appear to differ from tropocollagen α chains in containing a polypeptide extension of 20,000 to 25,000 molecular weight. This extra peptide differs greatly in composition from the α chains in possessing several cysteine residues, tryptophan, high amounts of glutamic and aspartic acids and serine, and little or no hydroxyproline (BORNSTEIN et al., 1972; DEHM et al., 1972; EHRLICH and BORNSTEIN, 1972; UITTO et al., 1972; VON DER MARK and BORNSTEIN, 1973) and cannot exist in a triple helical conformation. Conversion of procollagen to collagen appears to proceed by way of proteolysis by a specific protease, procollagen peptidase, (BORNSTEIN, EHRLICH, and WYKE, 1972). The enzyme appears to be absent from the skin of genetically defective dermatosparoxic calves, which is very fragile and contains disorganized collagen bundles. A high content of "procollagen", from which little or none of the extra peptide has been removed and which does not readily aggregate into fibrils, can be extracted from the dermis of affected animals (LENAERS et al., 1971; STARK et al., 1971). By contrast, only small amounts of high-molecular-weight α chains with heightened aggregation properties can be obtained from normal skin (CLARK and VEIS, 1972). The presence of extracellular α chains intermediate in size between pro-α and α suggests that at least part of the proteolytic conversion of procollagen takes place extracellularly (VEIS et al., 1972). Cartilage and basement membrane collagens are also synthesized and secreted in precursor forms (GRANT et al., 1972; DEHM and PROCKOP, 1973). Various suggestions have been made as to the functional role of the extra peptide in procollagen. These include: 1. alignment of chains and acceleration of helix formation, 2. intracellular translocation and secretion, and 3. control of fibril formation (B.D SMITH et al., 1972; SHERR et al., 1973; GRANT et al., 1973).

b) Hydroxylation

The vast bulk, and very likely all of the hydroxyproline residues in collagen is produced by the action of proline hydroxylase upon the appropriate proline residues. Since this enzyme acts readily upon denatured protocollagen and synthetic polypeptides as well as upon protocollagen in the triple-helix state, no absolute requirement for the triple-stranded conformation is indicated. The site (or sites) of proline hydroxlation within the cell have not been clearly established; it may be either on the nascent polypeptide chains bound to polyribosomes, or on a postribosomal, microsomal membrane, or on both. In plants, hydroxylation of cell-wall peptide takes place on completed polypeptide chains (SADAVA and CHRISPEELS, 1971b).

The same proline residues of protocollagen are hydroxylated by the enzyme *in vitro* as are synthesized *in vivo*, i.e. those residues preceding glycine. Thus, for example, only the proline residue in position 1 of synthetic $(\text{L-Pro} — \text{Gly} — \text{L-Pro})_n$ substrates is hydroxylated (NORDWIG and PFAB, 1969). The enzymic activity requires atmospheric oxygen and, as cofactors, α-ketoglutarate, L-ascorbate, and Fe^{++}. It can hydroxylate additional prolyl residues in collagen from various vertebrate sources and in elastin (UDENFRIEND, 1970). Differences in extent of hydroxylation of otherwise identical α1 chains of rat skin and tendon collagen are due to unknown cellular factors.

The *in-vivo* biosynthesis of collagen hydroxyproline in the intestinal helminth *Ascaris* also requires the action of a hydroxylase dependent upon atmospheric oxygen, although the metabolism of the worm is apparently predominantly anaerobic. The enzyme in muscle is similar to the hydroxylase of chick embryos in that it also requires the same cofactors, but it differs in being less active in normal atmospheric air than in a gas phase containing 1% oxygen (FUJIMOTO and PROCKOP, 1969). It also differs in several aspects from the related enzymic activities in *Ascaris* eggs (CAIN and FAIRBAIRN, 1971) and hypodermis (CHVAPIL, BOUCEK, and EHRLICHOVA, 1970). The muscle enzyme was capable of hydroxylating protocollagen from chick embryos and from *Ascaris* muscle layers. However, protocollagen from *Ascaris* cuticle was an acceptable substrate only after it had been denatured (FUJIMOTO and PROCKOP, 1968). This difference in the two protocollagens from *Ascaris* may be accounted for by the unusual structure of the cuticle collagen (see Sect. 6e). It is of interest that prolonged incubation of denatured cuticle collagen with excess enzyme increased the hydroxyproline content (Table 3.2) from 19 to about 80 residues per 1000 residues. The sites of excess hydroxylation, however, are unknown. This enzyme may be less specific than the avian enzyme, since it was found to hydroxylate the proline residues in both positions 1 and 3 in synthetic $(\text{L-Pro}—\text{Gly}—\text{L-Pro})_n$ substrates (NORDWIG and PFAB, 1969).

Selected lysine residues in protocollagen are converted enzymatically to hydroxylysine residues. The lysine hydroxylation reaction has the same requirements for oxygen and cofactors as does the proline hydroxylation reaction but requires a separate enzyme (MILLER, 1971; KIVIRIKKO and PROCKOP, 1972; POPENOE and ARONSON, 1972).

c) Glycosylation

D-galactosyl residues are transferred to specific hydroxylysine residues (Table 3.1). The glucosyl transferase that specifically catalyses the transfer of glucose from UDP-glucose to galactose residues in collagen has been isolated and purified by BOSMAN and EYLAR (1968). The exclusive localization of this enzyme in the plasma membrane of cells that synthesize collagen suggests that the attachment of carbohydrate to collagen may be a requirement for secretion of collagen (HAGOPIAN et al., 1968). The properties of this enzyme and of the corresponding galactosyl transferase been described (SPIRO, 1972). Both enzymes are also present in the plasma membrane of human platelets (BOSMANN, 1971) which, however, do not synthesize collagen. A relation to collagen-induced platelet aggregation is suggested (KATZMAN et al., 1973).

d) Aldehyde Formation

Data on the formation of aldehydes in collagen are meager. SIEGEL and MARTIN (1970) have partially purified an enzyme, lysyl oxidase, from chick cartilage which converted lysine residues to allysyl residues in both α1 and α2 chains. In the α1 chain, only the lysine residue at position 9 (Fig. 3.2) was oxidized. A metal cofactor, apparently copper, and molecular oxygen are involved with the enzymatic reaction forming allysine (SIEGEL et al., 1970). The same, or a related enzyme catalyzes a similar reaction with lysyl residues in elastin (PINNEL and MARTIN, 1968). It is possible that some aldehyde may be formed outside the cell.

Collagen formation has wide implications for cell biology. The biosynthesis of collagen-class proteins is a differentiated cellular function in tissues of most animals, from the most primitive diploblast porifera and coelenterates to the highest forms developing from three germ layers. Also, a large variety of different cell types including some of ectodermal origin (CHURCH et al., 1973; HAY and DODSON, 1973) as well as others of mesodermal derivation are involved in collagen production. Indeed, within the group of specialized secretory products, collagen must be regarded as extraordinary in regard to the great diversity of cell types involved in production of this protein. Perhaps it is in this general light that we should see to understand the findings of GREEN and GOLDBERG (1968) that collagen is synthesized, albeit at a low rate, by many cultured cell lines originating from cells not specialized for the production of collagen.

12. Immunology

Immunological methods offer possible approaches to the molecular structure and taxonomy of collagen. Although the antigenicity of collagen was first demonstrated about 40 years ago, this subject has been extensively investigated only within the last two decades. Results have been reviewed by SEIFTER and GALLOP (1966), O'DELL (1968), and KIRRANE and GLYNN (1968).

The antigenicity of collagen depends largely upon those regions of the α chains that are accessible to proteases, such as pepsin and trypsin, and are therefore nonhelical, N-terminal and C-terminal. However, antibody to tropocollagen also recognizes sites in the triple helical regions (PONTZ et al., 1970). These conclusions derive from studies with denatured collagens or gelatins, which exhibit wide species cross-reactivity. Major antigenic determinants are located in the C-terminal regions (STOLTZ, TIMPL et al., 1973) of both α1 and α2 chains (BECKER et al., 1972).

Using complement fixation and *in-vivo* immunofluorescence methods, ROTHBARD and WATSON (1965) demonstrated species-specificity of acid-soluble collagens from rat, mouse, guinea pig, chicken, carp, and man. Rat and mouse collagens were found to be indistinguishable and to cross-react with guinea-pig collagen. Cross-reactions also occurred between collagens of rat and man and chicken and man, whereas carp collagen showed no cross-reactions. These results are in general accord with taxonomy. Collagens from carp, buffalo fish, rats, calves, sheep, and humans were readily distinguished by complement-fixation tests with rabbit antiserums against the various collagens (DAVISON et al., 1967). The antiserums were species-specific except that they did not distinguish between ruminants and between collagens from different tissues of the same animal.

Most important, the complement-fixing activity was lost with denaturation of the collagen, but could be recovered by a renaturation process. This result indicated that the observed species-specific antigenicity of collagen was dependent upon the triple helical conformation. Studies by MICHAELI et al. (1969) with species-specific antibodies produced by injection of rat or guinea-pig skin collagen into rabbits showed that antibody is directed primarily against the α2 chains with little reaction with the α1 chain. The location of the species-specific determinant in the N-terminal region of the α2 chain is consistent with the apparent species variability of the amino acid sequence of this region in the α2 chain (see Fig. 3.7 and HIGHBERGER et al., 1971). The use of CNBr- and protease-derived peptides should permit a more precise characterization of antigenic sites.

The collagen component of basement membranes is revealed by immunological methods. Fluorescent-dye labelled antibodies to collagen, homologous or heterologous, are fixed in basement membranes of renal glomeruli and also at sites where collagen is present in liver and spleen of rat, mouse, guinea pig, and chicken (ROTHBARD and WATSON, 1965). The collagen components of sheep lens capsule and of dog glomerular basement membrane (KEFALIDES, 1973) produce antibodies in rabbits. Cross-reactions between these collagens and collagens of other basement membranes, both homologous and heterologous, have been demonstrated.

13. Evolution

a) General Considerations

It is apparent that collagens are a set of very similar proteins. However, the present data on amino acid sequences form a far from adequate basis upon which to establish detailed homologies and gene phylogenies according to the rigorous criteria discussed earlier (Chap. 1). Nevertheless, some tentative conclusions concerning molecular phylogeny may be drawn. At a less fundamental level, evolutionary variation of biological properties and functions may be correlated with differences in molecular structure. In order to do this, however, we need to identify the critical properties of collagen that are selected in evolution and their molecular-structural correlates.

b) Variation in Thermal Stability

An important parameter of the biological function of collagen fibers, tensile strength (HARKNESS, 1968), is greatly lowered when shrinkage is induced at elevated temperatures (see Sect. 4 for discussion of the associated indices, T_s and T_d). One might expect, therefore, that this characteristic of collagen would not only relate to molecular structure, but also have adaptive and hence evolutionary significance.

i) Correlation with Composition

The content of total imino acids of various collagens correlates positively with denaturation temperature, as shown in Fig. 3.8. JOSSE and HARRINGTON (1964) pointed out that T_d also correlates reasonably well with estimated frequences of triplets devoid of pyrrolidine residues, in support of a suggestion that it is primarily such residues (and to a lesser extent triplets containing one such residue) that are involved in the melting transition. Not only the total imino acid content but also the sequential distribution of imino acid residues in polypeptide chains may be an important factor influencing conformational stability (RAO and HARRINGTON, 1966). Still other structural aspects of collagen in the fibrous state and as a molecular dispersion in solution may contribute to stability. A specific contribution is attributed to hydroxyproline (Sect. 4).

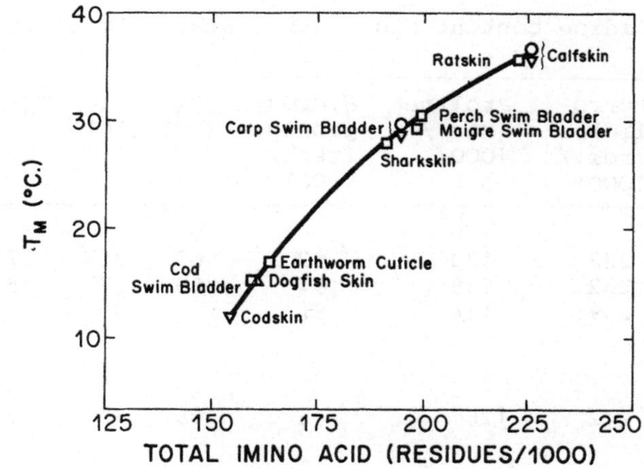

Fig. 3.8. Relation between total imino acid content and $T_d(T_m)$ of several collagens (see text) corrected to "equilibrium" conditions and constant solvent environment. (Redrawn with permission from VON HIPPEL and WONG, 1963)

The data in Tables 3.5 and 3.6, although obtained by different methods and under various ionic conditions, broadly reflect the strong relationship between imino acid content and thermal stability for different collagens. However, some collagens (particularly from invertebrates) appear to deviate to a considerable degree from this general rule. Unusually high values of T_s may be due to special linkages. Such linkages may also account for T_s - T_d values which fall outside the range of 20° to 30°, as in the case of shark elastoidin fibers where collagen is closely associated with noncollagenous proteins and cross-links involving tyrosine residues occur (see Sect. 8c). Here, the stabilizing influences are apparently intermolecular since the value of T_d is in the expected range. For frog-skin collagen, both T_s and T_d values appear high. A contribution of associated proteins to stability in mussel byssus ($T_s = 90^o$) was suggested by PIKKARAINEN et al. (1968); T_s of marine snail collagen is high in relation to the imino acid content.

Although the earthworm *(Lumbricus* sp.) cuticle collagen is of unusual structure (see Sect. 6d), the T_d of 22° correlates well with a pyrrolidine residue content of 173 per 1000 residues (JOSSE and HARRINGTON, 1964). It should be noted, however, that VON HIPPEL and WONG (1963) had obtained a T_d of 17° for preparations of earthworm (*Lumbricus* sp.) cuticle collagen for which MASER and RICE (1962) reported a total imino residue content of 163 residues per 1000. These values also conform to the correlation curve of Fig. 3.8. Differences between the two sets of data may be due to species variability.

The pyrrolidine content of *Ascaris* cuticle collagen is higher than that of any other collagen. Nevertheless, the data point for denaturation temperature may be placed on a reasonable extrapolation of the correlation curve (Fig. 3.8) for other collagens. In spite of a unique chain structure, including disulfide linkage (Sect. 6e), this molecule generally maintains the triple helical conformation characteristic of collagen. However, fiber shrinkage temperature under acid conditions (T_t, RIGBY, 1967a) is anomalously low. The difference of only 7° be-

Table 3.5. Pyrrolidine content and thermal stability of vertebrate collagen

Source of collagen	Pyrrolidine (res./1000)	Proline (res./1000)	Hydroxyproline (res./1000)	T_s (°C)	T_d (°C)	T_s-T_d	Approx. environ. temp. (°C)
Land mammals							
rat skin	222	130	92	63	36[a]	27	37
calf skin	232	138	94	65	37[a]	28	37
calf cartilage	209	114	95		39		37
Aquatic mammals							
dolphin skin	212	126	86	59	31	28	37
Reptiles							
python skin	221	119	102	57			warm
Amphibia							
toad skin	188	110	78	54			warm
frog skin	167	102	65	58	25	33	warm
Birds							
chicken skin	224	119	105	64	41	23	42
Warm water fish							
pike skin	199	129	70	55	27	28	
carp skin	197	116	81	54	29	25	24 to 28
shark skin	191	113	78	53	28[a]	25	
shark elastoidin	185	109	76	61	28	33	
lung fish skin	207	129	78	63			hot water
sturgeon cartilage[b]	209	121	88		33		
shark cartilage[b]	235	151	84		34		
Cold water fish							
cod skin	155	102	53	40	16	24	10 to 14
dogfish skin[c]	156	99	57	41	16[a]	25	
Ice-water fish							
T. eulepidotus skin	143	98	73	27	6	21	-1 to 3
R. glacialis skin	162	120	67	29	5.5	23	
Cyclostomes							
hagfish skin	154	92	62	35[d]	16	19	cold water
lamprey skin	173	107	66	35[d]	21	24	cold water

[a] Corrected values from VON HIPPEL and WONG (1963), other references in text; [b] MILLER and MATHEWS (unpubl.); [c] spiny dogfish, *Squalus acanthias*; [d] primary phase T_s (PIKKARAINEN, 1968).

Table 3.6. Pyrrolidine content and thermal stability of invertebrate collagen[a]

Source of collagen	Pyrrolidine (res./1000)	Proline (res./1000)	Hydroxyproline (res./1000)	T_s (°C)	T_d (°C)	T_s-T_d	Approx. environ. temp. (°C)
Coelenterata							
Sea anemone	167	75	92		26.6		warm water
Platylhelminthes							
liver fluke	206	111	95		36.6		37
Annelida							
Lumbricus cuticle	173	8	165	45	22[b]	23	15-20
Aschelminthes							
Ascaris cuticle	310	291	19	59	52[b,c]	7	37
Ascaris muscle					40		37
Arthropoda							
crab	180	101	79	52	32	20	warm water
Mollusca							
oyster	181	97	84	58	33	25	warm water
octopus	196	101	95	49	27	22	warm water
mussel byssus	147	82	65	90[d]			warm water
land snail	199	101	98	59			very warm
marine snail	177	98	78	59			very warm

[a] See text for references; [b] corrected value, see text;
[c] T_d (reduced and alkylated) = 33°; [d] mussel skeletal collagen T_s = 58°

tween T_s and T_d suggests that the unusual organization of monomers into fibrils (indicated by the absence of the usual banding pattern) results in a weak contribution to stability by intermolecular forces.

RIGBY (1967b) noted that both T_s and T_d correlated not only with the total of imino acid residues but also (negatively) with the number of serine residues. The data were best fitted by separate lines for fish and for nonfish collagens. However, a subsequent study (RIGBY, 1968b) showed that these correlations did not extend to ice-fish collagens; although these collagens have the same serine content as codfish collagen, their T_d is 10° lower.

A more general correlation that is based upon the genetic code has been suggested by RUDALL (1968). He noted that the four amino acids, serine, proline, threonine and alanine, are coded by four triplet sequences of bases in RNA, each triplet containing cytosine as the second

Table 3.7. The genetic code (WATSON, 1970)

First position	Second position				Third position
(5' end)	U	C	A	G	(3' end)
U	Phe	Ser	Tyr	Cys	U
	Phe	Ser	Tyr	Cys	C
	Leu	Ser	Term[a]	Term	A
	Leu	Ser	Term	Trp	G
C	Leu	Pro	His	Arg	U
	Leu	Pro	His	Arg	C
	Leu	Pro	GluN	Arg	A
	Leu	Pro	GluN	Arg	G
A	Ileu	Thr	AspN	Ser	U
	Ileu	Thr	AspN	Ser	C
	Ileu	Thr	Lys	Arg	A
	Meth	Thr	Lys	Arg	G
G	Val	Ala	Asp	Gly	U
	Val	Ala	Asp	Gly	C
	Val	Ala	Glu	Gly	A
	Val	Ala	Glu	Gly	G

[a]Chain terminating (formerly called "nonsense").

residue and differing from the others only in the first residue of the code (Table 3.7). For example, the codon for proline is formed directly by a point mutation of the first residue in the codons for any of the other three amino acids. If mutations leading to interchange of serine and proline are very probable, then one should observe an inverse relationship between the number of pyrrolidine and the number of serine residues, as was actually found by RIGBY (1967b). Also, the sum of residues coded by XCX, where X is one of the four bases, will remain relatively constant in evolution (see Sect. 13d).

ii) Correlation with Environmental Temperature

The T_s values of skin collagen from 25 species of fish, compiled by GUSTAVSON (1956) from his own studies and those of TAKAHASHI, form two main groups: those in the range 35 to 45° for fish living in cold (deep) water, and those in the range 54 to 57° for fish living in warm (surface) waters. Only a few values fall in an intermediate range. This correlation for fish collagen between T_s and environmental temperarure extends generally to the collagens of other animal species (see Tables 3.5 and 3.6).

The collagens of the homoiothermic mammals and birds have the highest values of T_s and T_d, values that are close to body temperatures. The vertebrates of other classes are poikilothermic and the T_d values of their collagens correspond approximately to the temperature of their habitats. Collagens of fish cartilage have unusually high total imino acid contents and T_d values (MILLER and MATHEWS, unpubl.). A similar correspondence exists between environmental temperature and T_d values of both cuticle and bodywall collagen of earthworms (RIGBY, 1968a). T_d is close to the upper limit of the preferred environmental temperature and hence to the upper lethal temperature (RIGBY, 1971). In addition, T_d values are identical for cuticle and bodywall collagen in the same organism, in spite of large differences in total imino acid contents. Differences in tissue organization of collagen account for the finding that values of T_s are lower for cuticle collagen than for bodywall collagen in the same organism. The T_d values of collagen in parasitic invertebrates, the liver fluke, and the roundworm *Ascaris* are likewise close to the body temperatures of their mammalian hosts.

Ther are a few instances where T_s is unusually high in relation to the pyrrolidine content, i.e. for shark elastoidin, mussel byssus and frog skin collagens. In these cases, physiological demands other than those related to temperature may have been superimposed upon the thermal requirement. The special adaptations appear to be achieved by means of ancillary components in association with collagen. A similar situation seems to obtain for some snail collagens (RIGBY and MASON, 1967). These proteins have T_s values from 53 to 59° that are lowered to 50° by treatment with glycanases. It is likely that basement membranes, when examined, will also yield high values of T_s.

iii) Biological Significance

GUSTAVSON (1956) had suggested that the differences in T_s values of collagen among various groups of fish are probably connected with the differences in habitat and stressed the importance of skin structure and its hydrothermal stability for vital functions. Although such an interpretation of relative values of T_s is supported by the observed correlation, the biological significance of absolute T_s values is not clear. Consideration of heat stability as an adaption to temperature does not *per se* explain the observations that the range of T_s values lies 15 to 30° above the maximum environmental temperature. These temperatures may be related to the metabolic turnover of collagen.

The biological significance of the thermostability of cells and proteins of poikilotherms has been extensively studied. This subject was reviewed by USHAKOV (1964) who subsequently published more recent data (USHAKOV, 1967). Interspecific variations in thermostability are characteristic for the main bulk of proteins from poikilotherms. Intraspecific differences between populations in different environments are uncommon, indicating that thermostability is a conservative characteristic. In closely related species, such variations for several sets of homologous enzymes tested correlated with reproductive and environmental temperature. However, other environmental factors (for example, the oxygen level) could be contributing influences.

The views of READ (1967) on protein thermostability differ from those of USHAKOV (1964). Heat stability may be of adaptive significance *per se* for certain proteins but not for others, in particular where denaturation or inactivation occurs only at temperatures well above those of the environment. This is the case for collagen and for several enzymes; for example, lactic dehydrogenase of chicken heart is stable at 60°, which is well above body temperature. Differences in the heat stability of this enzyme in different vertebrate classes reflect changes in primary polypeptide structure that have been selected in the course of evolution. However, heat stability and other properties also depend upon subunit composition, and it is the specific kinetic properties of the enzyme that are thought to be adaptive (KAPLAN, 1965; WILSON et al., 1964).

Since T_s values of collagen cannot be directly related to biological adaptation, alternative explanations of their significance are needed. It may be that T_s values correlate with another fiber property that undergoes transformation closer to maximum environmental temperatures. Such a property may be the glass transition, which occurs at a temperature (T_g) near, or slightly above, T_d. VEIS (1967) suggested a potentially large physiological significance for this transition, which results in irreversible deformation of stressed fibers (see also RIGBY, 1971). It is probable that tensile strength is a critical biological property of collagen fibers. However, there are no data available to relate this property directly to T_s, to the primary chemical structure, or to a specific mechanical role of collagen in tissues of different species. Furthermore, secondary chemical features, such as cross-links and glycosylation (MORGAN et al, 1970), may play a large part in modulating the mechanical properties of collagen fibers. High values for T_s may be necessary, but of incidental consequence, for the achievement of fibrous structures with high tensile strength.

A possible interpretation of the biological variation of T_s is that such values represent the summed contributions of triple helical stabilization represented by T_d, and the approximately constant contributions of intermolecular cross-linking and noncovalent interactions represented by T_s-T_d. It is T_d, then, that relates closely to a property of collagen adaptive to environmental temperature. BAILEY (1968) suggested that a certain difference between environmental temperature and T_d might be essential during synthesis and orientation of molecules into fibrils. Thus, it is likely that the molecular adaption is closely associated with the rate of collagen fold formation from α chains that contain the helical regions of pro-α chains (see Sect. 11). The rate constant for helix formation can be related on theoretical and experimental grounds to the experimental temperature, T, and to the denaturation temperature, T_d, (both in K) (see discussion by VON HIPPEL, 1967) by the following equation: $k' = A \exp(-B/kT\Delta T)$, where k' is the observed rate constant, k is the Boltzman constant, A and B are experimental constants, and $\Delta T = (T_d - T)$, the degree of undercooling of the gelatin sample. The form of the equation shows that the rate of helix formation is expected to increase with T to reach a maximum and then to decline to zero at $T = T_d$. In addition, it was found from renaturation experiments that a high degree of perfection of the resulting triple helix is achieved only at small values of ΔT, $\approx 2^{\circ}C$. Thus, as a consequence of the requirement for a high degree of crystallinity, it appears necessary for the optimum temperature for rate of collagen from α chains to lie close to T_d.

Experimental rates of helix formation vary with solvent environment, degree of cross-linking, and composition of chains. However, when data for various gelatins are compared on a normalized basis, the observed rates increase rapidly with total pyrrolidine content. Although *in vivo* conditions for triple helix formation are undoubtedly more complex and may be promoted by terminal regions of pro-α chains and include participation of other molecules as well as large cellular structures, it is possible that optimal rates for the process are attained biologically by synthesis of primary polypeptide structures with T_d values very close to environmental temperatures.

A highly speculative mechanism for thermal adaptation, involving base modification of mRNA, was offered by RUDALL (1968) who suggested that the pyrrolidine content of collagen might be regulated physiologically in response to environmental temperature. However, USHAKOV (1967) described experiments with frogs and leeches at different temperatures where the thermostability of proteins remained unaltered for prolonged periods of time. Unfortunately, collagen was not examined.

c) Fossils

The detection of amino acids in a variety of fossil bones by ABELSON (1956) suggested the possibility that data on collagen evolution could be derived from the fossil record (see review by WYCKOFF, 1972). Electron and X-ray diffraction methods confirm the presence of collagen in decalcified bone from the Triassic and the Devonian periods. Apparent fibril banding, somewhat reduced from 640 Å, was demonstrated in mammalian bone of the Pleistocene and Miocene epochs, and shown to be due to overlayering of apatite crystals. DOBERENZ and WYCKOFF (1957) observed loss of interband definition in electron micrographs of collagen in fossil bones of the Pleistocene when stained with uranyl acetate and phosphotungstic acid. They attributed this to stepwise degradation of the protein, a conclusion confirmed by chemical data that showed destruction of collagen with time. Thus, the amino acid content of bones from Upper Pleistocene deposits was consistent with the collagen pattern, but revealed a differential loss of certain amino acids that was attributed to differential stability. However, ARMSTRONG and TARLO (1966) found that the concentration of amino acids in the sediment immediately surrounding the specimen may exceed that in the fossil bone. They suggested that, until the processes of differential diffusion and destruction of amino acids are fully understood, interpretation of fossil data is highly uncertain. These views were echoed by MILLER II and WYCKOFF (1968) who reported that the amino acid content of bones from Cretaceous and Jurassic dinosaurs generally resembled those of Pleistocene bones, but in rather different relative amounts.

HO (1966, 1967a) compared amino acid analyses of bones and teeth from 21 different species of Pleistocene mammals from the Rancho La Brea tar pits with those of their modern relatives. The composition of the fossil collagens was typical for mammalian collagen. However, the minor differences between the collagens of each fossil species and its extant relative were considered significant. In particular, HO (1967b) used values of total imino acids to estimate the body temperatures of Pleistocene mammals from a relationship between measured body temperature and total imino acid content established for 18 living mammalian species. In all 9 cases studied, the estimated values for their modern relatives differed from the recorded values for their modern relatives by less than ± 3% (± 1.2^{o} or ± 6 pyrrolidine residues). From this result and indications that the high body temperature of mammals is an adaptation to environmental temperature, the author concluded that the Pleistocene mammals should have been able to survive in the relatively warm climate of recent time and that climatic change was not a causal factor in their extinction in North America.

Electron-microscopic studies of skeletal periderm of 450-million-year-old graptolites (Hemichordata) reveal a morphologically well-preserved fibrous material which resembles collagen in association with polysaccharide (see Chap. 10, Sect. 3c). The cortical fibrils, organized into lamellar plies, show a rough periodicity, a central "core" and helical structure, all of which can be related to extant collagens. Unfortunately, neither chemical nor X-ray evidence confirming the presence of collagen could be obtained from the carbonized fossil (TOWE and URBANEK, 1972).

d) Origins and Remote Ancestry

There are several groups of fibrous proteins from diverse animal phyla that show a general similarity in their mechanical functions as elements of extracellular supportive structures. In spite of differences in tertiary structure and biological context, they possess such striking chemical similarities as to suggest (RUDALL, 1968) the possibility of a common evolutionary ancestry. Data on the amino acid composition of selected representative proteins from a few of these groups (Table 3.8) show a glycine content of nearly one third of the total amino

Table 3.8 Amino acid composition of representative fibrous proteins (res./1000)

	Fibroin[a]	Silk collagen[b]	Resilin[c]	Brachiopod[d]	Conchiolin[e]	Tropoelastin[f]	Collagen[g]
Glycine	318	327	314	351	357	334	330
Proline	47	99	72	72	18	109	121
Hydroxyproline	0	0	0	0	nd	11	99
Hydroxylysine	0	37	0	0	nd	0	7
Lysine	2	15	9	16	0	48	28
Histidine	tr	6	9	9	0	0	3
Arginine	13	57	51	40	0	7	50
Aspartic acid	21	99	75	106	90	3	45
Glutamic acid	5	75	64	44	55	18	74
Serine	167	33	89	65	109	9	36
Threonine	18	39	34	27	15	14	19
Cystine (1/2)	0	0	0	13	0	0	<1
Tryptophan	nd	nd	0	nd	nd	nd	nd
Methionine	nd	1	0	7	nd	0	7
Alanine	249	120	135	96	272	218	116
Valine	52	32	31	33	22	121	18
Leucine	68	14	52	25	19	46	22
Isoleucine	35	17	19	14	18	19	10
Phenylalanine	2	2	28	62	24	28	12
Tyrosine	3	27	19	0		16	3
N(XCX)	481	291	330	260	404	361	391

nd, not determined; tr, trace; N(XCX), sum of proline, hydroxyproline, serine, threonine and alanine; [a] from *Galleria* (RUDALL, 1968); [b] from silk of the sawfly, *Nematus ribessi* (SPIRO, LUCAS, and RUDALL, 1971); [c] from locust, *Schictocerca* (ANDERSEN, 1971), including cross-linked residues derived from about 25 tyrosine residues; [d] whole calcite protein of the articulate *Notosaria* (JOPE, 1971); [e] from mollusc *(Nautilus)* mother-of-pearl (FLORKIN, 1966); [f] from porcine aorta (Table 4.1); [g] from calf skin.

acid residues, a high alanine content, and a roughly similar content of amino acids coded by nucleotide triplets (Table 3.7 and Sect. 13b) of the type XCX. It is, of course, possible that such similarities are the result of evolutionary convergence from a diverse ancestry (see Chap. 4, Sect. 5). However, the hypothesis of derivation from a common ancestor existing at the beginnings of metazoan evolution is attractive (Fig. 3.9).

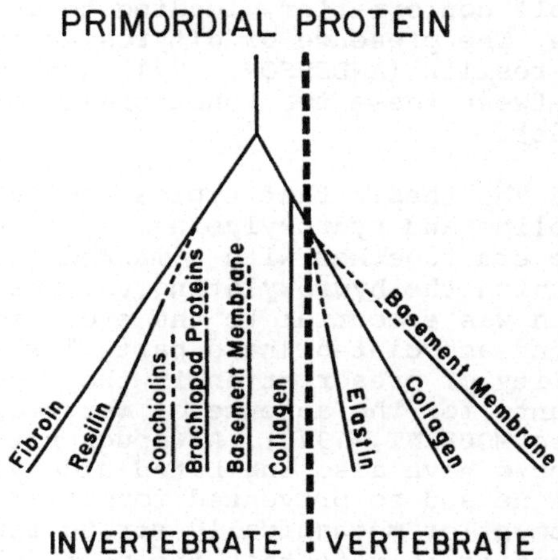

Fig. 3.9. Hypothetical origin of structural proteins

Hymenomonas, a primitive protist, possesses intracellular fibrils having a 200-Å periodicity, reminiscent of very narrow-diameter collagen fibrils (ISENBERG et al., 1965). The presence of hydroxyproline in its coccolith protein suggests the possibility that this unicellular organism may contain an ancestral form of collagen (see also Chap. 2, and Chap. 9, Sect. 1). A primitive form of collagen may also be present in nematocyst capsules of coelenterates (BLANQUET and LENHOFF, 1966).

Group 3 fibroins (Table 3.8) are characterized by a high glycine content (near 1/3 of the total amino acid residues), high serine, and no hydroxyproline. Hydroxylysine is usually absent and the X-ray pattern is that of a typical β structure. The single exception is the protein from the gooseberry sawfly which shows the typical wide-angle pattern of collagen and also contains hydroxylysine. This collagen-like protein forms regularly banded fibrils with an axial period of 550 Å (RUDALL and KENCHINGTON, 1971) but contains no carbohydrate (SPIRO, LUCAS, and RUDALL, 1971).

Resilin is a highly elastic (random-coil) protein that depends upon chemical alteration of tyrosyl side chains for its crosslinking.

The composition of brachiopod- and mollusk shell proteins suggests their inclusion in this list, despite the lack of adequate data on molecular structures. Similar proteins from inarticulate brachiopods contain both hydroxyproline and hydroxylysine (Sect. 6h, and Table 3.2). The composition of these proteins bears a general resemblance to that of enamel proteins (SCHIFFMAN, MARTIN, and MILLER, 1970).

Elastin and collagen show many points of resemblance, although they differ markedly in contents of acidic amino acids and residues bearing nonpolar side chains. Elastin has a significant, but small, amount of hydroxyproline. More interesting, perhaps, is the observation that both proteins share common initial steps of deamination and oxidation of lysyl residues leading to cross-links. These common features suggest (PIEZ, 1968) that the inextensible collagen molecule may have been the evolutionary precursor of elastin. The major alterations would have involved a loss of stabilization due to a decrease in imino acid content, and the introduction of a new system of cross-links between polypeptide chains in a random-coil conformation, leading to an insoluble, fibrous and elastic structure. The presence of dityrosine in elastin (LA BELLA et al., 1967) and in resilin (ANDERSON, 1971) suggests a remote evolutionary affinity between these two functionally similar proteins (see Chap. 4, Sect. 3e).

TOWE (1970) developed the thesis that typical collagen, i.e. collagen containing hydroxyproline and hydroxylysine, appeared near the beginning of the Paleozoic era together with a marked increase in atmospheric oxygen, upon which the hydroxylating enzymes depend. Abundant production of collagen was essential to the evolutionary development of larger metazoans and associated "hard parts," which brought a greatly increased probability of preservation in the fossil record. This hypothesis thus accounts for the absence of metazoan fossils of Precambrian age (see also MOURANT, 1971). Although not considered by TOWE, a low oxygen level would have also inhibited aldehyde production from lysine and hydroxylysine and so prevented formation of stabilized collagen and the development of mechanically strong fibrils. In this connection, it is interesting to note that the peptidylprolylhydroxylase of plants has the same cofactor requirements as the analogous enzyme of animals. In view of the cross-linking role of hydroxyprolyl residues in cell walls (Chap. 2), it is possible that the evolutionary appearance and proliferation of large multicellular plants may have also depended upon atmospheric oxygen tension.

A central assumption of TOWE's argument is that primitive organisms have already evolved the preadaptive ability to synthesize collagen. The hypothesis would establish collagen as a protein of ancient ancestry (see Chap. 2) and might also account for the apparent absence of a greater variety of structural proteins. We note here that the cuticle of *Ascaris lumbricoides* contains collagen with no hydroxylysine and very little hydroxyproline, although muscle collagen in the same species has relatively high contents of both amino acids (Table 3.2). It is likely that this condition is due to the anaerobic environment of the parasite and to tissue differences (CHVAPIL, BOUCEK, and EHRLICH, 1970). In both collagens, disulfide bonds are found to provide an alternate means of stabilization of molecular structure. Such bonds are also present in cuticle collagen of earthworms where, although over 95% of imino acid residues are hydroxylated, no hydroxylysine is present. However, in the case of the earthworm, the relationship between hydroxylation and availability of oxygen has not been investigated.

A somewhat different explanation of the origin of collagen was offered by RAFF and RAFF (1970). These authors proposed that the low oxygen concentration of the atmosphere at the time of evolution of early metazoans limited the absolute size attainable by these organisms for metabolic reasons and thus obviated the need for a firm body. There is no reason to believe that ancestral metazoans were limited to typical collagen as a primary structural material. If dictated by evolutionary pressure for a firmer body structure, a structural protein not dependent upon oxygenases for stabilization could have evolved.

e) Molecular Phylogeny

The data obtained to date show clearly that collagen has undergone structural and functional variation in the course of animal evolution. However, the relationships of such variations to evolutionary forces are still obscure. It is already apparent that data on collagen derived from a limited number of species, or restricted to a single tissue type or stage of development, furnish an insecure base upon which to erect hypotheses. Nervertheless, some reasonably sound generalizations and a few tentative speculations may be offered for their heuristic value.

The structural aspects of the collagen molecule in evolution may be roughly divided into three categories (Table 3.9): 1. those characters common to all collagen molecules that appear to be closely conserved; 2. variable structural features that have been slowly modified in the course of evolution in response to various selection pressures and are specifically adaptive; 3. secondary features of structure that change during development and growth and vary to some degree with physiological state. This last group rests upon mechanisms which act as modulators of the primary structure and, as such, are only indirectly related to the collagen genes.

Table 3.9. Structural features of evolving collagen molecules

Primarily conserved	Variable - specifically adaptive	Physiologically regulated
1. glycine in every third position	1. total imino acid content	1. aldehyde formation
2. minimum imino acid content	2. aldehyde formation	2. hydroxylation
3. α chain length	3. hydroxylation of proline	3. glycosylation
4. residues whose side chains are involved in fibrillogenesis	4. glycosylation of hydroxylysine residues	4. intra- and intermolecular cross-links
5. residues near cross-linking sites	5. special cross-links (e.g. disulfide bonds)	5. fiber diameter
6. residues near sites of hydroxylation of specific prolines and lysines	6. tissue-specific structures (e.g. byssus, cuticle, elastoidin, cartilage, vitrosin)	
7. residues near sites of glycosylation of hydroxylysine		

Two factors - the presence of glycine at every third position of the amino acid sequence and a minimum content of imino acids - are essential to the formation and stability of the triple helical conformation of collagen. Because the presence of glycine in every third position is essential to the triple chain packing, this feature of polypeptide sequence is strongly conserved. The total imino acid content varies in evolution and correlates with the environmental temperature of the species and the denaturation temperature of the collagen. Although contributions to stability are made by other structural aspects, such as hydrogen bonds, etc., it appears likely that there is a minimal content of imino acids. HIPPEL and WONG (1963), from extrapolation of their data (Fig. 3.8), suggested that 139 to 140 residues of imino acid are required for a T_d of 0° and that stable poly-L-proline-type helices should not be expected in proteins containing fewer imino acid residues.

Invariance of α chain length and of those amino acid residues that are involved in fibrillogenesis has been demonstrated by the identity of cross-striation patterns in segment long-spacing crystallites of collagen from species covering the whole phylogenetic range (Fig. 3.10).

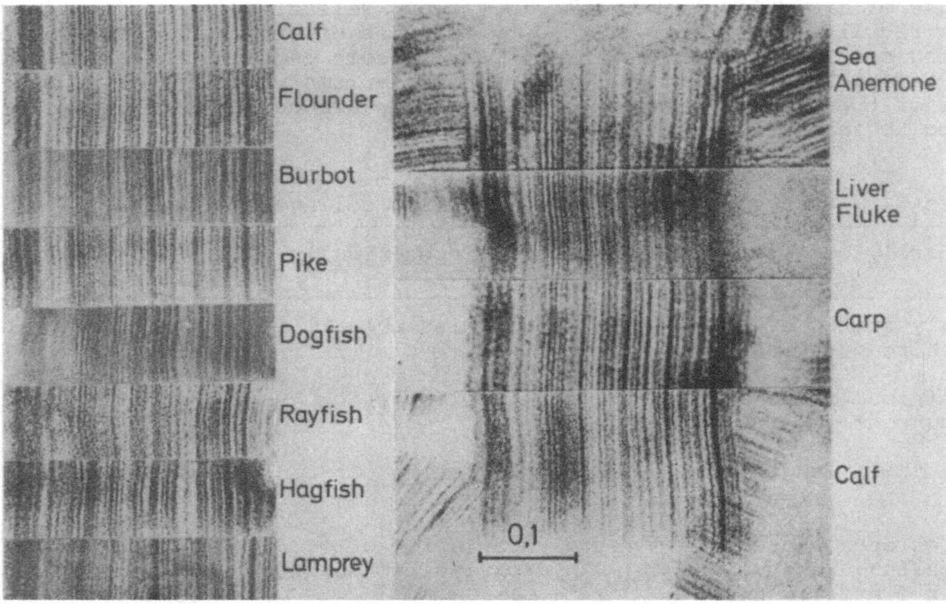

Fig. 3.10. Comparison of segments of collagen. Left, vertebrates; plate, gift of J. PIKKARAINEN and K. KUHN (unpubl.). Right, invertebrates and vertebrates; plate, gift of A. NORDWIG (NORDWIG and HAYDUK, 1969). Particular bands are correlated with known amino acid sequences (BRUNS and GROSS, 1973)

A similar identity of cross-striations, indicating the distribution of charged amino acid residues, was shown for carp, buffalo fish, rat, calf, sheep, and human collagens (DAVISON et al., 1967) and for collagens renatured from three identical chains, i.e. $(\alpha 1)_3$ and $(\alpha 2)_3$ for rat and calf (TKOCZ and KÜHN, 1969), and $(\alpha 1)_3$, $(\alpha 2)_3$, and $(\alpha 3)_3$ for codfish collagen (OLSEN, 1967). These results demonstrate a close homology of the α chains in collagens of all species so far examined. TRELSTAD et al. (1970) and STARK, MILLER, and KÜHN (1972) found small but distinct difference in the intensities but not in the positions of banding patterns for chick-cartilage collagen and chick-skin collagen (Sect. 8e), thus extending the homology to include cartilage collagen. Apparently the precise orientation of molecules in the native fibril, which is produced spontaneously from solutions of tropocollagen, results from highly specific side-chain interactions that are determined by the primary structure of polypeptide chains. A characteristic distribution of polar and apolar groups along the polypeptide chains of collagen appears essential for fiber formation and has remained fairly constant during evolution (HULMES et al., 1973).

SEGREST and CUNNINGHAM (1971) have considered the problem of the evolution of specific macromolecular structure in α chains leading to the observed three-dimensional packing of collagen molecules in quarter-stagger association. No clear choice is yet possible between hexagonal and nonhexagonal arrays. In both models, the evolutionary development of an α2 chain to allow formation of $(\alpha 2)_2$ α2 type collagen would increase the specificity of molecular association from that possible for a $(\alpha 1)_3$ type collagen.

Amino acid sequences about specific residues that undergo essential secondary transformations will tend to be conserved for catalytic reasons. Such sequences include those about lysine residues that are converted to aldehydes, in particular the lysine residues at the N- and C-terminal (nonhelical) ends of α chains (see Fig. 3.7). Other lysine and hydroxylysine residues also may be converted to aldehydes and utilized in subsequent cross-linking reactions. In addition, there may be a strong requirement for hydroxylation of specific prolyl and lysyl residues, in particular those lysine residues that are completely glycosylated, such as lysine 103 of the α1 chain (see Fig. 3.2).

Apart from the glycine content, it is not yet possible to define the structural features that are constant in all collagens. Some that are highly variable over the entire phylogenetic range may well be relatively fixed at a lower taxonomic level. Such constant features may be highly selected by evolutionary processes and be specifically adaptive. The most reliable example appears to be the sum of imino acids, discussed above. Some secondary chemical features, for example special cross-links, as well as various primary chemical structures in certain types of tissue seem also to be specifically adaptive.

The variation of primary structure with type of tissue in the same organism enlarges the task of identification of specifically adaptive collagen. Whereas chick collagens from skin and bone are apparently derived from the same structural genes, an additional gene, homologous in man, ox, and chicken, is necessary for the biosynthesis of the α1 type II chain in cartilage (MILLER, 1973). Tissue-specific genes are also involved in collagen production in cuticle and muscle of *Ascaris*, and in skin, elastoidin, and cartilage of sharks. In higher vertebrates additional genes are probably necessary for production of vitrosin and the collagen component of basement membranes.

Variations in secondary chemical features of a tissue-specific collagen may depend upon stage of embryonic development and growth, or occur in different physiological states. For example, the lysine residue in α1CB1 and α2CB1 of 3-week-old chick-skin collagen is not appreciably hydroxylated whereas this residue in homologous fragments of 3-week-old chick-bone collagen is partially hydroxylated, the degree of hydroxylation decreasing with increasing age (MILLER, 1973). Since this residue is involved in intra- and intermolecular cross-links, the hydroxyl group may have a functional role in development (PIEZ, 1968). Glycosylation is similarly dependent upon physiological conditions (BLUMENKRANTZ and PROCKOP, 1970; SPIRO, 1969).

No ready explanation is available to account for the selection of collagen molecules that are the product of the same genes for a group of such differentiated tissues as bone, skin, and tendon, while special collagen types are utilized in formation of another differentiated tissue, e.g. cartilage, in the same organism. It would seem that the capacity, or efficiency, of physiologically regulated mechanisms to modify the secondary chemical features of a single type of collagen molecule is inadequate to serve the varied requirements of all tissues.

In the living organism, collagen functions only as a component part of a whole tissue or organ containing many other kinds of molecules as well as cells. As such, its full physiological role can be elucidated only in terms of the dynamic, structured interactions of the integrated system. In particular, interactions of collagen with glycoproteins and acidic polysaccharide proteins may be of particular significance (see Chap. 10). Type-II collagen appears specifically adapted for function in cartilage, possibly with respect to enhanced ionic interactions with other components, or in calcification processes (MILLER, 1973).

i) Amino Acid Composition

The composition of collagen may vary considerably throughout the whole animal kingdom. Attempts to identify a particular type of composition with large taxonomic units and to identify broad trends in evolution encounter several difficulties. Thus, large differences in composition, which reflect specialized adaptations, appear frequently between different species of a small taxon and even between tissues of the same organism. Also, the composition of collagen reflects the summed contribution of multiple hereditary units, i.e. genes (each corresponding to one polypeptide chain), of unknown ancestry. Differences in composition cannot be utilized, as can differences in amino acid sequence, to infer the number and type of mutational changes that have occurred in structural genes. Yet, despite such severe limitations, compositional data reveal some general relationships that are of interest.

With respect to the sum of proline, hydroxyproline, serine, threonine and alanine residues, collagens may be divided into three groups in accordance with general taxonomy: invertebrates, lower chordates, and higher chordates (Table 3.10). $\Sigma N(XCX)$ for a collagen (for genetic significance, see Sect. 13b) appears to fall within a narrow range of values characteristic for the group, with the notable exception of the unusual cuticle collagens and vitrosin, and possibly the collagen of *Lingula*. Total imino acid residues show a much greater variability within each group. It appears that evolutionary changes in total imino acid residues, probably in response to changes in environmental temperatures, may have occurred to a large extent via mutations in position 1 of the nucleotide triplet XCX.

A more detailed analysis of compositional differences has been carried out by PIKKARAINEN and KULONEN (1969), who noted a marked increase in total imino acid content from poikilotherms to homotherms (mammals and birds), with a complementary decrease in serine and threonine. Also, vertebrate collagens as a group contain amounts of aspartic and glutamic acid much lower than those of invertebrates. Amphioxus (Table 3.3) occupies an intermediate position, somewhat closer to invertebrates than to vertebrates (Table 3.2). When differences in the contents of individual amino acids of several collagens were summed for every possible pair of species, invertebrates were clearly separate from vertebrates, with the homothermic species forming a subgroup within the vertebrates.

Table 3.10. Sum of proline, hydroxyproline, serine, threonine, and alanine [N(XCX)] in collagens and in α chains (res./1000)[a]

Source		Source	
INVERTEBRATES		LOWER CHORDATES	
Sponge	312	Amphioxus	362
Sea anemone	295	Hagfish	348
Sea anemone	309	Lamprey	374
Liver fluke	320	Dogfish	349
Earthworm (cuticle)	405	" (α1 chain)	360
Round worm	317	" (α2 chain)	326
Round worm (cuticle)	421	Shark[f] (α1 (II) chain)	386
Crab	300	Elastoidin	351
Mussel byssus	321	Codfish	358
Oyster	307	" (α1 chain)	365 (356[c])
Lingula	369	" (α2 chain)	356 (357[c])
Sea cucumber	355	" (α3 chain)	353 (361[c])
		Hake[b]	375
		" (α1 chain)	381
		" (α2 chain)	362
		Sturgeon[f] (α1(II) chain)	371
HIGHER CHORDATES			
Toad	378	Rat	388
Crocodile	399	" (α1 chain)	400
Chick	392	" (α2 chain)	364
" (α1 chain)	403	Guinea pig[c]	400
" (α2 chain)	370	" (α1 chain)	405
" (α1(II) chain)	376	" (α2 chain)	388
Calf	391	Human	385
" (α1 chain)	400	" (α1 chain)	396
" (α2 chain)	373	" (α2 chain)	363
" (α1(II) chain)	360[d]	" (α1(II) chain)	372[e]
" (vitrosin)	305	" (glomerular)	295

[a] Values computed from Tables 3.2 and 3.3, except for [b] PAZ, SALAZAR, and GAITE (1967); [c] NOLD, KANG, and GROSS (1970); [d] STRAWICH and NIMNI (1971); [e] MILLER (1973); [f] MILLER and MATHEWS (unpubl.).

MATSUMARA (1972) compared the amino acid composition of 97 different collagens according to number and type of residues. Three groups of residues were considered: a hydroxylated amino acid group including hydroxyproline, threonine, serine, tyrosine, and hydroxylysine; a hydrophobic amino acid group including valine, methionine, leucine, isoleucine, tyrosine, and phenylalanine; and a polar amino acid group including aspartate, glutamate, hydroxylysine, histidine, and arginine. The ratios of the number of residues in each of the three groups were used as parameters for locating each collagen by a vector in triangular co-ordinate space. The collagen preparations were classified into four groups according to their vector positions. The main group contained vectors of all collagen preparations reported to have regular fibrillar striations and none without such striations and included collagens from both ectodermal and mesodermal tissues. The other three groups, characterized by a high content of either polar, hydrophobic, or hydrophylic residues, contained only collagens of ectodermal derivation. Since the collagen preparations that fell into the main group are found throughout the phylogenetic range of the metazoa, it was suggested that these closely related collagens might have evolved from a "prototype of collagen." Collagens of the other three groups are specially differentiated and each has evolved along independent lines, although all derived originally from the same "prototype." The vectors of high-glycine proteins, including silks, resilin, and elastin, fell outside the range of any of the above groups.

Collagens from all the invertebrate species examined to date, as well as those from the lowest vertebrates, the Cyclostomata (hagfish and lamprey), appear to consist of only a single type of α chain. However, numerous higher-vertebrate collagens have been shown to contain two types of α chain. Three different α chains have been demonstrated conclusively only in codfish collagen. If we narrow our purview to a restricted group of tissues, i.e. skin, swim bladder, tendon, and bone of vertebrates other than cyclostomes, it is plausible to consider that the two different chains in these tissues arose as a result of duplication of a single, ancestral vertebrate gene. Subsequent independent evolution would have led to divergence in structure of $\alpha 1$ and $\alpha 2$ polypeptide chains. The $\alpha 3$ chain of cod collagen was the result of a second and later duplication of one of these two structural genes, an event which occurred after the bony fish were well established (PIKKARAINEN and KULONEN, 1969).

Support for the concept of a genetic relationship between the $\alpha 1$ and $\alpha 2$ chains of type-I collagen is suggested by the similarity of the magnitudes of $\Sigma N(XCX)$ for the respective chains shown in Table 3.10. The similarity extends to $\alpha 1$ type-II and $\alpha 3$ chains also. Analogous comparisons between $\alpha 1$ and $\alpha 2$ chains of type-I collagens were made by DEASY (1972) with like results for $\Sigma N(XAX)$, corresponding to tyrosine, histidine, glutamate, aspartate, and lysine (plus hydroxylysine) residues; for $\Sigma N(XGX)$, corresponding to cysteine, tryptophan, arginine, serine, and glycine residues; and for $\Sigma N(XUX)$, corresponding to phenylalanine, leucine, isoleucine, methionine, and alanine residues. Thus, most of the compositional differences between different α chains are accountable for by point mutation of the first nucleotide of the codon. Table 3.10 shows that the $\alpha 2$ chain generally has a lower content of $N(XCX)$ than the $\alpha 1$ chain of the same collagen molecule. Also, total imino acids (except in guinea-pig collagen) and alanine are consistently less in $\alpha 2$ than in $\alpha 1$ chains. These differences in composition between the two α chains are generally (with some exceptions) greater in higher than in lower vertebrates, suggesting independent and divergent evolutionary paths for these polypeptides.

The genealogy of collagen is both complex and highly branched. However, collagen probably had a simple beginning in early metazoan life as a single gene coding for a single type of α chain. As organisms grew larger and more complex and developed highly specialized tissues, gene duplication and divergent evolution led to separate types of collagen molecules. At this stage, each collagen molecule may still have consisted of three α chains of identical primary structure. Then, early in vertebrate evolution, further complexity was introduced by the appearance of collagen molecules composed of two different α chains.

The stage of vertebrate evolution at which collagens composed of heterologous α chains originated is unclear. According to KULONEN and PIKKARAINEN (1970), the presence of only a single type of α chain in collagen of cyclostomes, i.e. hagfish and lamprey of the Class Agnatha, and in collagen of cartilage fish, i.e. smooth dogfish and ray of the Class Chondrichthyes, indicates that duplication of the α chain to yield α1, α2, and α3 chains occurred only in the phyletic line leading to the Class Osteichthyes (Fig. 3.11). However, two different α chains

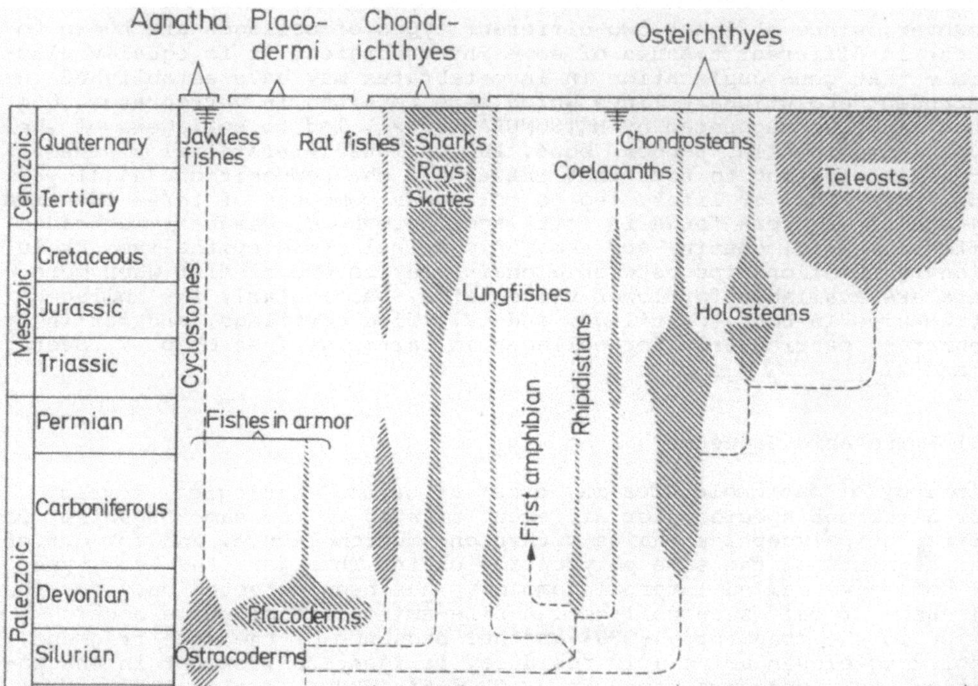

Fig. 3.11. Evolutionary relationships of fishes. The horizontal dimension of the shaded areas indicates the relative abundance of each group (adapted from MILLOT, 1955). The Acanthodii have been previously included with the Placoderms but are now placed in a separate class by NORMAN and GREENWOOD (1963) and in a subclass of the Osteichthyes by ROMER (1966)

are known to compose the skin collagen of the spiny dogfish (LEWIS and PIEZ, 1964). This nonuniversality of collagen composed of two different α chains among the Chondrichthyes suggests that α chain duplication may have taken place relatively late in the history of the class, possibly as recently as the Mesozoic era. The apparent restriction of the α3 chain to the codfish among the Osteichthyes similarly suggests a late occurrence of the corresponding gene duplication.

The finding that the collagen of chick cartilage is composed of three identical α chains led MILLER and MATUKAS (1970) to speculate that the α1(II) chain may be a direct descendent of a "primitive vertebrate α chain". Alternatively, cartilage collagen can be viewed as "embryonic" and hence neotenous in adults (see Chap. 7, Sect. 1). Furthermore, bone, which in extant species contains collagen composed of two different α chains, is a no less "ancient" type of tissue in vertebrate evolution than is cartilage. Thus, gene duplications may well have occurred very early in vertebrate evolution and continued to be expressed differentially in each tissue type.

However, since at least two different types of collagen are known to occur in different tissues of some invertebrates, it is equally plausible that gene duplication in invertebrates may have established independent evolutionary lines which were retained in vertebrates. One such line, as suggested by MATSUMURA (1972), led to collagens of the type found in skin, tendon, bone, and various interstitial tissues. It is interesting to note that these have the composition $[\alpha1(1)]_2\alpha2$. The other line, or lines, led to collagens composed of three identical α1 chains of types found in cartilage, notochord, basement membranes, vitrosin, skin, dentin, and embryonic spinal chord epithelium. Evolutionary relationships between α chains may become clearer when more data are available for lower vertebrates, particularly for collagen structures in bone, cartilage, and calcified cartilage, and for invertebrates, particularly for collagen in cartilage (see Chap. 7, Sects. 1 and 2).

ii) Amino Acid Sequence

Homology of macromolecules may exist at several biological levels: for different species, for different tissues in the same organism, for different polypeptide chains within one macromolecule, and for different segments of the same polypeptide chain. Thus far, the last type of homology, called internal homology, has been detected only for the α1 chain of calf-skin collagen. A nine-amino-acid sequence occurs twice in the fragment α1-CB6; another sequence of the same fragment, including eleven amino acid residues, is identical, except in one position, to a sequence found in α1-CB(4-5) (WENDT et al., 1972). However, the presently available data do not permit a decision as to whether gene doubling during evolutionary development of collagen was responsible for observed similarities in sequences.

Evidence for homology of α chains of collagens from different species derives not only from the electron-microscopic comparisons and compositional data discussed above, but also from compositional data on chemically defined peptides and from partial amino acid sequence data. Thus, the CNBr peptides from the α1 chains of chick-bone and human-skin collagens were shown to be homologous to those from the α1 chain of rat-skin collagen. Similarly, homology was demonstrated for the α2 chains of the respective collagens (MILLER, 1973). Homology of corresponding fragments has also been shown for α1-CB2 and α1-CB3 peptides from rat-skin collagen and human-skin collagen; the α2-CB2 peptides from the respective collagens were likewise found homologous (BORNSTEIN and KANG, 1970). Only three substitutions in a 30 residue sequence of α2-CB2 distinguish chick-skin collagen from rat-skin collagen (HIGHBERGER et al., 1971).

The homology of α1 and α2 chains is strongly suggested by the high degree of similarity of compositions and also by the marked similarity in distribution of charged residues, as shown by comparisons of the banding patterns of renatured molecules composed of only one type of chain (this sect., above). Support for this view was gained from the results of computer comparison of the α2-CB2 sequence of rat-skin collagen with all 30 possible amino acid sequences in the known structure of the α1 chain. The region in α1 that was found to possess a significantly greater degree of similarity than any other includes residues 106 to 135 in α1-CB8-HA2. This region corresponds precisely to α2-CB2 in its distance from the NH_2-terminus of the chain. Also, both sequences are almost identical in their distribution of charged residues (BALIAN et al., 1972).

The relative invariance of composition of the helical regions of collagen chains extends to the nonhelical, amino-terminal ends of α1 chains but not to the corresponding regions of α2 chains. This situation is illustrated in Fig. 3.7. The first 20 residues of the α1 chains of chick-skin, chick-bone, and rat-tendon collagens are identical; the corresponding peptide of human-skin collagen is surprisingly deviant. However, the amino terminal sequences of α2 chains of different species show a wide variation. These regions apparently contain the antigenic sites determining species-specificity (see Sect. 12).

iii) Secondary Chemical Structure

Aldehydes and the derived cross-linking structures, hydroxyl groups of hydroxyproline and hydroxylysine, and carbohydrate groups linked to hydroxylysine, are structural features common to all collagen molecules. The disaccharides linked to hydroxylysine residues appear to have the same structure in all collagens examined (KATZMAN et al., 1973). From the viewpoint of molecular phylogeny, however, they belong to the second level of phenotypic characters, i.e. the rank of "indirectly homologous" structures (see Chap. 1). The inference of "indirect homology" rests upon the plausible, but as yet undemonstrated "direct homology" of the enzymes involved in producing the secondary chemical transformations. However, evidence for enzyme homology, based upon comparisons of catalytic properties, is emerging slowly.

The precise contribution of modified residues in the α chains of the collagen molecule to fibril structure remains largely unknown (MORGAN et al., 1970). Secondary transformations are clearly of great significance and closely related to physiological adaptation of tissues for specific functions during development and adult life of organisms. A major contribution to our understanding of collagenous tissues can be made by comparative studies of the enzymes (and associated control mechanisms) responsible for producing the secondary chemical modifications of collagen molecules.

14. Development and Aging

Biological research is just approaching the threshold of meaningful investigation of the collagen molecule in differentiation, embryonic growth, maturation, and aging of connective tissues. The problems that arise are of great complexity and biological range (WEISS, 1961) and fall largely outside the scope of this monograph. However, certain temporal aspects of the structure of collagen molecules deserve brief mention here because of their critical importance. The interactions of collagen with other components of connective tissues, which appear to be dependent upon stage of embryonic development, growth, and aging, will be discussed in a subsequent chapter.

One of the primary consequences of differentiation of the connective tissue cell is apparently the biosynthesis of tissue-specific α chains. It is possible that the types of α chains produced vary not only with embryonic development but also with adult age in each tissue type. The variations in types of α chains referred to here relate to structural gene-determined primary structure. Such secondary chemical changes as aldehyde formation and hydroxylation of proline and lysine are disregarded. The biological significance of these and other structural variations with time is unknown. We have yet to understand the functional differences between various types of collagen molecules, for example, those in skin and in cartilage of the same organism (LINSENMAYER et al., 1973).

Shortly after secretion by the cell, collagen molecules become involved in formation of fibrils and fiber bundles as components of the connective tissue matrix. Subsequently, maturation results in a "normal" degree of covalent cross-linking of collagen, producing a general increase in stability toward external influences such as swelling, extractability, denaturation and enzymic attack. Students of aging of connective tissues have directed their attention largely to the time-dependent alterations of fibrillar collagen that occur in late life stages (see reviews by COMFORT, 1970; HALL, 1969; SINEX, 1968; CHVAPIL, 1967). Commonly, the maturation and aging processes of collagen are not clearly differentiated. Although the formation of intermolecular cross-links (as measured by the decrease in reducible cross-links) appears to be virtually complete in collagen from skin, tendon and cartilage of cows and humans at maturity, the cross-linking process in rat tissues, although extensive at 6 to 8 months of age, is still continuing even up to 3 years of age. This phenomenon is probably related to the continued biosynthesis of collagen well into the adult life of this animal. BAILEY and SHIMAKOKOMAKI (1971) concluded that this type of cross-linking must be an essential part of the growth process rather than an effect of aging.

Functional considerations have generally been confined to physical properties (VIIDIK, 1969); processes leading to "excessive" cross-linking have been emphasized primarily. However, PIEZ (1969) noted that the exclusive concern with increasing cross-linking obscured the consideration of other, possibly more important changes, such as the breaking of cross-links or peptide links with time. The slow accumulation of structural "defects" due to either type of process in such a protein of low metabolic turnover as collagen (and elastin) could lead to disturbances of function.

Chapter 4

Elastin

1. Identification

Elastin is a fibrous structure in connective tissues of higher vertebrates varying in appearance from thick fibers having a diameter of 4 to 5μ to fine fibrils near the limit of resolution in the light microscope (see reviews by AYER, 1964; BANGA, 1966; ROSS and BORNSTEIN, 1971; FRANZBLAU, 1971; ROSS, 1973). Elastin has been differentiated from collagen by its selective affinities for orcein, and other dyes, by the presence of branchings, and by the possession of a high degree of elasticity. X-ray diffraction patterns show only amorphous rings; no periodicity is evident in the electron microscope. The fibers are complex structures intimately associated with collagen and other tissue components.

Elastin is difficult to define. As the result of its chemical inertness, it is considered to be the protein material that remains after all other proteins have been removed. The methods used to remove contaminants have involved exposure of tissues to proteolytic enzymes, hot alkali, hot acids; and autoclaving temperatures. It seems likely that some degradation of the chemical structures occurred under such drastic conditions. Milder treatments, including the use of collagenase, are partially effective (GRANT et al., 1971).

JACKSON and CLEARY (1967) suggested that elastin be defined by a characteristic average amino acid composition. However, the recent isolation (SANDBERG et al., 1971; SMITH, BROWN, and CARNES, 1972; SYKES and PARTRIDGE, 1972) of a soluble protein precursor of insoluble elastin, for which the name "tropoelastin" has been suggested, offers hope that elastin will be defined less ambiguously in the near future. Tropoelastin is free from neutral sugars, hexosamines, tryptophan, cysteine, cystine, methionine and histidine.

2. Primary Structure

A description, or definition, of elastin that is based upon composition (see Table 4.1) reveals certain unusual features: 1. a great preponderance (about 95%) of nonpolar amino acids; 2. a very high glycine content near one third of the total residues (as in collagen); 3. the presence of 1 to 2 residues per 100 of hydroxyproline; and 4. the presence of two unique components, desmosine and isodesmosine. While there is generally good agreement between analytical values obtained by many investigators, the differences that exist indicate that the problem of the composition of insoluble elastin remains unsettled. It is not clear whether the difficulties are due to degradation or to contamination (JOHN and THOMAS, 1972).

Table 4.1. Amino acid composition of elastins (res./1000)

Amino acid	Porcine Aorta[a]	Tropo-elastin[a]	Bovine Aorta[b]	Bov. lig. Nuchae[b]	Bov. lig. nuchae[c]
Asp	8.8	2.9	8.2	5.0	5.4
Hypro	14.5	11.2	11.4	7.3	11.3
Thr	7.4	13.8	7.1	5.2	8.6
Ser	8.1	9.4	9.0	7.7	9.4
Glu	20.9	18.3	16.3	14.9	14.7
Pro	93.8	108.7	104.0	110.9	110
Gly	328.9	333.7	325.2	323.0	324
Ala	233.3	218.1	224.7	221.5	223
Cys (1/2)	0.0	0.0			5.0
Val	124.9	120.6	143.5	148.2	140
Met	1.7	0.0	tr	tr	
Ile	19.6	18.8	26.6	27.4	26.2
Leu	57.4	46.2	64.4	66.1	67.8
Tyr	16.9	15.6	8.7	11.7	8.2
Phe	32.3	27.9	25.6	28.2	30.9
Ide (1/4)	8.1	0.0	4.1	4.5	4.1
Des (1/4)	8.3	0.0	5.9	6.4	6.5
Lys	7.6	47.5	5.5	3.7	7.6
His	1.1	0.0	0.9	tr	0.7
Arg	6.6	7.1	5.2	4.4	5.2

[a] SANDBERG, WEISSMAN, and SMITH (1969)
[b] Alkali-treated preparation, STEVEN and JACKSON (1968)
[c] Enzyme-treated preparation of fetal tissue, ROSS and BORNSTEIN (1969)

Partial sequencing of tryptic peptides from tropoelastin reveals some unique features of the primary structure (FOSTER et al., 1973). Whereas the glycines in collagen (except in the short telopeptide regions) are restricted to a triplet repeating pattern (essential for forming the triple helical structure) those in elastin are not. Also, hydroxyproline is a real constituent of the elastin chain, occurring in the same local peptide environment as in collagen (-Gly-X-Hypro-Gly-), but not as part of an extended collagen-like structure.

The N-terminal sequence of tropoelastin is Gly-Gly-Val-Pro-Gly-Ala-Val-Pro-Gly-Gly-Val-Pro-Gly-Gly-Val-Phe-Phe-Pro- (GRAY et al., 1973). Several repeat amino acid sequences occur in tropoelastin (GRAY et al., 1972). One such sequence, -Ala-Pro-Gly-Val-Gly-Val, is repeated 5 times in consecutive order and the -Pro-Gly-dipeptide recurs more than 35 times in some 350 residues. URRY et al. (1974) have demonstrated that linear penta, deca- and pentadecapeptides of this sequence from stable β-turns with Pro-Gly dipeptides at the corners. They propose (URRY, 1974) that the single-chain conformation of tropoelastin consists of a series of α-helix and β-spiral segments. The basic fibrillar unit has up to six parallel-alligned chains held together primarily by cross-linking at the α-helical regions, by hydrophobic forces, and to a lesser extent by hydrogen bonding. In this connection it should be noted that elastin also contains peptide sequences that consist almost exclusively of alanine and that may occur in the α-helical conformation. Such sequences are present in the cross-link region (FOSTER et al., 1973).

3. Secondary Chemical Features

a) Cross-Links

The two polyfunctional amino acids, desmosine and isodesmosine, serve as the main cross-links between peptide chains in elastin (Fig. 4.1).

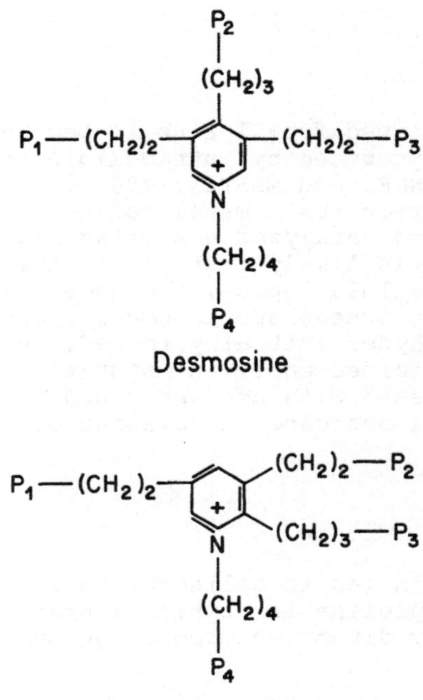

Fig. 4.1. Structure of cross-links in elastin. P = polypeptide chain

Elastin has been shown to contain also two kinds of intermediate cross-links, dehydrolysinonorleucine and dehydromerodesmosine as well as their naturally reduced derivatives, lysinonorleucine and merodesmosine (PAZ et al., 1971). Rather good evidence has been presented for the biosynthetic derivation of each of the cross-links from lysine residues that first undergo oxidation to aldehydes. These, in turn, react with other residues to form aldimine, aldol, or other types of condensation structures. Further transformations may then occur by oxidative processes, including dehydrogenation, as well as by reductive processes (GALLOP et al., 1972; FRANCIS et al., 1973).

Tropoelastin, has a molecular weight of about 68,000 and contains 38 lysine residues in a single polypeptide chain of about 800 residues. Since mature elastin contains only 6 unaltered lysines and only 12 to 14 lysine residues have been consumed in the formation of cross-links, there remain 18 to 20 lysine residues unaccounted for. At least some of these residues are converted into various intermediates of the desmosine cross-links and/or into independent cross-links. The lysine residues involved in desmosine and isodesmosine formation occur in sequences of the type, Ala-Ala-Ala-Lys and Ala-Ala-Lys and two such peptides occur in juxtaposition to yield sequences of the type Lys-Ala-Ala-Lys-Ala-Ala-Ala-Lys. Cross-links are formed from the lysines of two such longer sequences present in different tropoelastin molecules, but may also occur between sequences in the same molecule.

The possible presence of cross-links other than the desmosines and their intermediates was suggested by the presence in elastin preparations of dityrosine (LA BELLA et al., 1967). This compound is associated with glycoproteins (see Chap. 5) that are tightly bound to elastin (KEELEY et al., 1969; KEELEY and LA BELLA, 1972; GOTTE et al., 1963).

b) Aldehydes

The primary aldehyde derived from lysine in peptide linkage, α - aminoadipicsemialdehyde, is produced by intracellular reactions catalyzed by an amine oxidase (PINNEL and MARTIN, 1968). The enzyme is dependent upon molecular oxygen, requires a metal cofactor, probably copper SIEGEL et al., 1970), and catalyzes a similar reaction in collagen (Chap. 3, Sect. 11). It is likely that, after the cell secretes tropoelastin into the extracellular space, the sequence of condensations leading to cross-links proceeds spontaneously, apparently involving most or all of the aldehydes initially formed. Thus, a maximum content of the desmosines is attained early in maturation of mammals (PIEZ, 1969). However, an increase with age was found in rat skin (BARTOS and LEDVINA, 1971) and a decrease at advanced age in human skin (JOHN and THOMAS, 1972).

c) Hydroxylation

Hydroxyproline in elastin (as in collagen) derives primarily by an oxidation process from proline in peptide linkage and only in a very minor way, if at all, by direct incorporation of hydroxyproline.

4. Fibril Organization

Various views are held on the molecular organization of insoluble elastin and its derivation from precursor polypeptide chains. Elastin has long been thought of as a tight three-dimensional chain network (PARTRIDGE, 1962). Apparently, hydrophobic bonding is important in maintaining three-dimensional stability. Hydrophobic interactions are also involved in the reconstitution of solubilized elastin (α-elastin) and tropoelastin to filamentous structures similar to those of native elastin (COX et al., 1973, 1974; STARCHER et al., 1973). PARTRIDGE (1969) and collaborators attempted to visualize how the four lysine side-chains of separate tropoelastin molecules could be brought into precise conformation so as to form desmosine cross-links (Fig. 4.1).

They proposed that globular precursor molecules assembled first in a semicrystalline order which, upon cross-linking, formed a three-dimensional gel with a regular system of pores. This model was supported by electron micrographs of swollen elastin which showed an array of globular particles with open pores of 30 to 40 Å diameter.

However, the globular protein model does not appear to fit the data on X-ray diffraction, circular dichroism, and elastic properties of elastin (PIEZ, 1968; MAMMI et al., 1968; HOEVE and FLORY, 1958). PIEZ (1968) suggested, as a compromise of the conflicting viewpoints, that the elastin fiber, initially composed of a regular array of identical units, is converted during or after cross-linking by a process equivalent to denaturation into a randomly coiled, cross-linked network of chains. The proposed transformation is analogous to the process that occurs in fibrous collagen upon thermal denaturation and shrinkage and results in a great increase in elasticity of the structure. A "liquid drop" model, for which interfacial forces are the only restoring forces, was proposed by WEIS-FOGH and ANDERSEN (1970). However, the thermodynamics of stretching and the hydration of elastin are explained best by a general type of structure, which was termed an "oiled coil" by GRAY et al. (1973). The proposed molecular model, which was based upon sequence data, consisted of cross-linked polypeptide chains containing β turns as well as sequences in α-helical conformation.

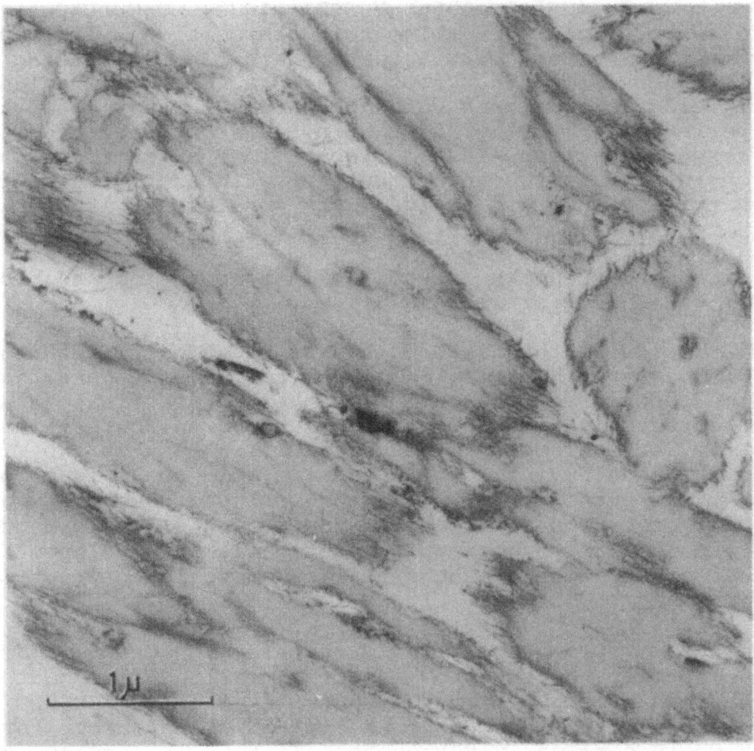

Fig. 4.2. Electron micrograph of a preparation of elastic fibers from ligamentum nuchae of term fetal calf (ROSS and BORNSTEIN, 1971; photograph by courtesy of Dr. R. ROSS)

The mature elastic fiber is a complex structure containing several components. During embryological development of elastic tissues a collagen fibril network appears first in the extracellular matrix. After 2 months of fetal development, the elastic fibers now present consist of small bundles of microfibrils that contain within their centers amorphous material (Fig. 4.2) destined to increase greatly in amount relative to the microfibrils. ROSS and BORNSTEIN (1970) and VARADI (1972) showed that the microfibrils were a second component of the elastic fibers, differing from both collagen and elastin in amino acid composition. Such additional macromolecular components are glycoprotein in nature, are intimately associated with elastin in the elastic fiber, and contribute to its mechanical properties. It should be noted also that the narrow-diameter spaces of the elastin network (PARTRIDGE, 1969) could entrap macromolecules mechanically and thus prevent their separation from elastin, except by chemically degradative means.

5. Comparative Biology

Elastin occurs in association with collagen in tissues where it functions as an elastic restraint. It is prominent in the nuchal ligament of the necks of ruminants and the aorta near the heart, but it is a minor component in most other connective tissues. Elastins from aortas of different mammals (GRANT, 1966; VARADI, 1972; MC CULLAGH et al., 1973) differ only slightly in amino acid composition, with the content of hydroxyproline, acidic amino acids and arginine increasing in the lower part of the vessel. The proportion of acidic amino acids in human aortic elastin increases with age. However, in view of the difficulty experienced in purifying elastin, it is possible that the reported differences between elastin preparations are not due entirely to variation in primary chemical structure with anatomical site and with age. Partial amino acid analyses are consistent with the presence of elastin in carp swim bladder (GROSS et al., 1958).

Aorta tropoelastin of chicks is similar in composition and properties to that of pigs (RUCKER et al., 1973).

HASS (1939) considered the distribution of elastin in animal species solely on the basis of histological observations and found that evidence for the occurrence of elastin in invertebrates is contradictory. Elastin was detected in tissues of all vertebrates, but stained weakly and atypically in the lamprey. These identifications in the microscope require confirmation by isolation and amino acid analysis of purified preparations. The presence of the desmosines could serve as an important secondary criterion for elastin. However, the development of a satisfactory molecular phylogeny of elastin will depend upon the isolation and characterization of the specific monomeric polypeptides, i.e. tropoelastins.

The evolutionary origin of elastin is obscure (see Chap. 3, Sect. 13d). An analogous elastic protein from insects, resilin, also contains dityrosine, but has a composition quite unlike elastin (ANDERSON, 1964). Our present meager data indicate that elastin was probably formed for the first time in some early vertebrate. If this is correct, the new vertebrate protein may have derived from some related, but functionally different, invertebrate antecedent. Strong similarities between collagen and elastin (such as high contents of glycine, alanine, and proline, the presence of hydroxyproline, common initial steps leading to cross-links, and the acquisition of elastic properties by collagen fibers upon thermal shrinkage) suggested to PIEZ (1968) that collagen may have been the evolutionary precursor of elastin (see Fig. 3.9). Also, amino acid differences between the two proteins could be due to substitutions resulting from point mutations for residues coded by nucleotide triplets of the type XCX (see Chap. 3, Sect. 13b). This is suggested by the observation that Σ(XCX), i.e. the sum of residues of proline, hydroxyproline, serine, threonine and alanine (see Table 3.8) is nearly the same in mammalian collagen (391) and in elastin (361). However, the partial data on amino acid sequences (Sect. 2) indicated that a genetic relationship between the two proteins cannot be accounted for by this simple point-mutation hypothesis. It is likely that the biochemical similarites are the result of evolutionary convergence from a diverse ancestry. A phylogenetic association between silk fibroin and elastin appears more plausible in view of the close relationship of repeating sequences in these two proteins (GRAY et al., 1973).

Chapter 5

Structural Glycoproteins

A great many proteins that are widely diverse in chemical structure, have been loosely classified as glycoproteins solely by virtue of a content of covalently bound carbohydrate residues (SPIRO, 1970). However, only a small number of such proteins are known to be present in significant amounts in the extracellular phases of tissues that have structural functions. These proteins fall into three large groups with unique characteristics that require consideration in individual categories. One such group includes the collagens described in Chap. 3; the acid polysaccharide-protein group will be described in subsequent chapters. This division reduces considerably the list of heterogeneous macromolecules that may be considered "structural glycoproteins" (ROBERT et al., 1972).

The chemistry, metabolism and functions of glycoproteins have been reviewed by HUNT (1970) for invertebrates, and by SPIRO (1970), MARSHALL and NEUBERGER (1970), MARSHALL (1972), SCHMID (1972), and GOTTSCHALK (1972) for vertebrates. Summary descriptions of glycoproteins in specific connective tissues have been given for blood-vessel walls (MUIR, 1965), cartilage, tendon and intervertebral disk (SCHUBERT and HAMERMAN, 1965), skin (PEARCE, 1965), kidney basement membrane (SPIRO, 1965a), eye (DISCHE, 1970a,b) ear (BALAZS, 1965), and bone and teeth (BALAZS and ROGERS, 1965; HERRING, 1968a, 1972).

1. Characterization

Incomplete knowledge of the highly complex structures of glycoproteins makes the tasks of characterization and definition very difficult and any present classification necessarily tentative. To increase the likelihood that the macromolecules to be discussed will fall into homologous groups, it is prudent to limit consideration to those glycoproteins present in specific tissues of clear structural character. However, in doing so, we necessarily omit a large number of glycoproteins which serve primarily lubricatory, protective, enzymatic, hormonal, and other functions. A few of the latter glycoproteins may well be either homologous to, or even identical with, some structural glycoproteins.

a) Definition

The structural glycoproteins are proteins having one or more heterosaccharide chains. The protein moiety is quantitatively the most important constituent. Each heterosaccharide chain, of molecular weight from 500 to 4000, contains up to seven different monosaccharide residues, often linked in a highly branched manner. The most common constituent is hexosamine and either N-acetyl-D-glucosamine, N-acetyl-D-galactosamine, or both. In addition, some or all of the neutral monosaccharides, D-mannose, D-galactose, L-fucose, D-xylose, L-fucose,

D-xylose, L-arabinose, and less commonly D-glucose, may be present. A sialic acid is an important component of glycoproteins, always linked glycosidically in a terminal position of the heterosaccharide chain. Ester sulfate may be present also. The glycoproteins are distinguished from the acid polysaccharide-proteins, which contain much smaller proportions of protein. Also, the latter possess linear polysaccharide side-chains with polymeric structures constructed of disaccharide repeating units that commonly contain a uronic acid, a hexosamine, and an ester sulfate residue.

Keratan sulfate-peptides of cartilage present a problem in classification. Although their polysaccharide side-chains consist principally of disaccharide repeating units of sulfated galactosyl-N-acetylglucosamine, they also contain other neutral sugars, N-acetylgalactosamine, and sialic acid. For traditional, and other reasons, these molecular fragments are grouped with the acid polysaccharide-proteins and hence will be considered in subsequent chapters. However, some workers prefer to include them among the glycoproteins.

b) Structure

The difficulties of isolation, purification, physical characterization, and determination of primary structure of glycoproteins are greatly intensified by the polymorphism of polypeptide sequences and the microheterogeneity of carbohydrate chains. As a consequence, although a large number of glycoproteins have been prepared in a high state of purity, few have been investigated extensively, and none has had its structure completely elucidated.

A glycoprotein molecule consists of a polypeptide chain to which may be attached many carbohydrate side-chains. In many specific instances the oligosaccharide side-chains are not alike in structure. Two types of glycan-protein linkages occur characteristically in glycoproteins: the N-(β-aspartyl) glycosylamine link, and the O-seryl glycoside link. Whereas all proteoglycans contain the latter link, with the exception of keratan sulfate-protein, they do not contain the former link. Microheterogeneity of the oligosaccharide chains is also a common occurrence. This phenomenon is principally due to variation in the number and type of monosaccharide composing each oligosaccharide chain, but may also result from positional isomerism of glycosidic linkages without compositional changes. Internal homologies may occur in the polypeptide chain (SCHMID, 1972).

In many of the structural glycoproteins, as in the general class of glycoproteins found in vertebrates, sialic acid is an important component. Because the pk value of the average residue is about 2.6, an acidic character is imparted to the molecule. When, as in the case of submaxillary glycoprotein, the content of sialyl residues is very high (25%), electrostatic repulsion of ionized carboxyl groups near neutral pH results in distention of the average conformation of the molecule. This asymmetric conformation causes a large increase in viscosity of aqueous solutions; enzymatic removal of sialyl residues greatly reduces the viscosity increment due to the glycoprotein. Thus, these glycoproteins may be classed as polyanionic polyelectrolytes; they appear similar in general architecture, and possibly in some aspects of physiological function, to the chondroitin sulfate-protein macromolecules (see Fig. 6.3). It may be noted that sulfate groups in a glycoprotein enhance the polyanionic character of the molecule.

c) Biosynthesis

In glycoproteins the amino acid sequence of the polypeptide chain is determined by the usual template mechanism of protein synthesis. However, biosynthesis of the carbohydrate groups appears to be of the nonribosomal type and occurs subsequent to synthesis of the polypeptide. It is directed by a series of glycosyltransferases that transfer sugar residues from their nucleotide derivatives to specific acceptors. The precise structure of the oligosaccharide side-chains is determined by the specificities of the enzyme transferring the initial saccharide to the appropriate amino acid residue of the polypeptide chain and of the enzymes catalyzing oligosaccharide chain elongation and branching (SPIRO, 1970).

GOTTSCHALK (1969) suggested that microheterogeneity of the oligosaccharide moieties is due to small deviations from absolute specificity of the transferases. Other factors, such as enzymatic degradation, may also contribute to this phenomenon.

d) Immunology

Glycoproteins isolated from cornea, sclera, skin, tendon, cartilage, and other connective tissues resemble each other in composition and exhibit immunological cross-reactivity (ROBERT, ROBERT, and ROBERT, 1970; ROBERT, DARREL, and ROBERT, 1970). For these reasons, L. ROBERT and collaborators suggest that such molecules form a natural grouping, or family, of structural glycoproteins. However, because chemical and physical characterization is still far from complete, the precise nature of the relationships between these glycoproteins remains unclear.

e) Biological Function

Various physiological roles have been suggested for structural glycoproteins in different tissues. Glycoproteins may be involved in calcification processes of vertebrates (HERRING, 1969) and may have regulatory functions in morphogenesis of connective tissues (ROBERT, 1966).

2. Invertebrate

The structure of very few glycoproteins of the general class (and none of those that are clearly associated with a structural function in connective or supportive tissues of invertebrates) have been adequately investigated. In a comprehensive review of the available data, HUNT (1970) notes that the invertebrate glycoproteins are generally similar to their vertebrate counterparts except for the usual absence of sialic acid and a greater frequency of occurrence of xylose and arabinose.

Although sialic acid appears to occur universally in the vertebrates, it is found quite irregularly among invertebrate phyla, appearing only in cephalochordates, hemichordates, echinoderms, crustaceans, molluscs and platylhelminths. This sugar is frequently a component of a sialopolysaccharide-protein complex that also contains ester sulfate. A recently well-characterized preparation from sea urchin *(Pseudocentrotus depressus)* eggs contained 70% of sialic acid (mainly N-glycolylneuraminic acid), minor amounts of fucose, hexosamine, glucose, mannose, and galactose, and about 4% protein (HOTTA et al., 1970). The small proportion of protein to carbohydrate excludes this compound from the class of glycoproteins as defined here. A large number of similar substances, some with much higher proportions of protein or with little or no sialic acid, have been isolated from the egg jelly coats of other echinoderms (HUNT, 1970).

Collagen from the sea anemone *(Metridium dianthus)* apparently contains a glycoprotein-like region integral with the molecular structure (Chap. 3, Sect. 6b). Collagenase digestion leads to isolation of a homogeneous fragment containing a heteropolysaccharide consisting of fucose, xylose, mannose, galactose and glucose linked glycosidically to a threonine residue of a large polypeptide with characteristic collagen composition (KATZMAN and JEANLOZ, 1970c). These findings indicate the desirability of more careful investigation of minor carbohydrate components in collagens from other invertebrate phyla.

3. Vertebrate

The available data on structural glycoproteins are insufficient to permit informative comparisons on a systematic basis. Since most studies have been concerned with mammalian tissues, it is preferable at this time to consider structural glycoproteins grouped according to tissue source.

a) Aorta

Glycoproteins have been isolated from aorta of man (BARNES and PARTRIDGE, 1968; BERENSON et al., 1966; SRINIVASAN et al., 1971), cow (RADHAKRISHNAMURTHY and BERENSON, 1973), pig (WAGH and ROBERTS, 1972), horse, sheep (ROBERT, ROBERT, and ROBERT, 1970), elephant (MC CULLAGH et al., 1973), and chicken (KEELEY et al., 1969). The chicken glycoprotein is unusual in containing dityrosine residues. It is firmly associated with elastin (KEELEY and LaBELLA, 1972a).

b) Ligament and Tendon

The glycoproteins of bovine ligamentum nuchae and tendon are closely associated with elastic (see Chap. 4, Sect. 4) and collagenous structures and, hence, are difficult to characterize (GOTTE and SERAFINI-FRACASSINI, 1963; GOTTE et al., 1963; KEELEY et al., 1969; ANDERSON and JACKSON, 1972). They are stabilized principally by disulfide bonds (ROSS and BORNSTEIN, 1971) and contain dityrosine residues (KEELEY and LaBELLA, 1972b).

c) Skin

Glycoproteins of roughly similar amino acid composition (Table 5.1) have been obtained from skin of cattle (BOWES et al., 1958; WOLFF et al., 1971), rat (FURTHMAYER and TIMPL, 1970), and rabbit (TIMPL et al., 1969; BOREL et al., 1970). Glycoproteins of similar chemical composition, which are also associated with collagen, are present in *Ascaris*.

Table 5.1. Amino acid composition of structural glycoproteins (moles/1000 total residues)

Source	Human aorta[a] (BARNES and PARTRIDGE, 1968)	Pig aorta (ROBERT et al., 1970)	Chick aorta (KEELEY et al., 1969)	Bovine ligament[a] (GOTTE et al., 1963)	Bovine ligament (ROSS and BORNSTEIN, 1971)	Bovine ligament (KEELEY et al., 1969)	Bovine skin[a] (BOWES et al., 1958)	Rabbit skin (TIMPL et al., 1969)	Human amyloid fibrils (COHEN, 1970)	Bovine bone (HERRING, 1972)	Bovine cornea (ROBERT et al., 1970)
Glycine	84.0	87.9	115	242.5	110	103	89.8	69	100.0	122	105.3
Proline	69.6	52.1	61	55.2	63.5	87	58.6	44	54.0	49	56.6
Hydroxyproline		0	0	28.7[b]	0	0	5.9[b]	0	0		0
Hydroxylysine		1.3			0			0			11.4[c]
Lysine	52.8	57.2	54	38.8	45.0	62	54.4	50	59.4	29	44.2
Histidine	19.7	22.3	16	12.6	15.4	17	17.0	15	18.9	13	15.9
Arginine	48.9	45.8	40	35.7	45.2	37	49.8	30	59.4	11	59.8
Aspartic acid	99.3	89.8	103	73.9	114	102	119.2	91	94.6	173	102.1
Glutamic acid	112.4	99.5	125	93.9	114	139	112.2	158	108.1	228	113.2
Serine	60.7	56.5	60	57.3	62.2	51	67.7	56	63.5	80	59.3
Threonine	51.1	54.4	37	33.2	55.9	35	52.4	59	43.2	111	50.9
Cystine (1/2)	11.9	26.7	12	5.8	48.2	8		2	18.9	12	15.2
Tryptophan	13.9							19		11	
Methionine	20.6	0.7	12	13.2	15.6	13	8.0	29	23.0		10.8
Alanine	70.9	80.8	85	109.8	65.1	75	72.1	92	87.8	40	70.1
Valine	64.4	73.9	69	62.4	56.3	73	66.9	54	56.7	30	86.3
Leucine	85.8	76.7	93	65.5	68.6	81	102.7	98	81.1	29	96.5
Isoleucine	50.0	43.4	43	33.9	47.7	38	53.5	60	40.5	25	46.9
Phenylalanine	41.8	31.4	39	27.3	37.7	42	37.2	40	44.6	11	37.7
Tyrosine	42.2	28	29	10.4	36.0	31	32.7	34	35.1	22	17.4
Σ (XCX)	252	244	243	256	247	248	250	251	249	280	236.9

[a] recalculated; [b] not included in Σ (XCX), see text; [c] includes 3.8 residues allohydroxylysine.

d) Cartilage

Sialic acid is a common constitutent of glycoproteins as well as of keratan sulfate-proteins of vertebrate cartilage (SCHUBERT and HAMERMAN, 1968; ENOMOTO et al., 1966). Such glycoproteins have been isolated in good purity from ox cartilage (JANADO and DUNSTONE, 1972; VITTUR et al., 1972a, b). They aggregate readily in solution and bind cations strongly (Chap. 10).

e) Bone, Enamel, and Dentin

The organic matrix of bone (see review by HERRING, 1972) contains several different sialoproteins. One of these (Tables 5.1 and 5.2) is closely associated with a chondroitin sulfate proteoglycan. Glycoproteins of tooth enamel matrix are poorly characterized. The phosphoglycoproteins of dentin matrix that are covalently bound to collagen were discussed earlier (Chap. 3, Sect. 9f). Developmental aspects are discussed by A. WEINSTOCK (1972).

Table 5.2. Composition of bovine bone sialoprotein (HERRING, 1972)

	% (w/w)	mole/mole of BSP
Sialic acid	20.5	15.3
Galactose	8.2	10.5
Mannose	2.5	3.1
Fucose	0.7	1.0
Xylose	0	0
Galactosamine	4.6	5.9
Glucosamine	4.6	5.9
Uronic acid	0.0	0.0
Phosphate residues	1.4	3.4
Sulfate residues	0.0	0.0
Amino acid residues	47.9	87.6

f) Cornea

Sialoglycoproteins of complex composition are present in corneal stroma (DISCHE, 1970a, b). The bovine preparation (Table 5.1) contained sulfate as well as hydroxylysine residues. The physicochemical properties of fish cornea differ from those of mammalian cornea (Chap. 10, Sect. 4b). These differences are related to glycoprotein chemistry (ROBERT and SCHILLINGER, 1967; MOCZAR et al., 1969; MOCZAR and MOCZAR, 1970).

g) Brain

Rat brain contains an abundance of glycoproteins that are associated with membranous elements and difficult to solubilize (DiBENEDETTA et al., 1969). An unusual sulfated glycopeptide resembles keratan sulfate-protein in composition (MARGOLIS and MARGOLIS, 1970).

h) Other Sources

Amyloid is a class of glycoproteins that occur extracellularly in connective tissues in disease states (BENDITT and ERIKSEN, 1972) but may also be a normal constituent. The fibrous amyloid proteins of man (COHEN, 1970) resemble those of mouse in morphology and chemical composition (GLENNER et al., 1971).

An unusual glycoprotein precipitates from notochordal fluid of the coelacanth (Chap. 7, Sect. 4c) as fibrils of 100 Å diameter (MATHEWS, unpubl.).

A sialic acid-rich glycoprotein occurs in the extracellular matrix surrounding the rod and cone cell layers of the retina (BACH and BERMAN, 1970).

4. Conclusions

A cursory view of the glycoproteins in Table 5.1 shows a broad similarity in the amino acid composition of their protein moieties (in particular, a high content of acidic amino acids) with the possible exception of those from bone and cornea. The relative constancy of the sum of residues coded by the triplet XCX (serine, threonine, alanine, proline, hydroxyproline; see Chap. 3, Sect. 13b), about one fourth of the total residues, is also consistent with a genetic relationship, albeit remote, between these molecules. Given the present state of our knowledge, one may only hazard a guess that some of these molecules, in particular those associated with the fibrous proteins of aorta, ligament, and skin, will be shown to be homologous by future investigations.

Homologies may extend beyond the group of glycoproteins selected here. Significant biological relationships between macromolecules are probably obscured by the somewhat arbitrary nature of a classification based upon presumed structural functions. In this respect, it is worth mentioning a general group of vertebrate epithelial mucous secretions including, for example, dermal mucins of cyclostomes (LEHTONEN et al., 1966) and bony fish (FLETCHER and GRANT, 1968), and mucins from gastric (PAMER et al., 1968; SLOMIANY and MEYER, 1972), intestinal (NEMOTO and YOSIZAWA, 1969), and colonic (INOUE and YOSIZAWA, 1966) mucosa. The glycoprotein components of such mucins resemble the structural glycoproteins with respect to the types of carbohydrate present and, in addition, contain appreciable ester sulfate. Yet, these mucins occur externally or within body cavities and appear relatively short-lived by comparison with skeletal elements. However, they do have a high viscosity and therefore may be considered to possess a "structural" character, albeit of a somewhat fluid nature.

The structural glycoproteins present challenging opportunities for biochemical investigation because of their complexity and heterogeneity of structure and their unknown biological roles. In addition, recent observations by SRINIVASAN et al. (1971) suggest the possibility of fine genetic analysis of these macromolecules. These workers studied glycoproteins of different organs in sets of both identical and nonidentical twins of cow, sheep, and man by electrophoretic and immunological methods. The same glycoprotein was present in the homologous tissues of homozygotic individuals but not in those of dizygotic individuals. However, the glycoproteins present in the different organs of each individual differed from each other. Thus, the structural glycoproteins appear to be both individual and organ-specific.

Chapter 6

Polyanionic Proteoglycans

In 1925 our total knowledge of the molecular structures of polysaccharide components in connective tissues was presented in a few pages by LEVINE (1925). Chemical data were limited largely to a single substance, chondroitin sulfate. Since then the list of polysaccharides that have been identified chemically and characterized structurally has grown rapidly, particularly within the last two decades.

Initial hopes that the intensive efforts of many workers would soon lead to complete structural characterization of the polyanionic proteoglycans were gradually replaced by a growing realization of previously unsuspected complexities. The macromolecules studied were not only difficult to isolate in a native state in purified condition, but also exhibited types of chemical structural variability that fell outside the range accessible to characterization by available methods. Their present description must therefore encompass an extensive heterogeneity (see especially MEYER, 1970). In order to describe them, it is essential to revise older terminology and to introduce novel descriptive concepts.

Some years ago this chapter might have been entitled "acid mucopolysaccharide-proteins" However, the recent introduction and widespread acceptance of a new nomenclature (now coexisting with the old one) may lead to confusion unless our terminology is clarified. The term 'proteoglycan' as employed here refers to a molecule in which the glycan, or polysaccharide portion, is the major constituent. Such molecules are distinguished roughly from glycoprotein molecules, which contain principally protein. Also, unlike the glycoproteins discussed in Chapter 5, the proteoglycans possess glycan chains having a carbon backbone that consists predominantly of repeating disaccharide periods, usually of a single structural type.

The term "glycosaminoglycan" is equivalent to the older term "mucopolysaccharide" as originally applied to protein-free polysaccharides of animal origin. The new term seems preferable since the hexosamine component is specified explicitly. In view of their content of uronic acid and/or ester sulfate, the polysaccharides from vertebrate connective tissues have been referred to commonly as "acidic glycosaminoglycans." However, at physiological pH the prevailing counterions (cations in this case) are not H^+, but Na^+, K^+, Ca^{++}, etc. Substitution of 'polyanionic' for 'acidic' gives emphasis to the ionic state of the molecule and to the presence of a multiplicity of negative charge sites; such molecules are members of the class of anionic polyelectrolytes (see BETTELHEIM, 1970). Finally, the terms 'polyanionic glycan' and 'glycan' will be used, when necessary, to refer to polysaccharides without reference to the presence of amino sugars.

Complete details of structure for any of the polyanionic proteoglycans of animal connective tissues are not yet available. Nevertheless, we may profitably compare these macromolecules in a limited fashion with respect to specific portions of their structure, such as the region of linkage between carbohydrate and protein, amino acid composition, general features of molecular architecture, immunological properties, and chemical structure of the glycan moiety. We shall limit the discussion here to the broad features and definition of the glycan structures themselves, reserving for subsequent chapters detailed comparisons of these molecular moieties in different species.

1. Characterization as Parent Types

In order to trace evolutionary history from molecular structural data it is essential either to demonstrate or to assume that the particular set of macromolecules studied are genetically related, i.e. are homologous (see Chap. 1). Such relatedness between macromolecues can be established more readily for macromolecules that differ very little among themselves than for those that are highly variable in chemical structure and function. In the latter case, which applies particularly to proteoglycans, it becomes necessary to invent appropriate archetypal models, or *parent* types. When this is done, many obscuring details due to chemical features that are variable with such factors as developmental stage, physiological state, etc., are removed. We may than compare "directly homologous" features and postpone for the moment detailed comparisons of "indirectly homologous" features.

a) Glycosaminoglycans

The best known polyanionic glycans, i.e. those present in vertebrate connective tissues, are all glycosaminoglycans. With the possible exception of hyaluronate (hyaluronic acid), these polysaccharides are present in tissues as segments of larger macromolecules, the proteoglycans, which consist of carbohydrate chains covalently linked via a linkage region to protein. Protease treatment of tissues permits the release of glycosaminoglycan molecules which also contain the linkage region and a small amount of residual peptide. Methods of isolation, analyses and charcterization are described in detail by RODÉN et al. (1972a). The chemical structures of these polysaccharide segments have been reviewed extensively by MEYER (1969), JEANLOZ (1970), RODÉN (1970), and LINDAHL and RODÉN (1972). The reader is referred to their articles for fuller details and bibliography. Although various elements are associated in trace amounts with glycosaminoglycans, with the exception of silicon (SCHWARZ, 1973) their contribution to covalent structure has been neglected.

A molecule of a polyanionic glycosaminoglycan may be generally described as a linear-chain polymer (although a small degree of branching cannot be rigorously excluded) consisting principally of a sequence of disaccharide repeating units. One component of the characteristic disaccharide is a hexosamine (with amino group acetylated or sulfamated); the other component may be a neutral sugar or a hexuronic acid. At least one anionic group is present, either carboxyl or sulfate.

The best-known classes of polyanionic glycosaminoglycans of vertebrate connective tissue are listed in Table 6.1. It is evident that within each class considerable variation occurs in monosaccharide components, O-sulfate, N-sulfate, and N-acetyl groups. For this reason, it is essential to define a class-type polysaccharide. This may be done by considering a polymer constructed entirely of one of the types of re-

Table 6.1. Composition of polyanionic glycosaminoglycans of vertebrate connective tissues[a]

Type	Tissue	Repeating period monosaccharides	Sulfate M/period	Intrinsic viscosity range[b]	mol.wt. range (X 10^3)	Remarks
1. hyaluronate	vitreous humour, synov. fluid, umbil. chord, skin, aorta	D-glucuronic acid, D-glucosamine	0	0.2-48	4-8,000	
2. chondroitin 4-sulfate	cartilage, bone, cornea, notochord, skin	D-glucuronic acid, D-galactosamine	0.1-1.0	0.2-1.0	5-50	
3. chondroitin 6-sulfate	cartilage, umbil. chord, skin, aorta	D-glucuronic acid, D-galactosamine	0.2-1.3	0.2-1.8	5-50	
4. dermatan sulfate	skin, tendon, aorta	L-iduronic acid, D-galactosamine	1.0-2	0.5-1.0	15-40	also contains D-glucuronic acid
5. corneal keratan sulfate	cornea	D-galactose, D-glucosamine	0.9-1.7	0.2-0.5	4-19	
6. skeletal keratan sulfate	cartilage, nucleus p., annulus	D-galactose, D-glucosamine	1.1-1.8	0.2-0.5	8-12	also contains D-galactosamine, sialic acid, and methyl pentose
7. heparan sulfate	aorta, beef lung liver amyloid	D-glucuronic acid D-glucosamine	0.1-2.8	-	-	also contains L-iduronic acid, variable in N-acetyl, N-sulfate, O-sulfate residues
8. heparin	lung, liver, skin	D-glucuronic acid D-glucosamine	2-3	0.1-0.25	7-16	also contains L-iduronic acid; variable in N-sulfate, O-sulfate residues

[a] See references to reviews.
[b] See Table 10.1 for relationships between intrinsic viscosity and molecular weight of each glycan.

Chondroitin 4- sulfate
Ch4S (CS-A)

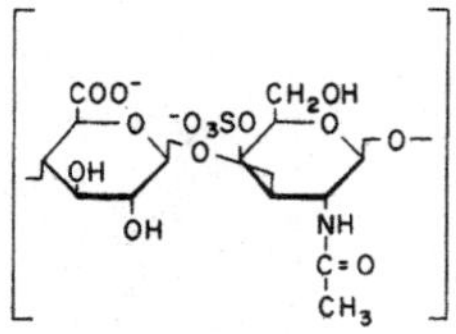

Chondroitin 6 - sulfate
Ch6S (CS-C)

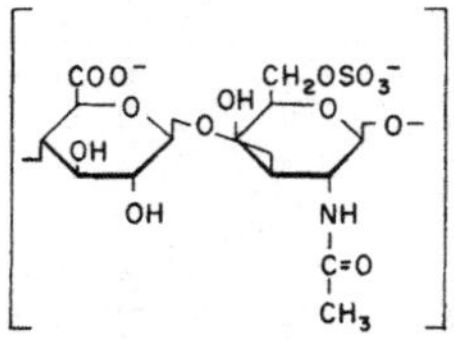

Dermatan sulfate
DeS (CS-B)

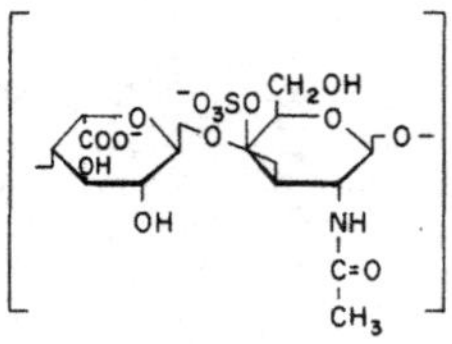

Hyaluronate
Hya

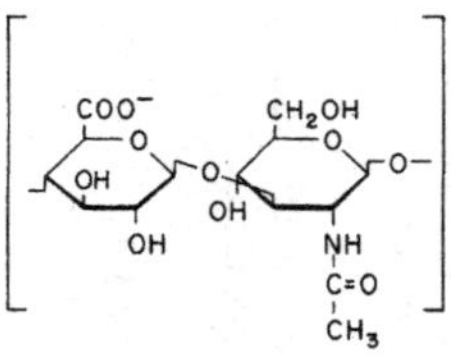

Fig. 6.1. Disaccharide repeating periods of *parent* type polyanionic glycosaminoglycans. Extra ester sulfate groups may exist on the *parent* (glycan) repeating period of Ch 6-S either at positions C-2 or C-3 of the glucuronic acid residue, leading to the repeating period, CS-D, or at position C-4 of the galactosamine residue, leading to the repeating period, CS-E. Similarly, an extra ester sulfate group may occur on the *parent* (glycan) repeating period of dermatan sulfate either at the C-2 or the C-3 of the iduronic acid residue (MALMSTROM and FRANSSON, 1971) to produce the repeating period, Derm-D, or at position C-6 of the galactosamine residue, to yield the repeating period, Derm-H (SUZUKI et al., 1968; ANNO et al., 1971). The occurrence of both types of sulfate substitution leads to an as yet unnamed repeating period with a total of three sulfate groups (AKIYAMA et al., 1971; SENO et al., 1972). A disaccharide repeating period that differs from that of chondroitin 4-(or 6-) sulfate only in lacking ester sulfate is present in chondroitin (Ch). It may occur to a variable extent together with *parent* repeating periods of chondroitin 4-(or 6-) sulfate in the same polymer chain. The unsulfated homopolymer, consisting only of Ch periods, has been isolated from squid skin (Chap. 9, Sect. 1) but its presence in vertebrate tissues is uncertain (KERN and BRASSIL, 1967; BERMAN, 1970; HOPWOOD, 1972).

Although the repeating period and the related *parent* homopolymer are designated here by the same term, they are always distinguishable by the context. Explicit designations for disaccharide periods appear in Table 7.1. The older, trivial name, hyaluronic acid, is replaced by the name, hyaluronate

peating periods of Figs. 6.1 and 6.2 to be a parent-type macromolecule (the linkage region is omitted from this definition).

Keratan Sulfate
KeS

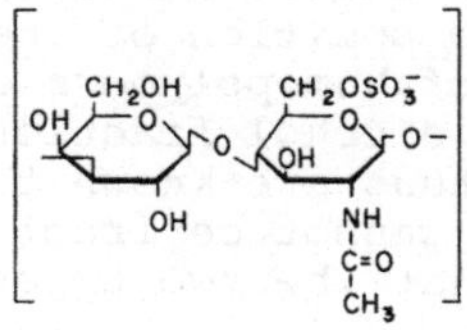

Heparan Sulfate
HeS

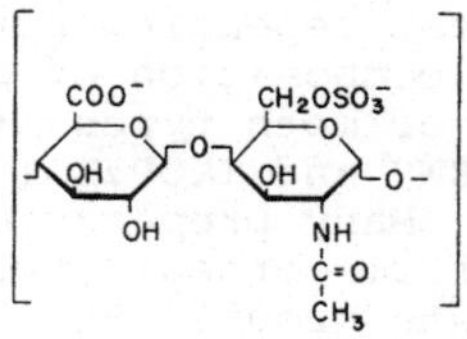

Fig. 6.2. Disaccharide repeating periods of additional *parent*-type polyanionic glycosaminoglycans (see text for details)

Actual molecules will rarely, if ever, contain only a single type of repeating period. Thus, it has become necessary to classify generally the actual polysaccharide preparation according to the predominent type of disaccharide repeating period present. Usually, for convenience, small (and hence difficult to detect) differences in the proportions of different repeating periods between molecules are overlooked. However, even if all the polymer molecules in a preparation had identical sizes and also contained identical proportions of two different repeating periods, extensive polymorphism could arise from differences in the ordering of repeating periods. This complexity also is usually neglected.

i) Hyaluronate

Hyaluronate is in many respects the simplest of the glycosaminoglycans. Preparations from a variety of sources (including bacterial) appear to consist virtually entirely of the parent-type disaccharide repeating period. A linear structure of the polymer is supported by electron microscopy (FESSLER and FESSLER, 1966) and by physicochemical studies of dilute solution properties (CLELAND and WANG, 1970).

ii) Chondroitin 4-Sulfate, Chondroitin 6-Sulfate

It is notable that all the parent-type repeating periods shown in Fig. 6.1 are similar with respect to the glycosidic linkages, i.e. the glycosyluronic acid linkage is β-(1 → 3) and the hexosaminyl linkage is β-(1 → 4). It is not surprising, therefore, that these linear polyelectrolytes exhibit generally similar polymer chain behavior (MATHEWS, 1959), with hyaluronate molecules possessing the least chain flexibility (CLELAND, 1970; BETTELHEIM, 1970). Such comparisons, however, may be valid only for the homopolymers, i.e. those having a single type of disaccharide repeating period.

Chondroitin 4-sulfate and chondroitin 6-sulfate periods are isomers, differing only in the position of the ester sulfate group. Although artificial mixtures of homopolymers of the two types of periods can be separated well by alcohol fractionation, actual preparations of chondroitin sulfate that are known to contain both types of disaccharide periods usually cannot be fractionated in this respect. It seems likely, therefore, that the two types of periods generally occur in the same molecule.

A particular polysaccharide preparation known to be composed only of the two types of parent repeating periods may then be characterized by an average molecular composition of such periods. Methods of assay of the relative amounts of both types of period have been devised by MATHEWS (1958), MATHEWS and INOUYE (1961), SAITO et al. (1968) and SUZUKI et al. (1968). Many preparations, particularly those from embryonic tissues and from corneal stroma, contain repeating periods devoid of ester sulfate groups. Such periods (designated Ch or D-CS) are detectable qualitatively (MATHEWS and DECKER, 1968a) and determined quantitatively by the just mentioned quantitative methods.

It should not be assumed that the three types of repeating period, Ch 4-S, Ch 6-S, and Ch, are either present in identical proportion in every molecule of a given preparation or even randomly distributed in every molecule. Thus, considerable polymorphism may be expected for the polysaccharide chains, although the actual extent of variability is uncertain. For a preparation containing equal proportions of Ch 4-S and Ch 6-S in molecules having 50 repeating periods, the number of random permutations is very great, approximately 1.2×10^{14}. If, in addition, periods of Ch are introduced to replace an equivalent number of Ch 4-S and Ch 6-S in a molecule of the same size, the number of possible permutations increases greatly and may exceed the number of molecules in a milligram of preparation. However, the degree of randomness is modified, at least partially, leading to preferred distributions of ester sulfate groups along the polysaccharide chains (WASTESON and LINDAHL, 1971).

A further complication in structure results from the occurrence of ester sulfate on C-2 or C-3 of the uronic acid residue of Ch 6-S periods as well as on the galactosamine residue (see Legend, Fig. 6.1).

iii) Dermatan Sulfate (Chondroitin Sulfate B)

Dermatan sulfate differs in structure from chondroitin 4-sulfate in containing L-iduronic acid, which is the C-5 epimer of D-glucuronic acid. While L-iduronic is the major uronic acid present, D-glucuronic acid is also an integral component of the same chain, a circumstance which results in a hybrid, i.e. copolymer molecule. Dermatan sulfate parent repeating periods form hybrid polysaccharides with Ch 4-S parent repeating periods and with Ch 6-S parent repeating periods in preparations from pig skin and from human umbilical chord, respectively (FRANSSON, 1968, 1970). The distribution of parent-type periods may occur in blocks, or series of like-repeating periods, in the glycan chain (FRANSSON and MALMSTROM, 1971). Usually, the content of D-glucuronic acid does not exceed 20% of the total uronic acid content of the preparations. It should be noted that, according to convention, the L-idosyluronic acid linkage is an α glycosidic bond.

The presence of more than one type of parent repeating period in dermatan sulfate molecules results in polymorphism, as noted above for chondroitin sulfates. In addition, 'extra' ester sulfate groups may be present on both the uronic acid and the galactosamine residues (see Legend, Fig. 6.1).

iv) Keratan Sulfate

The parent disaccharide repeating period and its variants that are shown in Fig. 6.2 are major components of two types of macromolecule, viz. corneal keratan sulfate and skeletal (cartilage and nucleus pulposus) keratan sulfate. The major differences between these two classes of keratan sulfates arise from the presence of carbohydrate residues other than D-galactose and N-acetylglucosamine and in the structure of the linkage region. Sialic acid, fucose, mannose and additional galactose are found in small and variable amounts in corneal preparations and in considerably greater amounts in preparations from skeletal tissues. The sialic acid from whale cartilage keratan sulfate was identified as N-acetylneuraminic acid (TODA and SENO, 1970). Galactosamine may comprise 10-20% of the total amino sugar in skeletal keratan sulfate preparations but is absent from corneal preparations. Cartilage keratan sulfate chains have a complex, branched structure and have multiple links to the peptide moiety.

The average sulfate content of different preparations is highly variable. Sulfate groups may be lacking from the C-6 position of the glucosamine residue and extra sulfate groups may be present on the C-6 position of the galactose residue. Low molar ratios of sulfate to hexosamine (less than 1.0) are common in corneal preparations while molar ratios up to 1.8 may be found in fractions from skeletal preparations.

Keratan sulfate, particularly in proteoglycan preparations from skeletal tissues, shows a closer resemblance to glycoproteins than to the other polyanionic glycosaminoglycans with which it is generally included. It is the only member of the class having no uronic acid residues. Also, the presence of sialic acid, fucose, mannose, and glycosidic linkages to asparagine (see Sect. 2b) are features characteristic of glycoproteins and absent from other vertebrate connective tissue glycosaminoglycans.

Fractions of keratan sulfate from whale cartilage are heterogeneous in composition with sialic acid content inversely related to sulfate content and molecular weight (NAKAGAWA and SATAKE, 1971). This is understandable since most of the sialic acid appears to be at the nonreducing terminus of the chain (UENO and KITAMIKADO, 1972) while the neutral sugars are concentrated at the linkage region to peptide (HIRANO and MEYER, 1971) and are probably little sulfated.

v) Heparan Sulfate

The structure of the repeating periods of heparan sulfate (Fig. 6.2) is not firmly established. A number of variations in structure are known to occur. Various preparations of heparan sulfate may have N-acetylglucosamine contents between 40 and near 100% of the total hexosamine, with the remainder of the hexosamine accounted for by N-sulfate groups. The N-acetylglucosamine residues are located primarily in the vicinity of the carbohydrate-protein linkage. L-iduronic acid may also be present and some variations in molar ratio of sulfate to glucosamine residues are found (LINKER and HOVINGH, 1973; TAYLOR et al., 1973).

vi) Heparin

Although heparin preparations contain less than 40% N-acetylglucosamine and a high proportion of iduronic acid residues they are assigned the same parent repeating period as heparan sulfate. This selection may be justified further on biosynthetic grounds (see later). Heparin is virtually completely sulfated at the C-6 and the N-positions of glucosamine residues. In addition, partial sulfation of the C-2 position of uronic acid residues occurs. However, it appears that the majority of these ester sulfate groups are not bound to glucuronic acid residues but to iduronic acid residues which partially replace glucuronic acid residues in heparin (LINDAHL and AXELSSON, 1971). Indeed, iduronic acid residues may represent the major uronate constituents of some heparin preparations (PERLIN et al., 1970). A large portion, possibly all, of the glucosyluronic acid linkages have the β-D configuration (HELTING and LINDAHL, 1971).

b) Proteoglycosaminoglycan of Cartilage and Notochord

i) Parent Type

The polyanionic proteoglycan of bovine nasal septum cartilage has been more thoroughly studied than any other proteoglycosaminoglycan of connective tissue. A model structure for the basic molecular unit was proposed several years ago by MATHEWS and LOZAITYTE (1958). A more recent study (MATHEWS, 1971) has added further details and also extended the validity of the bovine cartilage model to the polyanionic proteoglycans of cartilage and notochord of a wide phylogenetic range of vertebrate species. The parent type structure proposed is shown in Fig. 6.3 in a generalized and schematic form. Electron-microscopic observations by SERAFINI-FRACASSINI and SMITH (1966), ROSENBERG et al. (1970) and WELLAUER et al. (1972) are consistent with this model structure (Fig. 6.4). Further confirmation has been obtained by degradative studies (HASCALL et al., 1972). More compex structures consisting of multiple polypeptide chains have been suggested by SERAFINI-FRACASSINI (1968). The polypeptide core apparently lacks any secondary structure (EYRING and YANG, 1968a). However, a secondary structure may be present in the polysaccharide chains (STONE, 1969).

A parent repeating period of the proteoglycan of Fig. 6.3 has been tentatively suggested (MATHEWS, 1971). According to the schematic model of Fig. 6.5 it consists of a polypeptide chain of about 45 amino acid residues divided into a long sequence and a short sequence to which two chains of polysaccharide are linked. More detailed specifications are not yet available, since the amino acid sequences vary with species while the length of the polysaccharide chains varies with both species and tissue source, as well as with topological region of the tissue, development, and aging.

Other repeating sequences are likely. Thus, trypsin digestion of the proteoglycan of bovine cartilage yields, in part, molecular fragments, or units, consisting of five or six glycan chains linked to peptide (BAKER and JOHNSON, 1973; V. HASCALL, priv. commun.). It is evident that the linear distribution of glycan chains along the polypeptide backbone is not uniform in a single molecule of proteoglycan and may

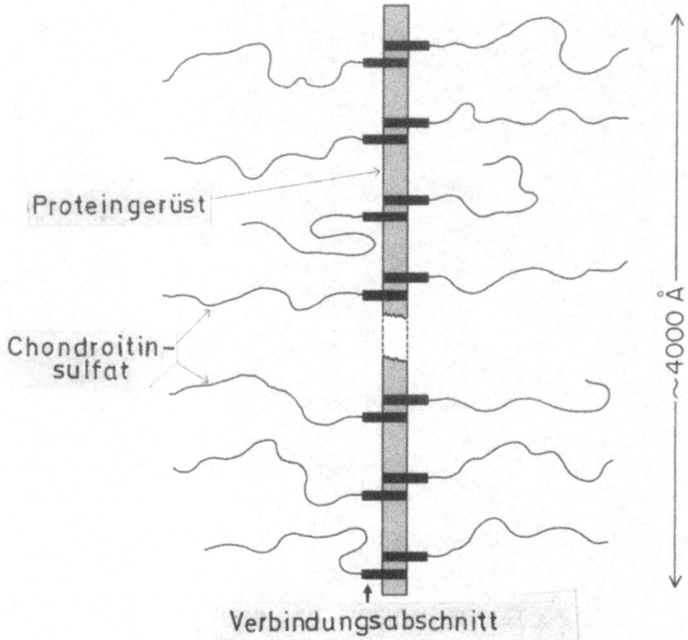

Fig. 6.3. Schema of *parent* -type polyanionic proteoglycosaminoglycan of vertebrate cartilage and notochord (MATHEWS and LOZAITYTE, 1958; MATHEWS, 1971). To a linear polypeptide consisting of 2000 to 3000 amino acid residues, paired sets of side-chains of chondroitin 4 (or 6)-sulfate are attached at regular intervals. Variants with alternate patterns of chain attachment may occur. Neither the exact size, the number of polymorphic variants, nor the amino acid sequence of the polypeptide can be specified at this time. Each polysaccharide chain is joined glycosidically to a serine residue via an oligosaccharide sequence of the linkage region (see Fig. 6.6).

The number of disaccharide repeating periods of *parent* and variant types from 10 to 100 periods per chain, dependent upon species and tissue, is not specified and hence the molecular weight of the parent proteoglycan is not stated. The approximate molecular length indicated was obtained by light-scattering measurements on proteoglycan of bovine nasal cartilage (compare Fig. 6.4)

also differ among homologous molecules in different tissues. It is of particular interest, therefore, that one or more tissue-specific proteochondroitin sulfates are produced by cartilage cells (PALMOSKI and GOETINCK, 1972; LEVITT and DORFMAN, in press; KIMATA et al., 1973).

An important variation in the structure indicated in Fig. 6.3 results from incomplete addition of the chondroitin sulfate side chains during biosynthesis. As a consequence, either one, or both, of the polysaccharide chains in the repeating period shown in Fig. 6.5 may be deleted in the final product. Various heterogeneities in composition and molecular parameters of proteoglycans may be due to differences in biosynthetic activities among cells at different sites within a tissue sample, to the presence of molecules at various stages of degradative turnover, and to preparative artifacts (BRANDT and MUIR, 1971; WELLAUER et al., 1972; HOPWOOD, 1972).

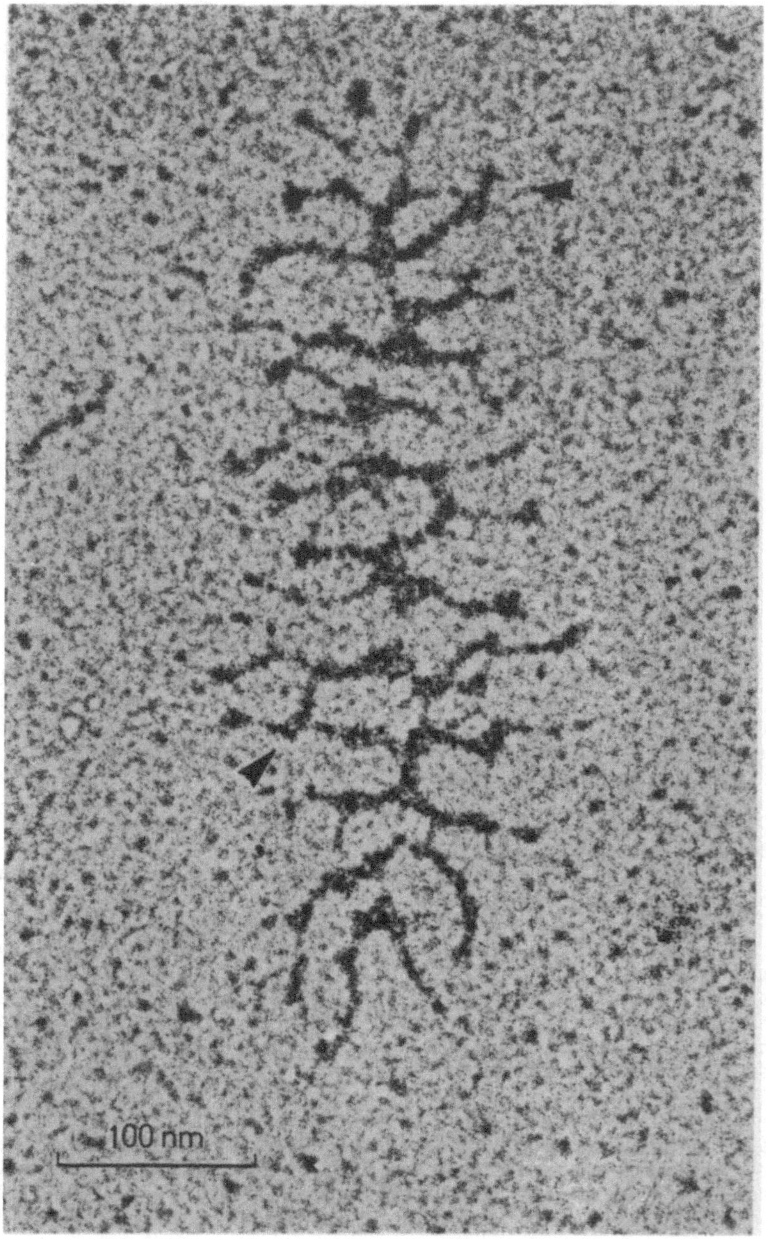

Fig. 6.4. Electron micrograph of proteoglycosaminoglycan of bovine nasal cartilage (ROSENBERG et al., 1970; courtesy of Dr. ROSENBERG and permission of the Journal of Biological Chemistry)

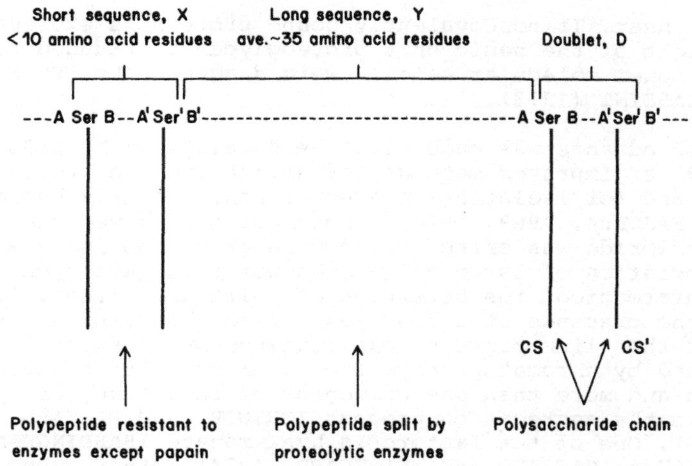

Fig. 6.5. Possible *parent* repeating period of model proteoglycan of Fig. 6.3, consisting of a short peptide sequence and a long peptide sequence (about 45 amino acids, total) and two polysaccharide side-chains (MATHEWS, 1971; with permission of the Biochemical Journal)

ii) Relationship to Skeletal Keratan Sulfate

The structure of the actual proteoglycans is more complicated than is suggested by Fig. 6.3. Preparations from cartilage (particularly from aged adults) but not those from notochord, may be closely associated with significant (over 5% by weight) amounts of skeletal keratan sulfate. The available evidence indicating that the minor glycosaminoglycan component is covalently linked to the same proteoglycan molecule that bears chondroitin sulfate chains, was confirmed by HASCALL and RIOLO (1972). Whether all of the skeletal keratan sulfate of a preparation of proteoglycan of cartilage is associated in this manner is still unclear.

iii) Aggregates

Molecular aggregation is a factor that frequently greatly complicates the chemical characterization and the investigations of structure of proteoglycans. The problems involved are exemplified in the particular case of the proteoglycan of bovine nasal septal cartilage. Early physical chemical studies (MATHEWS and LOZAITYTE, 1958; MATHEWS, 1962) showed that most preparations of the proteoglycan were polydisperse in molecular size. A very small proportion of the material had molecular weights below 3×10^6 and may have been either 'native' or the resultant of the action of lysosomal proteases upon larger molecules. However, the greater proportion of various preparations, with molecular weights up to 50×10^6 could be disaggregated by concentrated urea or potassium sulfocyanate solutions or by mild heating in dilute salt solution. These aggregates retained a portion of their protein by noncovalent bonds that were susceptible to hydrogen bond-breaking agents and to heat. If noncovalently bound protein is excluded, the molecular weight of the basic unit proteoglycan is reduced to about 3.5 million. Lower molecular weights were suggested by WELLS and SERAFINI-FRACASSINI (1973).

A considerable advance was made with the development by SAJDERA and HASCALL (1969) of improved methods for extracting the proteoglycan from tissues and for isolating it free of noncovalently bound protein (HASCALL and SAJDERA, 1969, 1970). These authors showed that 4 M guanidinium chloride was effective in disaggregation and that maintenance of association of large aggregates was dependent upon native molecular conformation, the retention of sulfhydryl or disulfide groups, and the presence of a fraction termed the "link protein". Separation of the "link protein" was achieved by density-gradient centrifugation and by chromatography. The "link protein" fraction is heterogeneous and more than one component of this fraction appears to be involved in the aggregation process (GREGORY and HASCALL, 1972; GREGORY, 1973). One of the factors is hyaluronate (HARDINGHAM and MUIR, 1972, 1973a; HASCALL and HEINEGÅRD, 1974). Other types of molecular interactions (Chap. 10) that occur in solutions of cartilage extracts may obscure the role of specific aggregation factors.

The proteoglycan "sub-unit" is polydisperse in molecular weight with an average molecular weight of 2.5×10^6 (as the guanidinium salt; by sedimentation velocity). When due allowance is given to the considerable differences in methodology among the different procedures used for molecular weight determination, this last value appears to be in substantial accord with the value of 3.5×10^6 obtained by light-scattering methods.

Other types of aggregates, also containing basic units of proteoglycan, have been separated from cartilage (SCHUBERT and HAMERMAN, 1968). Most such preparations contain significant amounts of collagen and, very commonly, large proportions of other matrix components (CAMPO, 1970). They may be generally regarded as fragments of intercellular matrix and are discussed in Chap. 10. However, some proteoglycan may be covalently linked to collagen (KOBAYASHI and PEDRINI, 1973).

c) Other Proteoglycosaminoglycans

The data obtained for proteoglycosaminoglycans, other than those characteristic of cartilage and notochord, are as yet too meager to justify proposals of parent structures. However, in all cases the presence of covalently bound amino acids indicates an association with peptide or protein components at some stage of biosynthesis.

i) Skeletal Keratan Sulfate

As noted above, this type of keratan sulfate is bound to the proteoglycan of cartilage. It has not been possible to isolate, even from such tissues as whale nucleus pulposus (HASHIMOTO and LUDOWIEG, 1968) that contain skeletal keratan sulfate as the major glycosaminoglycan component, a keratan sulfate proteoglycan that is free of chondroitin 4 (or 6)-sulfate.

ii) Corneal Keratan Sulfate

Corneal stroma contains a proteoglycan of keratan sulfate, which has not been well characterized because of difficulties in separating a homogeneous macromolecule from accompanying chondroitin sulfate-protein (KERN and BRASSIL, 1967; STUHLSATZ et al., 1971). The possible covalent association of keratan sulfate chains with chondroitin sulfate chains in the same proteoglycan molecules has not been fully explored.

iii) Dermatan Sulfate

Proteoglycans containing dermatan sulfate as the sole glycosaminoglycan component have been isolated from mammalian tissues but require further structural characterization (TOOLE and LOWTHER, 1968; PRESTON, 1968; ÖBRINK, 1971).

iv) Hyaluronate

The conclusion that hyaluronate exists in animal tissues as a proteoglycan rests principally upon the retention of a very small (less than 1%) content of amino acids by highly purified preparations from various sources. In the opinion of LINDAHL and RODÉN (1972), who have reviewed the available evidence, more convincing data are needed. However, covalent association of hyaluronate and peptide seems indisputable (SCHER and HAMERMAN, 1972).

v) Heparan Sulfate

JANSSON and LINDAHL (1970) showed clearly that preparations obtained from bovine aorta contain heparan sulfate in a complex molecular structure consisting of several glycan chains and a polypeptide moiety bound together as a proteoglycan unit. Further structural studies are needed.

vi) Heparin

The presence of a covalently linked peptide moiety in highly purified preparations of heparin suggested the possibility that this glycan may occur in tissues as a multichain proteoglycan. LLOYD et al. (1967) and SERAFINI-FRACASSINI et al. (1973) reported the isolation of a heparin proteoglycan from mast-cell granules and ox-liver capsule, respectively. However, the presence of the proteoglycan in ox-liver capsule was disputed by LINDAHL (1970a), who claimed that heparin occurred essentially as single glycan chains in this tissue. In explanation of these seemingly disparate results, LINDAHL and RODÉN (1972) suggest that heparin, which may be synthesized as a multichain proteoglycan, is hydrolyzed by glycosidases (ÖGREN and LINDAHL, 1971) and proteases intracellularly to yield a heterogeneous population of degradation products. It may be noted in this connection that heparin is the only glycosaminoglycan that is contained wholly within cells and is not a normal component of extracellular matrices. The structural problem is complicated further by the possibility of branched polysaccharide structures. HORNER (1971) isolated a heparin of molecular weight about 10^6 from rat skin. Surprisingly, the macromolecule appeared to be constructed of many short heparin chains (each about 30,000) joined to a common core polysaccharide by bonds sensitive to splitting by ascorbic acid.

2. Structure of the Linkage Region

A linkage region of a polyanionic proteoglycosaminoglycan consists of an amino acid residue with a directly linked monosaccharide unit, and the succeeding oligosaccharide structure that leads to the first disaccharide repeating period. The lengthier and more precise term would be *carbohydrate-protein linkage region*. It is generally assumed that, in each proteoglycan, all glycan chains with the same type of repeating period have identical oligosaccharide linking structures, although more than one kind of amino acid residue may be involved in the linkage to protein. Characterization of the linkage region structure has required the following steps: 1. isolation and purification of the proteoglycan, 2. hydrolysis of the proteoglycan by proteases and polysaccharidases, 3. isolation of the resulting oligosaccharide-peptides, and 4. determination of the chemical structure by standard procedures of organic chemistry. Details, as well as complete references for the discussion that follows, are given in reviews by RODÉN (1970), LINDAHL and RODÉN (1972), and MEYER (1970).

a) Proteoglycans of Chondroitin Sulfate Type

i) Vertebrates

The linkage-region structure of the proteoglycan of bovine cartilage was determined largely through the efforts of RODÉN and coworkers. A description of the methodology and detailed results have been given by LINDAHL and RODÉN (1972). Fig. 6.6 shows the established structure.

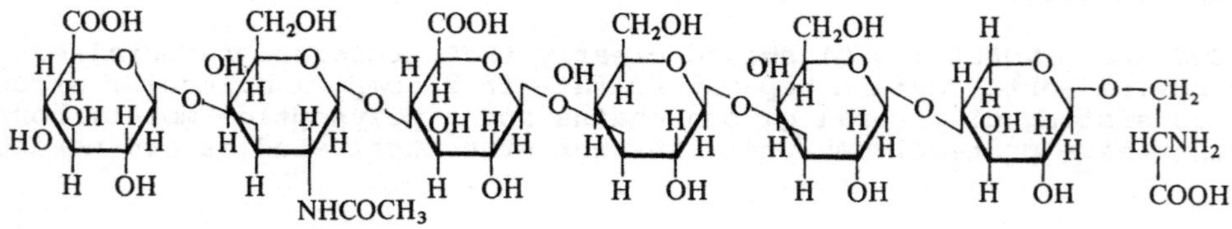

Fig. 6.6. Structure of the linkage region of chondroitin 4- and 6-sulfate (the sulfate group on the galactosamine residue is omitted). The ordered sequence of neutral sugars, beginning at the linkage to serine, is: residues of D-xylose, D-galactose and D-galactose. The next sugar is a residue of D-glucuronic acid, which is followed by residues of N-acetyl-D-galactosamine and D-glucuronic acid. The last two residues comprise a disaccharide repeating period and appear to be added to the carbohydrate chain by the same enzymes as are responsible for biosynthesis of the remaining repeating sequences of the polysaccharide chain (not shown here).

The first glucuronic acid residue is attached to D-galactose by a glucosyl uronic transferase that differs from the one responsible for the addition of the subsequent glucuronic acid residues in the chain. On account of its biosynthetic distinction, this glucuronic acid residue is taken together with the initial sequence of neutral sugars to comprise the carbohydrate linkage group of the proteoglycan. (From LINDAHL and RODÉN, 1972; with permission of Elsevier Publishing Co.)

This proteoglycan contains three times as many 4-sulfated repeating periods as 6-sulfated repeating periods. The identical linkage region has been demonstrated in the proteoglycan of chondroitin 6-sulfate from human umbilical chord (HELTING and RODÉN, 1968), which has a ratio of 6-sulfated periods to 4-sulfated periods greater than 10 to 1. The chondroitin sulfate chains of the proteoglycan in rat and rabbit brain and in bovine nasal cartilage are linked exclusively to serine (MARGOLIS et al., 1972; ROBINSON and HOPWOOD, 1973). Although there have been indications that a small proportion of the glycan chains may be bound to protein by other types of linkages (DANISHEFSKY and ABRAHAM, 1971; BRANDT and MUIR, 1971; SIMPSON and DAVIDSON, 1972), substantiating evidence has not been forthcoming. Fewer sulfate groups are present in the vicinity of the glycan-protein linkage region than in the distal repeating periods of the chain (WASTESON and LINDAHL, 1971).

The complete structure of the linkage region sequence has not been determined for chondroitin sulfate proteoglycans of animal species other than the ox. Available comparative data are restricted to determination of the carbohydrate composition of the linkage region. SCHMIDT et al. (1966) isolated linkage-region glycopeptides from enzymatic hydrolysates of cartilage proteoglycans of a variety of vertebrates. For preparations from ox and man they found that the molar ratios of galactose:xylose:serine were very close to 2:1:1, in accord with the now established structure (Fig. 6.6). Very similar results were obtained with preparations from shark *(Prionace glauca, Squalus acanthias, Scylliorhynchus canicula)*, ray *(Dasybatis centrura)*, salamander *(Amblystoma tigrinum)* and turkey *(Meleagris gallaysavo)*.

Analyses of chondroitin 4- and 6-sulfate preparations obtained by proteolysis of cartilage and notochord from the lamprey *Petromyzon marinus* and from the sturgeon *Acipenser fulvescens* show similar ratios for galactose, xylose, and serine (MATHEWS, unpubl.). In additon, number-average molecular weights indicate that each glycan chain contains one residue each of serine and xylose, and two residues of galactose. However, otherwise comparable preparations from cartilage of the hagfish *Myxine glutinosa* contain approximately equimolar proportions of galactose, xylose and serine. Thus, it appears likely that, with the possible exception of the hagfish, vertebrates contain chondroitin sulfate proteoglycans with chemically similar, and possibly identical, carbohydrate linkage groups.

ii) Invertebrates

The chondroitin sulfate preparations from cartilage of two species of squid *(Loligo pealii* and *Architeuthis dux)* contain near equimolar amounts of xylose, galactose and serine, whereas similar preparations from cartilage of the horseshoe crab *Limulus polyphemus* contain much more xylose than either galactose or serine (MATHEWS et al., 1962; MATHEWS, unpubl.). Comparable preparations from squid *(L. opalescens)* skin yielded a mixture of glycopeptides. One of these consisted of serine, xylose, galactose, glucose and glycine in a molar ratio of 1.00:0.98:4.83:1.14:1.58 and was probably derived from the linkage region. RADHAKRISHNAMURTHY et al. (1970) concluded that the probable linkage of glycan to peptide was through galactosylxylosylserine. Also, the same peptide chain appeared to carry an additional oligosaccharide chain that contained glucose, galactose and glucosamine, probably linked in an alkali-stable linkage to aspartic acid. However, the isolation and characterization by HABUCHI et al. (1971) of a glucose-containing hexasaccharide derived from an enzymatic digest of chondroitin sulfate-peptide of squid cartilage leaves little doubt that glucose is an integral component of the linkage region. The chondroitin sulfate proteoglycans of squid cartilage resemble their counterparts of vertebrate cartilage with respect to the xylosylserine linkage and to the presence of galactose in the carbohydrate sequence. However, they may differ with respect to the number of galactose residues per linkage group, the presence of glucose, and other structural features of the linkage region.

b) Proteoglycans of Other Types

i) Dermatan Sulfate

Proteoglycans of dermatan sulfate type from mammalian tissues have been demonstrated to contain a linkage region identical in structure to that (Fig. 6.6) of chondroitin sulfate type proteoglycans (STERN et al., 1971). The position next to galactose is largely, if not entirely occupied by a D-glucuronic acid residue, although the complete polysaccharide chain contains predominantly L-iduronic acid residues.

ii) Corneal Keratan Sulfate

The keratan sulfate proteoglycan of bovine corneal stroma contains a carbohydrate-protein linkage similar to that found in glycoproteins. A bond between an asparagine residue and N-acetylglucosamine was established (GREILING et al., 1970; BAKER, CIFONELLI, and RODÉN, unpubl.). BAKER et al. (1969) showed that the linkage region is contained in a glycopeptide consisting of mannose, galactose, glucosamine and aspartic acid in molar ratios of 2:1:2:1.

Aspartic acid is the predominant amino acid residue in preparations obtained by proteolytic digestion of corneas from the chicken, whitefish *(Coregonus)* and a shark *(Carcharchinus leucas)*. Only aspartic acid or asparagine was present in amounts sufficient to account for the linkage of the polysaccharide to peptide (MATHEWS and CIFONELLI, 1965; MATHEWS, unpubl.). The linkage region of this proteoglycan may be homologous in the vertebrate subphylum.

iii) Skeletal Keratan Sulfate

The structure of the principal linkage region of keratan sulfate from mammalian cartilage and nucleus pulposus appears to be quite complex. In preparations from aged human rib cartilage a majority of the bonds between carbohydrate and protein moieties are O-glycosidic linkages to serine and threonine. Available data suggest that the sugar residue involved is N-acetylgalactosamine and that this residue is substituted not only at C-1 but also at C-3 and C-6, indicating a branched structure (BRAY et al., 1967; BALDUINI et al., 1970). Similar data were obtained by SENO and TODA (1970) for preparations from whale cartilage. Further complexities are found. In addition to the above linkages that are split by treatment with alkali, a small proportion of unidentified linkages of carbohydrate to protein are stable to alkali. A linkage to glutamic acid (glutamine) is possible (HEINEGÅRD, 1972). Variable amounts of other sugars including mannose, fucose and sialic acid are commonly present. Mannose appears to be located at, or near, the linkage region (HIRANO and MEYER, 1971). Also, many preparations show a molar excess of galactose over glucosamine. The location of such extra galactose residues is unclear. HOPWOOD (1972) suggested that an N-acetylneuraminylgalactose disaccharide is linked to the C-3 position of the linkage region N-acetylgalactosamine residue and that the remainder of the keratan sulfate chain is linked to the C-6 position of the same residue. Two more such disaccharide units, as well as fucose and mannose, are linked to the keratan sulfate chain at positions distant from the linkage region.

Data on preparations from cartilage of vertebrates other than mammals are meager. Galactosamine appears to be a structurally significant component, since it is characteristically present in preparations from mammals as well as from shark *(C. leucas)*, coelacanth *(Latimeria chalumnae)*, chicken and garfish *(Lepidosteus)* (MATHEWS and CIFONELLI, 1965). However, it is evident that no clear conclusions can be drawn at this time as to the extent of chemical homologies of skeletal keratan sulfate fragments.

iv) Hyaluronate

A considerable amount of work has been directed toward demonstrating carbohydrate-peptide linkages in preparations of hyaluronic acid from animal tissues (see LINDAHL and RODÉN, 1972). Linkages to serine and threonine have been suggested. WARDI et al. (1969) proposed that arabinose and glucose have a role in the linkage region of hyaluronate analogous to that of xylose and galactose in chondroitin sulfate proteoglycans. However, the presence of arabinose has been questioned. Although the present evidence indicates that hyaluronate exists in tissues as a proteoglycan, detailed supporting data are lacking.

v) Heparin

The carbohydrate-protein linkage group of mammalian heparin is identical to that of chondroitin sulfate proteoglycans (Fig. 6.6; see LINDAHL and RODÉN, 1972). About one half of the small proportion of repeating periods of heparin that contain acetylated glucosamine residues is concentrated near the linkage region end of the polysaccharide chain (CIFONELLI and KING, 1972).

vi) Heparan Sulfate

The details of the structure of the carbohydrate-protein linkage group of heparan sulfate proteoglycans has not yet been determined. However, the presence of galactose and xylose in heparan sulfate preparations and consideration of the close structural similarities and biosynthetic relationships (see Sect. 4, below) between heparan sulfate and heparin suggest that the two glycans are likely to have identical linkage-group structures. In the brain proteoglycan, the linkage of glycan to peptide appears to be exclusively to serine *via* an O-glycosidic bond (MARGOLIS et al., 1972).

3. Comparative Macromolecular Structures of Proteoglycans

Comparative studies of proteoglycans are seriously hampered by difficulties of purification and characterization. The major problems arise largely from the very high molecular weights and expanded conformations but also from the commonly wide range of heterogeneity and polydispersity of these macromolecules (see Sect. 1b). The types of variability in the covalent structure of proteoglycans from a single tissue source that may be encountered include:

1. Polymorphism of polypeptide structures
2. Variation in type, number, and ordering of glycan molecules linked as side-chains to polypeptide
3. Variation in size of glycan side-chains
4. Variation in sulfation of side-chains

It is evident that the available data for proteoglycans permits only a description of *average* characteristics of a population of macromolecules in a given preparation.

This section will deal mainly with general features of molecular architecture and with the polypeptide moiety of proteoglycans. Details of glycan structure will be treated in the Chap. 7. The reader may refer to Fig. 3.5 and 3.11 for phylogenetic orientation and to Tables 6.2 and 6.3 for data on the molecular parameters and chemical composition.

a) Cyclostomes

The proteoglycans from lamprey *(Petromyzon marinus)* cartilage and notochord have been described. The glycan moieties are chondroitin 6-sulfate in the cartilage preparation and chondroitin 4-sulfate in the notochord preparation. They exhibit a wide range of heterogeneity with respect to the glycan/protein ratio (very little keratan sulfate is present) and to the ordering of glycan chains along the polypeptide core (see Fig. 6.5). In addition, it is likely that the polypeptide core is polymorphic in structure.

b) Chondrichthyes

A number of species of sharks and rays have been utilized as source material for investigations of proteoglycans of cartilage. In all cases, the glycan consists predominatly of repeating periods of the chondroitin 6-sulfate and CS-D types (see Fig. 7.1). Skeletal keratan sulfate is also present. Molecular parameters (MATHEWS, 1962) for preparations from the tiger shark *Prionace glauca*, the seven-gill shark *Notorynchus maculatum* and the stingray *Dasyatis* sp. are given in Table 6.2. Similar high molecular weights are indicated for preparations from cartilage tissues of the spiny dogfish *Squalus acanthias*, the bull shark *Carcharhinus leucas*, and the ratfish *Hydrolagus collei*.

The molecular weights and sizes of chondrichthye preparations are similar to the values given for proteoglycan preparations from bovine cartilage when measurements are made in dissociating solvents, such as 2 M KSCN. Aggregates of very high molecular weight are present in 0.4 M NaCl solutions of preparations from both vertebrate classes. However, unlike the otherwise similar bovine preparations, the preparations from chondrichthye cartilage do not disaggregate in the presence of 8 M urea or upon mild heating (MATHEWS, 1962). It is possible that this stability reflects an adaptation to the high urea concentrations in chondrichthyan tissues (see Chap. 3, Sect. 9b, discussion of elastoidin). A similar adaptation to internal environment has been suggested by ALLISON and KAPLAN (1964) for the lactic dehydrogenase of dogfish, which is more stable to urea and to heat than are the homologous enzymes of other vertebrates (WILSON et al., 1964, 1967).

c) Osteichthyes

The proteoglycans of cartilage of a chondrostean *(Acipenser fulvescens)* and of two teleosts *(Salmo gairdneri* and *Oncorhynchus garbuscha)* contain chondroitin 6-sulfate as the predominant glycan component. The proteoglycan of notochord of the same species of sturgeon has an average composition of 30% of protein and 70% of chondroitin 4-sulfate. The macromolecule has a weight-average molecular weight of 1.3×10^6.

d) Amphibia

The proteoglycan isolated from cranial cartilage of the bullfrog *Rana catesbiana* consists predominantly (96%) of chondroitin 4-sulfate. Molecular parameters resemble those of the proteoglycan of bovine cartilage.

e) Reptiles

No studies of proteoglycan from reptile tissues have been reported.

f) Birds

Molecular parameters for proteoglycans of the chondroitin sulfate type from 14-day chick epiphyseal cartilage and from xiphoid cartilage of 4-month-old chickens (Table 6.2) resemble those for the proteoglycan from bovine cartilage. NEVO et al. (1972) isolated comparable proteoglycans from 14-day chick epiphyseal cartilage and from the culture medium of chondrocytes grown from the same tissue.

A proteoglycan was isolated from xiphoid cartilage of young adult chickens by SHULMAN and MEYER (1970) and fractionated on gel columns under conditions that permit aggregation to persist. Amino acid analyses, which may reflect the presence of noncovalently associated protein, demonstrate that the protein compositions of the fractions differ from each other and also from the amino acid composition of chick-embryo proteoglycan given in Table 6.3. A preparation of proteoglycan of xiphoid cartilage of 4-month-old chickens, isolated and purified according to methods of HASCALL and SAJDERA (1969), had an amino acid composition different from that of the above-mentioned preparations from adult chicken and from embryo chick (MATHEWS, unpubl.). Proteoglycans of bone have been studied also (CANDLISH and HOLT, 1971; HERRING, 1972).

g) Mammals

Sufficient work has been done with proteoglycans from mammalian tissues to justify subdivision of this section into separate classes of proteoglycans, characterized by the predominant type of glycan component present (Table 6.4). This classification is somewhat arbitrary, since it is occasionally possible to isolate a small fraction that consists mainly of a minor glycan of the tissue.

A detailed discussion of the structure of the proteoglycan from bovine nasal cartilage has already been given (Sect. 1b). SERAFINI-FRACASSINI et al. (1971) suggested that a fragment of the proteoglycan, isolated by mild hydrolysis with HCl, was a subunit of the native macromolecule. However, the amino acid composition of their preparation was quite

Table 6.2. Molecular parameters of proteoglycans of cartilage (light-scattering results)

Preparation	Protein %	[η]	$M_w(10^6)$	S. Å	L. Å	M_n/L_n Avog./Å	Shape Model
Lamprey	42	0.83	0.34	450			stiff coil
Tiger shark	12	3.7	~25	~2,800			
(in 2*M* KSCN)			4.0	1,050			stiff coil
Seven-gill shark	6	4.5	12.5	2,060			stiff coil
(in 2*M* KSCN)			3.0	1,010			stiff coil
Stingray	5	4.3	~20	~2,500			
Trout	14	1.90	2.5	740			stiff coil
Rock sturgeon	8	2.65	2.7	850			stiff coil
(in 2*M* KSCN)	7		2.9	1,060	3,600	890	Rod
Frog	8	2.25	3.8	930			stiff coil
Chick embryo	8	2.30	2.4	740			stiff coil
Chicken (4 mo)	10	2.00	1.8	700	2,400	950	Rod (Polyd.)
Ox (40 cm fetus)	9	2.73	6.7	1,800			stiff coil
(in 2*M* KSCN)			3.1	1,280	4,400	1,250	Rod
Ox (calf)	22	3.20	~50	~3,500			
(in 2*M* KSCN)	10		3.5	1,070	3,700	940	Rod

Explanations:

[η] = intrinsic viscosity in 0.4*M* NaCl at ave. gradient of 100 cm^{-1}

Molecular parameters in 0.4*M* NaCl and in dissociating solvent, 2*M* KSCN, by light-scattering methods (MATHEWS and LOZAITYTE, 1958): S = mean radius of gyration; L = mean length of theoretical rod; M_n/L_n = average linear density of theoretical rod in avograms/Å.

different from that given in Table 6.3. TSIGANOS, HARDINGHAM, and MUIR (1971), and BRANDT and MUIR (1971) noted that proteoglycan preparations from pig laryngeal cartilage were polydisperse in size as well as heterogeneous in composition and concluded that the preparations contained proteoglycans having different core proteins. However, polymorphism of the core protein was not found by HASCALL and SAJDERA (1970) and by HASCALL and RIOLO (1972) for the bovine proteoglycan. The structure and molecular parameters of proteoglycans from porcine costal cartilage were extensively investigated by DAVIDSON and coworkers. The preparations obtained were polydisperse in molecular weight, ranging from 37,000 upward. Neither cysteine nor methionine appeared to be present in the protein moieties (HRANISAVLJEVIĆ et al., 1972; SIMPSON and DAVIDSON, 1972). Chondroitin sulfate chains were nearly homogeneous in molecular weight (13,000 w.ave.); variation in molecular size of the proteoglycans was attributed to differing levels

Table 6.3. Amino acid composition of proteoglycans of chondroitin sulfate type (res/1000)

Source	Lamprey cartilage (MATHEWS, 1971)	Lamprey notochord (MATHEWS, 1971)	Stingray cartilage (MATHEWS, 1971)	Sturgeon cartilage (MATHEWS, 1971)	Sturgeon notochord (MATHEWS, 1971	Frog cartilage (MATHEWS, unpubl.)	Chick (emb.) cartilage (NEVO et al. 1972)	Ox cartilage (HASCALL and SAJDERA, 1970)	Ox heart valves (TOOLE and LOWTHER, 1968)
Serine	134	71	85	116	85	123	133	127	67
Glycine	100	54	131	131	98	114	99	123	106
Glutamate	190	214	131	125	165	122	160	150	145
Aspartate	87	80	83	77	85	117	63	67	110
Proline	67	91	110	48	92	54	77	109	75
Threonine	58	104	51	51	71	69	86	59	59
Alanine	65	66	68	65	71	68	84	69	75
Valine	65	32	59	60	51	59	59	52	52
Isoleucine	41	70	80	102	32	78	50	32	35
Leucine	44	71	51	40	67	50	64	74	91
Tyrosine	0	28	17	71	26	32	17	19	22
Phenylalanine	15	20	25	23	26	36	36	34	32
Lysine	27	22	38	28	45	42	19	20	61
Histidine	61	40	11	17	28	14	15	21	23
Arginine	40	31	40	28	39	21	37	33	42
Cystine (1/2)	6	9	19	17	18		(20)[a]	6	
Methionine								3	4
%carbohydrate	54	48	95	93	70	92	90	93	40

[a] From MATHEWS (1971).

Table 6.4. Proteoglycans of mammalian tissues[a]

Tissue	Glycan[b]	Protein (%)	$M_w (10^5)$	Remarks
Nucleus pulposus	chondroitin 6-sulfate	30	1.7	ROSENBERG, SCHUBERT, and SANDSON, 1967[c]
Nucleus pulposus	keratan sulfate	40	85	HASHIMOTO and LUDOWIEG, 1968; LUDOWIEG et al., 1972[d]
Aorta	chondroitin 6-sulfate	20	20	KRESSE, HEIDEL, and BUDDECKE, 1971[e]
Umbilical chord	chondroitin 6-sulfate	8		DANISHEVSKY and ABRAHAM, 1971
Heart valves	chondroitin sulfate	6-32	0.4-0.7	LOWTHER, PRESTON, and MEYER, 1970
Heart valves	dermatan sulfate	50-70	12-22	TOOLE and LOWTHER, 1968; PRESTON, 1968[f]
Skin	dermatan sulfate	58	29	ÖBRINK, 1972[f]

a Except cartilage (Sect. 1b).
b Predominant type present.
c Fractions having keratan sulfate predominating present also; compare ROSENBERG, SCHUBERT, and SANDSON (1967), LOWTHER and BAXTER (1966), and HEINEGARD and GARDELL (1967).
d Preparation from whale reduced by dithiothreitol to M_w of 6.6×10^5.
e Also contains dermatan sulfate; compare FRANEK and DUNSTONE (1968).
f Possibly contains heparan sulfate also.

of glycan substitution in the polypeptide core or to catheptic proteolysis (WOODWARD et al., 1972). These proteoglycans isolated from porcine cartilage appear to be of extremely low molecular weight and too small to function physiologically in cartilage tissue (Chap. 10, Sect. 4d). Similarly low-molecular-weight proteoglycan fractions were isolated from rabbit-ear cartilage by SERAFINI-FRACASSINI and STIMSON (1971).

Proteoglycans have been isolated from a variety of tissues in addition to those mentioned in Table 6.4. These include bone (Chap. 5, Sect. 3e), chondrosarcoma (CHOI et al., 1971), Kurloff cells of spleen (Chap. 9, Sect. 4), and leukocytes (OLSSON, 1969a-d). Proteoglycan fractions obtained from bovine corneal stroma (STUHLSATZ et al., 1971; SALITERNIK-GIVANT and BERMAN, 1970) varied in protein content from 5 to 60%, in ratio of keratan sulfate to chondroitin sulfate from 0.06 to 15, and in molar ratio of sulfate to hexosamine from 0.5 to 1.7. Information on other proteoglycans (Sect. 1c) such as hyaluronate proteoglycan, heparan sulfate proteoglycan, and heparin proteoglycan is still meager.

h) Invertebrates

Proteoglycans of invertebrate tissues have been accorded scant attention. The glucan sulfate-peptide component of mucin from the marine snail *Buccinum undatum* was studied by HUNT (1970) and collaborators. This proteoglycan of molecular weight about 200,000 consists of 2.2% peptide and 97.8% of a sulfated glucan with a cellulose-like structure. Thus, the polypeptide chain has a maximum molecular weight of about 4,000. Four amino acids, serine, glutamic acid, glycine and alanine, comprise over 50% of the total residues. On extensive proteolysis, however, the residual residues are mainly glutamic acid, serine, glycine, and aspartic acid. Since the proteoglycan is stable to alkali, the linkage region is unlikely to contain the type of O-glycosyl bond to serine found in the proteoglycosaminoglycans of vertebrates.

A proteoglycan containing a chondroitin sulfate type glycan and about 10% protein was obtained in low yield from L. *opalescens* 'cranial' cartilage (MATHEWS, unpubl.). The preparations was polydisperse and had an average molecular weight of only 300,000. Since the single glycan chains obtained by proteolyses of the tissue had a molecular weight of about 30,000, it appeared from these preliminary experiments that an average-sized proteoglycan molecule contained about 10 glycan chains. Virtually complete extraction of the glycan is effected by the action of trypsin on the cartilage. In this case, the major product consists of molecules in which two glycan chains are linked to the same small peptide.

4. Biosynthesis

Investigations of the mechanisms of biosynthesis of proteoglycans are complicated by the necessity to consider the formation of molecules containing both polypeptide and heteropolysaccharide moieties. Because of the inherent difficulties of such studies, available results have been limited largely to a single type of proteoglycan, i.e. the one from vertebrate hyaline cartilage. The rather limited data relating to other proteoglycans suggests that general modes of biosynthesis are probably common to most proteoglycans and that many of the findings for the cartilage proteoglycan may be of broader validity. These beliefs are encouraged by the apparent universality of distribution of nucleotide sugars and the likelihood of the general utilization of such intermediates by animal cells for glycan biosynthesis. Reviews by STOOLMILLER and DORFMAN (1969), RODÉN (1970), RODÉN et al. (1972b) and by KIMATA et al. (1973) may be consulted for details.

a) Precursors

i) Vertebrates

D-glucose appears to be the biosynthetic origin of all of the component monosaccharides of the glycans. A series of transformations of this hexose leads to the various nucleotide sugars. The latter substances are the more immediate precursors and are directly involved in the transfer of glycosyl groups to growing polysaccharide chains of both glycoproteins and proteoglycosaminoglycans. The introduction of ester sulfate residues into carbohydrate groups is performed by enzymes (sulfotransferases) that utilize 3'-phosphoadenosine 5'-phosphosulfate. This activated precursor is formed in a two-step process from inorganic sulfate and adenosine triphosphate.

ii) Invertebrates

The ubiquitous occurrence of nucleoside pyrophosphate sugars in plants and animals, which was noted in Chap. 2, suggests that many of the pathways of biosynthesis of acidic glycans may prove to be common to both invertebrates and vertebrates. Although data on biosynthesis are very meager for invertebrates compared to those on vertebrates, the initial findings do not conflict with this hypothesis. Thus, PUYEAR (1967) showed that the UDP-glucose-to-glucuronic acid pathway in the blue crab *Callinectes Sapidus* Rathbun is similar to that in mammals. The uridine pyrophosphates of glucose, N-acetylglucosamine, and N-acetylgalactosamine have been identified in the slug *Limax maximus* Linné and in other invertebrates (WHEAT, 1960; LUNT and KENT, 1961). 3'-Phosphoadenosine 5'-phosphosulfate is utilized as an agent of sulfate transfer to glycans by a clam (DE MEIO et al., 1967) and by a snail (YOSHIDA and EGAMI, 1965). A unique sulfotransferase of the squid, also utilizing 3'-phosphoadenosine 5'-phosphosulfate as sulfate donor, catalyzes the introduction of sulfate into position 6 of an N-acetylgalactosamine residue already bearing a sulfate residue at position 4 (HABUCHI et al., 1971). A different enzyme is responsible for sulfation at position 4. HUNT (1970) gives a more detailed discussion of these studies of invertebrate metabolism.

b) Assembly of Biosynthetic Units in Vertebrates

The synthesis of proteoglycans involves the coordinated synthesis of two major, covalently linked components, protein and glycan. Each of these types of biopolymer is formed by the addition of subunits to the growing chain in a sequential fashion. The routes followed are essentially those already established for simple proteins and polysaccharides separately. In addition, however, special features appear owing to the covalent union of protein and glycan. Since most of our understanding of proteoglycan synthesis derives from studies of the formation of chondroitin sulfate-protein by cartilage cells, the following discussion deals mainly with the biosynthesis of this macromolecule as a model for proteoglycan synthesis in general, other proteoglycans being considered only insofar as problems specific for the individual compounds arise.

i) Chondroitin Sulfate Type

Since the biosynthesis of proteoglycans is a coordinated process of formation of both protein and glycan, one of the first questions to be asked relates to the temporal relationship between protein and polysaccharide synthesis. The best answer that can be given at this time is that the synthesis of the conjugated molecules most probably proceeds by the prior synthesis of protein followed by stepwise addition of monosaccharide units to form the glycan chains. It is possible, however, that monosaccharide addition may be initiated and that chain growth proceeds to some degree before the polypeptide chain is completed. Polysaccharide chain synthesis may be initiated while the polypeptide chain is still attached to the rough endoplasmic reticulum, proceeding to some further extent while the nascent molecule moves through the cisternal spaces of the smooth endoplasmic reticulum, and be completed in the vesicles of the Golgi apparatus. The finished molecule is then extruded from the cell by exocytosis.

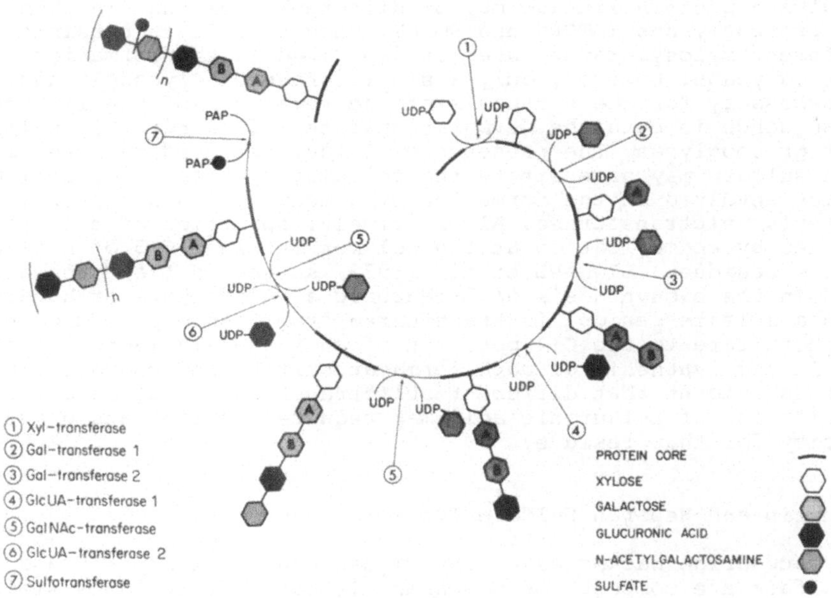

Fig. 6.7. Biosynthesis of the chondroitin 4-(6-) sulfate chain (RODÉN, 1970; with permission of Academic Press). Attachment of the polysaccharide chain and assembly of the sugar residues requires the sequential action of six different glycosyl transferases. A sulfotransferase catalyzes the transfer of active sulfate to the C-4 positions of N-acetylgalactosamine residues. Another sulfotransferase is responsible for sulfation at C-6 positions to yield chondroitin 6-sulfate

The biosynthesis of chondroitin 4-sulfate chains requires the sequential action of six distinct glycosyltransferases and a specific sulfotransferase, as illustrated in Fig. 6.7. The specificities of both acceptor and donor molecules are such as to ensure initially the addition of a D-xylosyl residue to the appropriate serine residue in the polypeptide chain (MATHEWS, 1971; BAKER et al., 1972), and then the addition of sugar residues in proper sequence. Acceptor specificity of the growing polysaccharide chain may be determined by the nature of the terminal nonreducing sugar residue. Thus, there exist two different galactosyltransferases corresponding to the two acceptors, the xylosyl residue and the galactosyl residue, and two different glucuronyluronictransferases corresponding to the two acceptors, the galactosyl residue and the N-acetylgalactosamine residue of the disaccharide repeating unit. It is likely that sulfation at C-4 and at C-6 of acetylgalactosamine residues is catalyzed by two different sulfotransferases with identical donor and acceptor specificities. The sulfation of glucuronic acid residues in proteoglycans of chondrichthyes particularly, and of a few other vertebrate species, probably requires a third sulfotransferase.

ii) Dermatan Sulfate Type

Further biosynthetic complexities are presented by the dermatan sulfate proteoglycans. Since the sequence of amino acid residues adjacent to the glycan-protein linkage may be different from that in chondroitin sulfate proteoglycans (STERN and RODÉN, unpubl.), it is possible that two different xylosyltransferases are involved in the formation of the two proteoglycans. However, only a single group of glycosyltransferases may be necessary for the synthesis of the remainder of the linkage-region sequence in both the dermatan sulfate and chondroitin sulfate-types of proteoglycan. The presence of L-iduronic acid residues in dermatan sulfate may necessitate the cellular synthesis, or activation, of enzymes involved in the formation of a donor precursor as well as an iduronyluronictransferase. Alternatively, formation of this residue may proceed by epimerization at the polymer level of C-5 of D-glucuronic acid residues. LINDAHL et al. (1972) suggested that such a step occurred in the biosynthesis of L-iduronic acid residues of heparin. Although a sulfate residue is transferred from 3'-phosphoadenosine 5'-phosphosulfate to the C-4 position of an N-acetylgalactosamine residue in the synthesis of both dermatan sulfate and chondroitin sulfate, it is evident that different sulfotransferases are involved. Also, sulfation of L-iduronic acid may require a sulfotransferase with specificity for that residue.

iii) Heparin and Heparan Sulfate Types

The polysaccharide backbones of the glycan chains of heparin and heparan sulfate are composed of the same disaccharide residues similarly linked and have been assigned the same parent disaccharide repeating period (see Sect. 1c). Also, both glycans contain L-iduronic acid residues as well as O-sulfate, N-sulfate, and N-acetyl groups, although in different proportions. These similarities in glycan structure, however, are insufficient grounds for considering that the corresponding proteoglycans are also related. Such a hypothesis had been proposed by several workers (SILBERT, 1970; CIFONELLI, 1970; LINDAHL, 1970b; TAYLOR et al., 1973). Convincing support may derive from comparative metabolic studies of both proteoglycans.

Biosynthesis of the glycan chains of heparan sulfate and of heparin requires the sequential action of enzymes following the initial addition of D-xylose to the peptide chain. The enzymes then introduce the monosaccharides forming the linkage region and finally D-glucosamine and D-glucuronic acid in alternating order to complete the chain. L-iduronic acid appears to be formed by C-5 epimerization of D-glucuronic acid residues in the polymer (LINDAHL et al., 1972). The glycosidic linkages formed in the repeating polymer portion are presumed to be of the α-D type and, hence, involve transferases specific for donor, acceptor, and type of glycosidic linkage. The glucosamine residues, according to SILBERT (1970), are transferred as N-acetylderivatives to both glycans. One or more enzymes are involved in partial exchange of the N-acetyl group for an N-sulfate group, a process which proceeds to a much greater degree with heparin than with heparan sulfate. Additional enzymes are responsible for the transfer of sulfate groups, primarily to C-6 of glucosamine and to L-iduronic acid residues (LINDAHL, 1970b). It is possible that sulfate groups may be transferred to D-glucuronic acid residues also, although no evidence for this pathway has been adduced. The data, although still meager, suggest that heparin and heparan sulfate proteoglycans share a common intermediate glycan (possibly proteoglycan) structure in biosynthesis.

iv) Keratan Sulfate Types

Incomplete knowledge of the very complex and heterogeneous structures of the proteoglycans of skeletal and corneal keratan sulfates has inhibited investigations of the enzymatic mechanisms of biosynthesis of these molecules. In addition to transfer reactions involving the saccharides of the parent repeating unit, viz. galactose and N-acetylglucosamine, one must also include reactions transferring sialic acid, fucose, mannose and, in the case of skeletal keratan sulfate, N-acetylgalactosamine. Also to be considered are the reactions transferring sulfate groups to galactose and N-acetylglucosamine residues.

v) Hyaluronic Acid

The biosynthesis of hyaluronic acid has been actively investigated in two systems: human fibroblasts in tissue culture, and *Streptococci* (STOOLMILLER and DORFMAN, 1969). The possibility that hyaluronic acid is linked to protein intracellularly and that its synthesis is dependent upon prior protein synthesis is supported by observations that general inhibitors of protein synthesis also prevent hyaluronic acid synthesis in human fibroblasts. However, similar effects of inhibitors of protein synthesis were not detected in streptococcal systems. Mechanisms of glycan chain elongation for hyaluronic acid in mammals are probably analogous to the mechanisms involved in chain elongation of other glycosaminoglycans.

c) Assembly of Biosynthetic Units in Invertebrates

Apart from the above-mentioned implications of sulfotransferases utilizing 3'-phosphoadenosine 5' phosphosulfate as donor, no reports dealing with mechanisms of assembly of subunit groups in polyanionic glycan synthesis in invertebrates have been published to date.

d) Regulation of Biosynthesis

Regulation and control of the biosynthesis and secretion into the intercellular matrix of proteoglycans is achieved by the interaction of a multiplicity of factors, intrinsic and extrinsic to the connective tissue cell. The following remarks will be limited to some observations on possible mechanisms of control at the enzymic level with a more comprehensive discussion postponed to Chapter 10.

Several mechanisms have been described which operate at the level of precursors. The first involves the inhibition of UDP-glucose dehydrogenase by UDP-xylose and probably involves the synchronization of polysaccharide and protein synthesis in proteoglycans. Thus, a decrease in the normal rate of synthesis of the protein acceptor would result in an accumulation of UDP-xylose which, in turn, would effectively decrease the rate of formation of UDP-glucuronic acid and consequently also the rate of synthesis of UDP-xylose itself. An indirect consequence of the inhibition of glucose dehydrogenase is the accumulation of UDP-glucose. This may result, as BALDUINI, BROVELLI, and CASTELLANI (1970) have shown with surviving corneal stroma, in an increase in the formation of UDP-galactose and hence an increased rate of biosynthesis of keratan sulfate.

The second mechanism involves the inhibition of fructose 6-phosphate: glutamine transamidase by UDP-N-acetylglucosamine and results in control of the concentration levels of both UDP-N-acetylglucosamine and UDP-N-acetylgalactosamine, since these two nucleotide sugars are readily interconvertible.

5. Immunology

The results of studies on the antigenicity of proteoglycans have been plagued with confusion and dubious interpretations in large measure because of the uncertain character of the preparations employed. The many discordances between results of different investigators point to the crucial importance of extensive purification of antigens (DI FERRANTE, 1970). In spite of many complexities (BRANDT et al., 1970), two general conclusions with regard to proteoglycans of cartilage appear likely to stand the test of time: 1. the proteoglycans of vertebrate cartilage contain common cross-reacting determinants and 2. the same preparations also contain one or more species-specific determinants. A brief resumé of the principal results of immunological studies follows, with primary emphasis on correlation with chemical structure of proteoglycans.

Both precipitating and agglutinating antibodies may be induced by proteoglycans of mammalian cartilage. Although the main glycan component of these proteoglycans is chondroitin sulfate, variable amounts of keratan sulfate, and possibly other unidentified components as well, may be present. The antigenic determinants appear to be associated with, or dependent upon, the protein components of the preparations since the glycan components themselves, when freed from peptide linkage have very little or no capacity to act as antigens. LOEWI and MUIR (1965) found that the precipitating antibody to a proteoglycan preparation from porcine cartilage, which was produced by immunized rabbits, cross-reacted strongly with proteoglycan preparations from porcine cartilage and bovine cartilage but weakly with cartilage proteoglycans of man and guinea pig. These authors speculated that the cross-reacting determinant might reside near the linkage region of glycan and protein. SANDSON et al. (1966), using similar preparations, demonstrated a minimum of two antigenic determinants, one species-specific, and one common to pig, ox, and man.

The chemical-structural basis for one common antigenic determinant was elucidated by the studies of SANDSON et al. (1970). Proteoglycans of cartilage and notochord were exhaustively digested with trypsin and chymotrypsin, the resulting glycan-peptide was purified and the glycan moiety digested with testicular hyaluronidase. In addition to the peptide sequence, these 'truncated doublets' included the linkage region sequence of glucuronic acid-galactose-galactose-xylose, bound glycosidically to serine, as well as one or two repeating disaccharide periods of chondroitin sulfate. These preparations inhibited the agglutination reaction of red blood cells coated with porcine proteoglycan by antiserum to bovine proteoglycan and also prevented the precipitation reaction between proteoglycans and antiserum. Thus, it was likely that the 'truncated doublets' contained a common antigenic determinant structure located near the juncture of the polysaccharide chains to the peptide sequence of the "doublet" (Sect. 1b and Fig. 6.5).

The presence of a common antigenic determinant was demonstrated also for proteoglycan preparations from cartilage of pig, cow, frog *(Rana catesbiana)*, chicken, trout (*Salmo* gairdneri), sturgeon (*A.* fulvescens), shark (*P.* glauca), lamprey (*P.* marinus), and squid (*L.* pealii), and from notochord of sturgeon and lamprey. In a subsequent study, BAXTER and MUIR (1972) showed that the sugars of the linkage region did not contribute significantly to the expression of the species-common determinant, thus pointing to the amino acid sequence of the doublet peptide as the principal determinant structure. One or more common antigenic determinants are associated with the amino acid sequence near the points of attachment of keratan sulfate side-chains to the core protein (KEISER and SANDSON, 1973).

One or more additional species-specific antigenic determinants are connected with the "glycoprotein link" (DI FERRANTE et al., 1970; KEISER et al., 1972), a component of cartilage proteoglycan preparations that is non-covalently associated in aggregates of the basic molecular units of proteoglycan of bovine cartilage (see Sect. 1b).

The hyaluronate-peptide of human synovial fluid contains two antigenic determinants, one specific for this compound and one that cross-reacts with antiserum against the proteoglycan of human cartilage (HAMERMAN et al., 1966). The cross-reaction may be due to either an undetected contaminant or some chemical structural feature common to both macromolecules. However, it appears unlikely that the same contaminant would be present in preparations of hyaluronate from Group-A hemolytic streptococci. To explain the cross-reactivity of all three preparations, HAMERMAN and SANDSON (1970) suggested a similarity of carbohydrate linkage region. However, hyaluronate does not contain xylose, which is a normal unit of the linkage region of the proteoglycan of cartilage.

Relatively crude preparations of proteoglycan from nucleus pulposus of man, ox, and dog contain at least two antigenic determinants common to all three species and, in addition, at least one species-specific determinant. The common determinants are closely related, if not identical, with determinants in similarly obtained preparations from cartilage (PANKOVICH and KORNGOLD, 1967).

6. Evolutionary Relationships

Since the chemical structural information, especially of amino acid sequences, is still insufficient to permit definitive conclusions to be drawn, we are limited to the consideration of various hypotheses regarding homologies (MATHEWS, 1967a; see Chap. 1).

a) Proteoglycans of Cartilage and Notochord

i) Macromolecular Architecture

The evidence presented in this chapter indicates that the proteoglycan molecules of cartilage and of notochord in vertebrates are probably constructed according to identical architectural plans. The parent type molecule (Fig. 6.3) consists of a polypeptide backbone with chondroitin sulfate side-chains grouped in pairs (Fig. 6.5). The unaggregated molecules of phylogenetically broad range of vertebrates, with the apparent exception of the lamprey, are similar with respect to average molecular weight and size (Table 6.2). The differences in molecular parameters may be attributed to differences in length of the glycan chains, which are variable over a ten-fold range, and in linear density of substitution of glycan chains in the polypeptide. However, small differences in peptide chain length are not excluded. Some differences may arise because of artifacts of preparation, natural processes of catabolism, and incomplete biosynthesis in respect to glycan chain initiation (MATHEWS, 1971). Also, genetically determined polymorphism of the protein backbone may contribute additional complexities. Nevertheless, although the proteoglycan of various species differ considerably in secondary structural details, they possess a similar general molecular morphology.

ii) Polypeptide Structure

The polypeptide chain that forms the backbone of the proteoglycan molecule appears to consist in part of "short" repeating polypeptide sequences of less than 10 amino acid residues. These short sequences contain at least one glutamic acid, two glycine, and two serine residues, the latter residues being responsible for linkage to glycan chains (Fig. 6.5). A "long" polypeptide sequence of approximately 35 amino acid residues is associated with the "short" sequences. Probably because of its role in biosynthesis of the linkage region, the "short" sequence shows a high degree of structural constancy in evolution (Sect. 5). In contrast, less rigid functional requirements seem to have permitted a greater latitude in amino acid composition of the "long" sequence, as indicated by the considerable variation in amino acid compositions of the proteoglycans of Table 6.3. It is possible that internal homologies within the total amino acid sequence occur in the proteoglycan. Such homologies have been shown for α1-acid glycoprotein of serum (SCHMID et al., 1972). The duplicated sequence might be as short as the sequence that comprises the "short" sequence and the "long" sequence. The sequence of "short" and "long" sequences may not be regularly repeating and either sequence may exist in block polymer repeats.

The polypeptide moieties of the proteoglycans of cartilage and notochord appear to be specific for connective tissue type (Table 6.3). In addition, the differences in amino acid composition between cartilage proteoglycans of embryo and adult chicken suggest a relationship of molecular structure to developmental stage. Polypeptide polymorphism may occur also within the same tissue (MATHEWS, 1971; BRANDT and MUIR, 1971). The existence of separate genes coding for different polypeptide backbones of chondroitin sulfate-proteins of different tissues was suggested by MATHEWS (1967b) as an interpretation of observations on nanomelic chicks showing a defective synthesis of chondroitin sulfate in cartilage but not in skin (see also Sect. 1b).

iii) Glycan Structure

As noted above (Sect. 2) the linkage-region carbohydrate components of chondroitin sulfate chains appear to be the same for all vertebrates with the exception of hagfish. The remainder of the glycan chain consists of repeating disaccharide units. In addition, chains of skeletal keratan sulfate are also present in the proteoglycan molecules, commonly in highly variable proportions of the total glycan present. In view of the likelihood that the corresponding transferases involved in assembly of the glycan chains in different vertebrates will prove to be homologous proteins, it is reasonable to assume that the glycan chain portions of the proteoglycans are "indirectly" homologous.

b) Other Proteoglycans

Comparative biochemical data on proteoglycans other than those in cartilage and notochord are very meager. Results relating to homology of polypeptide structures are limited to observations on residual amino acids bound to glycans after extensive proteolytic treatment. Similarities in composition suggestive of homology were found in preparations of skeletal keratan sulfate from several species of shark, coelacanth *(Latimeria chalumnae)*, garfish *(Lepisosteus spatula)* chicken, ox, and man. Residual amino acids were also similar in preparations of corneal keratan sulfate from ox, chicken, and whitefish *(Coregonus* sp.). In all the latter preparations, aspartic acid was predominant (MATHEWS and CIFONELLI, 1965).

The similarities in linkage-region sugars and amino acids among proteoglycans of chondroitin sulfate, dermatan sulfate, heparan sulfate, heparin and possibly hyaluronic acid suggest that homologies are not restricted to proteoglycans characterized by a single type of glycan chain.

c) Perspectives

The concept of molecular homology has contributed greatly to our understanding of a wide range of biologic phenomena. Applied to the polyanionic proteoglycans, it serves as a unifying principle, helping not only to organize seemingly diverse structural data, but also to illuminate aspects of physiologic properties and functions. Difficulties arise because of great chemical complexity.

The proteoglycans, like the other structural macromolecules of connective tissue matrices (with the apparent exception of elastin), possess carbohydrate as well as a protein moieties. Both types of segments of the macromolecule exhibit an array of secondary chemical variations. The immediately following chapters will treat the secondary chemical features of the glycan moieties, such as site of sulfation, extent of sulfation, minor constituents, etc., and attempt to assess their biologic significance. Subsequent chapters will be concerned with secondary, tertiary, and higher order structures. However, before proceeding further, it is desirable to summarize briefly and place in evolutionary perspective our present understanding of the structure of polyanionic proteoglycans.

The basic, chondroitin sulfate proteoglycan molecules of vertebrate cartilage have a similar molecular architecture but differ in many structural details (see Table 6.5). Their glycan chains are identical

Table 6.5. Conservative and relatively variant structural features of chondroitin sulfate proteoglycans in vertebrate phylogeny - a tentative listing

Conservative	Relatively variant
1. Molecular architecture: polypeptide with glycan side chains glycosidically linked to serine residues	1. "Long" sequence
	2. Polypeptide polymorphism
	3. Glycan chain length
2. Polypeptide chain length	4. Sulfation of glycan
3. "Doublet" chain substructure	5. Degree of glycan addition to polypeptide
4. "Short" sequence	
5. "Linkage region" sequence	6. Degree of keratan sulfate addition to polypeptide

with respect to the parent repeating disaccharide and the linkage-region carbohydrate sequence. The tentative classification of these molecules within a single homologous group is strengthened by their association with homologous skeletal structures, i.e. with elements of the chondrocranium, elements of the axial skeleton, etc. However, firm establishment of the hypothesis of homology awaits completion of the formidable task of complete structural determination. In addition, the question of polymorphism of proteoglycans of cartilage requires elucidation. Notochordal proteoglycans resemble their cartilagenous counterparts in the same organism very closely, but appear to differ in their protein moieties. It is possible that the proteoglycans of these two tissue types diverged from a common ancestral molecule soon after the evolutionary emergence of the earliest vertebrates. Also, structural gene relationships for chondroitin sulfate proteoglycans of other tissues remain to be explored. In the absence of sufficient data, it may be unwarranted to assume (as many do implicitly) an identity of the polypeptide moieties of proteoglycans of the same glycan type from different types of connective tissue of the same organism.

The noted similarities in linkage region between proteoglycans of dermatan sulfate, heparan sulfate, heparin, and chondroitin sulfate suggest some degree of homologous relationship between the molecules. The polypeptide moieties of these proteoglycans in the same organism may be very similar, or possibly even identical. Thus, the type of glycan chain introduced may depend in part upon the polypeptide chain structure and, in part, upon the types and activities of the glycosyltransferases present in the cell.

Although present data are very meager, it appears reasonable to anticipate that homologies may extend beyond the phylogenetic boundaries of the vertebrates. This expectation is based principally on the presence of serine, galactose, and xylose in preparations of chondroitin sulfate-type glycans in cartilage of the squid and the horseshoe crab. Also, the general similarity of sulfotransferases and glycosyltransferases of invertebrates to their counterparts in vertebrates with respect to specificities for nucleotide donors and glycan acceptors suggests that the analogous enzymes may prove to be homologous. However, the homologies may be remote in view of the apparently great evolutionary separation of many invertebrate phyla from the line of chordate evolution (see Fig. 2.2).

Chapter 7

Polyanionic Glycans of Cartilage, Bone, and Notochord

"Cartilage, like pornography, is usually not defined since it is easily recognized by experts." ANON.

There has been general agreement among biologists for a long time that bone is a uniquely vertebrate tissue. However, conflicting views have been held of the taxonomic distribution of cartilage. The earlier arguments in favor of the presence of "true" cartilage in invertebrates were summarized by NOWIKOFF (1912). Opinions, still widely prevalent, that cartilage is confined to vertebrates have been expressed by SCHAFFER (1930), LUBOSCH (1938), HYMAN (1940), ROMER (1942), and others. Recently, the entire question has been reopened and indisputably resolved in review articles by PERSON and PHILPOTT (1967, 1969a). To the older morphological and histological observations were added newer data including results from electronmicroscopic and chemical studies. Collagen as well as polyanionic glycans (discussed below) were clearly identified as essential components. PERSON and PHILPOTT (1969a, b) suggested that cartilage may have originated in invertebrates but recognized that this tissue type could also have evolved independently in both invertebrates and vertebrates in response to similar needs for structural support.

1. Biological Significance

a) Cartilage and Bone

An interpretation of the comparative biochemical data on vertebrate cartilage and bone must necessarily consider the evolutionary history and functional relationships of these two intimately associated tissues. For this reason, and also because there are divergent views of the origin of cartilage and bone in vertebrates, it is desirable to briefly summarize current opinion. In particular, the relative primitiveness of cartilage and bone has been widely debated.

BERRILL (1955) proposed a plausible theory of the phyletic origin of vertebrates from the free-swimming larval stage of precambrian tunicates. He suggested that cartilage evolved as a secretion of connective tissue cells, serving primarily a mechanical function, and that bony material formed as an outer layer, partly from a tendency for calcium phosphates to be excreted from the skin and partly as a development of impermeability. SMITH (1961), a proponent of the view that the earliest vertebrates, the ostracoderms, evolved in fresh water, concluded that the characteristic bony armor of the ostracoderms served primarily for osmotic insulation in an environment of relatively low salt content. URIST (1962a) and MOSS (1964) believe that bone evolved primarily as part of the homeostatic mechanism for control of calcium, phosphate, and other ions. However, DENISON (1963) considers this speculation to be unlikely. ROMER (1963) suggested that the hard dermal tissue of the ostracoderms served mainly as protection against predaceous freshwater eurypterids. On one point, nevertheless, the paleontological evidence appears clear: bone is an ancient tissue, probably dating from the Ordovician period, a basic and primitive characteristic of the earliest vertebrates (ROMER, 1968).

If, as has been argued convincingly by DENISON (1956, 1963), ROBERTSON (1957) and TARLO (1964), the first vertebrates were marine, then neither osmotic nor defensive functions could have been primary factors in the early development of dermal bone. HALSTEAD (1968) held that acellular aspidin, a primitive calcified tissue of vertebrates, functioned first as a phosphate store and later as protection. Only subsequently in evolution did bone penetrate below the dermal surface to become associated with, or replace, cartilage and take on an endoskeletal supporting function. However, HALSTEAD (1969) now accepts the view propounded by DENISON (1967) and ØRVIG (1967, 1968) that the earliest calcification in vertebrates occurred in cartilage. He takes the position that bone is indeed a direct evolutionary derivative from cartilage in the endoskeleton (but not in the dermal skeleton).

Thus, it is possible that some cartilage tissues in adult vertebrates may represent a primitive condition. URIST (1962a) and DENISON (1963) have suggested that the ancestral vertebrate was soft-bodied and that the chondrichthyes may have descended from some wholly cartilaginous ostracoderm and not from the ossified placoderms by reduction of bone (see Fig. 3.1). However, since cartilage does not readily fossilize, paleontological support for this view is lacking. Yet, extensive loss of endochondral bone with replacement by cartilage has evidently occurred in many phyletic lines of fish and amphibia, although the general direction in vertebrate evolution has been toward increasing endochondral ossification. A further interpretation of the significance of cartilage in these instances has been offered by DE BEER (1958) and by ROMER (1942, 1968). Cartilage may be regarded primarily as an embryonic adaptation essential for the orderly development of an ossified internal skeleton. The presence of cartilage in adult vertebrates therefore indicates neoteny, or the retention of an embryonic structure in an adult stage. BERRIL (1955), however, disagrees with the conclusion that cartilage could never have been an important component of a fully adult vertebrate skeleton.

Thus, while the view of cartilage as an embryonic adaptation would appear to be widely justified, it is probably not applicable to certain specific cartilagenous structures of particular descendent lines (for example, vertebral centra of chondrosteans, in which the appearance of ossification has been comparatively slow). Accordingly, I shall accept here the eclectic position that describes cartilage as a primitive endoskeletal support tissue which became associated with bone at many anatomical sites in adult vertebrates early in evolutionary history and subsequently was restricted mainly to embryonic stages of most vertebrates. From this broader viewpoint, a particular cartilagenous structure is primitive if the homologous element in ancestral forms has remained continuously unossified. Alternatively, if ossification has occurred in some ancestral form, the descendent cartilagenous structure may be regarded as neotenous.

Dermal or membrane bone must be distinguished from "endochondral" or "cartilage-replacement" bone on the basis of differences in origin, development, anatomical site and evolutionary history. It is not preceded in the embryo by cartilage, and its absence (or reduction) from descendents of the armored ostracoderms is a secondary condition. It should also be borne in mind that in evolution cartilage and bone are closely interwoven. MOSS (1964) broadly considers that the group of skeletal tissues, including cartilage and bone and all intermediate tissues, forms a series connected by common embryonic inductive mechanisms under the influence of extrinsic factors of mechanical function. For example, SCHAEFFER (1961) has indicated that the distribution and density of cartilage and bone in the fish skeleton are adaptively responsive mainly to factors of mechanical stress, equilibrium, buoyancy, and protection.

b) Notochord

The notochord, an ancient structure which vertebrates apparently (but see Sect. 3) share with such lower chordates as amphioxus and tunicates, has an origin in embryogeny of vertebrates distinctly separate from the bulk of mesoderm (DE BEER, 1958; ROMER, 1964). In vertebrates it is essentially a highly cellular elongated tissue surrounded by a cylindrical thickened sheet, which contains collagen and anionic glycosaminoglycans. The source of the acellular sheath is obscure but may derive from the interior cells, which are epithelial in origin. In all fish, most fossil and all living amphibia, except anurans, and in some fossil reptiles, the notochord remains continuous, though constricted. In higher forms the notochord is limited to intervertebral bodies composed mainly of other elements, including cartilage. The notochordal sheath is penetrated conspicuously only in chondrichthyes, chondrostei, and dipnoi by embryonic, mesenchymal cells which contribute a matrix in the formation of centra (GOODRICH, 1958; SMITH, 1960). Cyclostomes alone show no trace of centra. Histochemical studies are briefly reviewed by LEESON et al. (1961).

The axial skeleton of ostracoderms remains unknown. However, a tapering notochordal canal in the cranial region of these fossils indicates the presence of a persistent and well-developed notochord (STENSIÖ, 1958). If, as suggested by BERRIL (1955), ROMER (1970), and others, vertebrates arose from some larval Precambrian tunicate that possessed a notochord, one might expect to find biochemical evidence of affinity between notochordal structures in ancestral and descendent forms. Unfortunately, such data are not yet available.

c) Taxonomic Interpretation

The interpretation of taxonomic data for evolution has led to the widespread use of a variety of terms, some of which were introduced in Chapter 1. To avoid the possibility of misinterpretation, the definitions and descriptions given by SIMPSON (1961) will be used.

"The concepts of primitive and specialized are relative and are meaningless unless definitely related to a particular taxon, lineage, or phylogeny."

"Within a taxon, the characteristics of the common ancestry are primitive and others are more or less specialized in proportion to their departure from the ancestral condition."

"Within a single lineage, characteristics occurring earlier are more or less primitive and those appearing later more or less specialized in proportion to their times of appearance."

"Within a phylogeny (which is divided into taxa and consists of branching lineages) characteristics of any one common stem are more primitive than different characteristics of its descendant branches; within single branches they are more specialized in accordance with earlier or later appearance; and between branches they are more primitive or more specialized in accordance with less or more departure from the stem condition."

"*Homology* is resemblance due to inheritance from a common ancestry."

"*Homoplasy* is resemblance not due to inheritance from a common ancestry."

"*Parallelism* is the development of similar characters separately in two or more lineages of common ancestry and on the basis of, or channeled by, characteristics of that ancestry."

"*Convergence* is the development of similar characters separately in two or more lineages without a common ancestry pertinent to the similarity but involving adaptation to similar ecological status."

"*Analogy* is functional similarity not related to community of ancestry. Convergent characters are analogous insofar as the similarity can be related to function, which is usually and perhaps always the case."

SIMPSON (1961) has also commented on the difficulty or practical impossibility of distinguishing parallelism from homology on the one hand and convergence on the other. "The three processes, distinct enough in their characteristic manifestations, do intergrade through marginal cases. Fortunately, it is improbable that pure parallelism in a whole organism and at equal rates in all lineages ever occurs. It is always accompanied by homologies and perhaps also always, certainly usually, by divergent specialization in each lineage. It also commonly occurs at different rates in different lines. The lines may be sorted out on their divergent characters, and any convergence involved can usually be recognized by criteria already discussed. Remaining resemblances are then either homologous or parallel. The distinction between homology and parallelism can then frequently be made from the fact that homologous characters are primitive for the group but parallel characters are not."

It should be noted that data on only the glycan portion of proteoglycans are presented in this chapter. Also, presumed homologies between glycans of similar chemical structure necessarily fall in the category of "indirect homology" as defined in Chapter 1. Furthermore, it is important to bear in mind that cartilage tissue of a given species of organism may vary greatly in content and chemical structure of its glycans with development and with aging. Therefore, comparisons between species are biologically meaningful only if the tissues considered correspond to similar stages of development and aging in the life cycle of the individual organisms. In the absence of complete data covering the entire life cycle, an acceptable compromise is made by considering only results for organisms at an approximately similar early adult stage. The problems involved in comparative studies of cartilage are discussed further in Sects. 4 and 6 of this chapter; a detailed treatment of developmental and aging changes is reserved for the following chapter.

2. Invertebrates

The morphology and biochemistry of invertebrate cartilage have been summarized by PERSON and PHILPOTT (1967, 1969a), and by HUNT (1970).

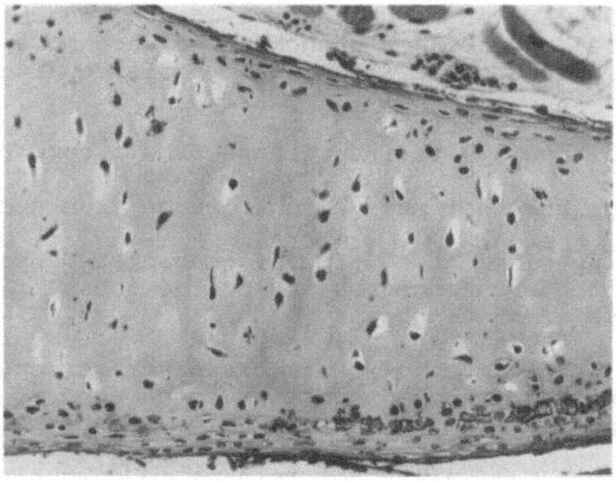

a

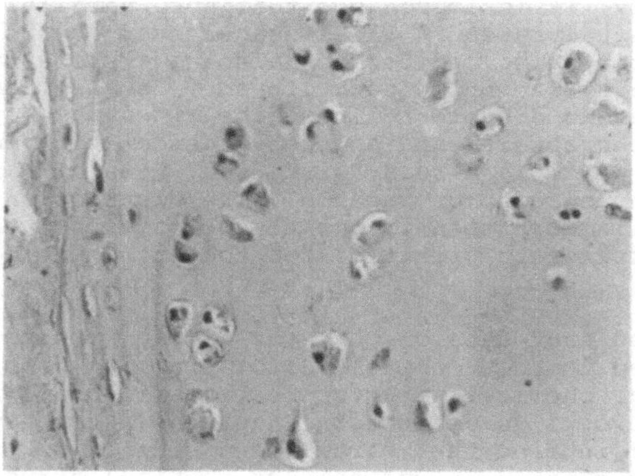

b

Fig. 7.1a-e. Invertebrate cartilages: (a) Dogfish *(S. acanthias)* gill cartilage, 170×; (b) squid *(L. loligo)* head cartilage, 170×; (c) horseshoe crab *(L. polyphemus)*, 8 in. diameter, gill cartilage, 250×; (d) feather-duster worm *(E. polymorpha)* endoskeletal complex, 200×; (e) marine snail *(B. canaliculatum)* odontophore, 700×. (Courtesy of PERSON and PHILPOTT, 1963, 1967, 1969b)

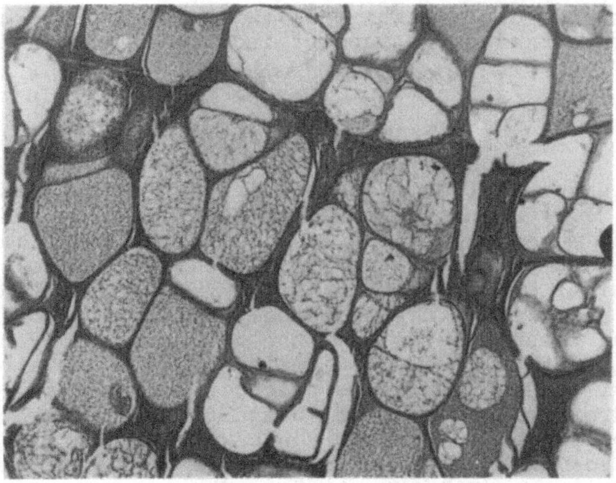

c

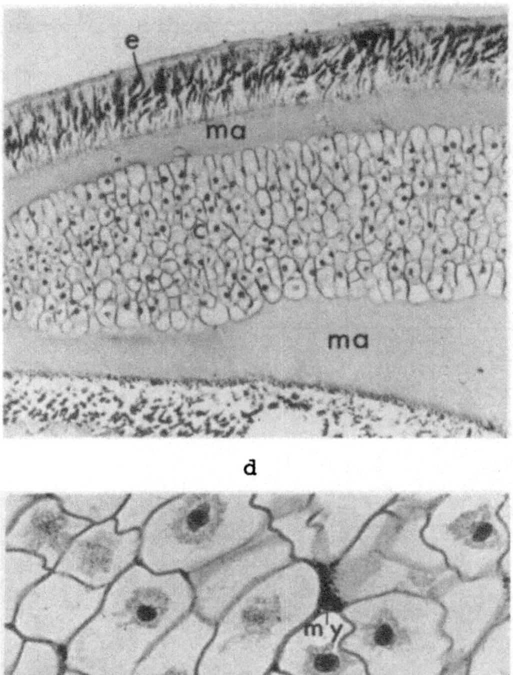

d

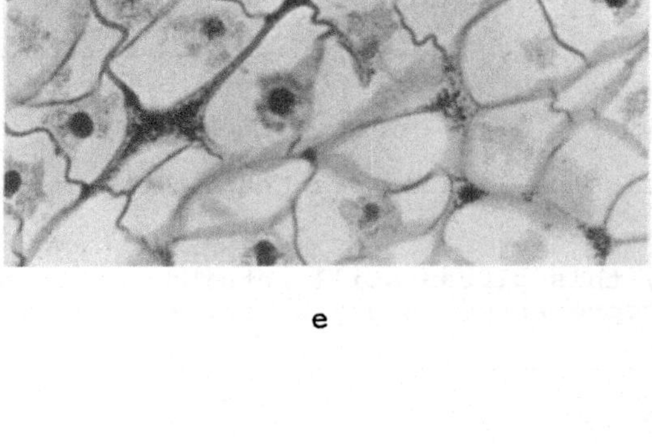

e

HUNT (1970) has also dealt extensively with polyanionic glycans of other invertebrate tissues, a subject included in the following chapter.

a) Annelida

Endoskeletal tissues derived from mesoderm and resembling cellular cartilage (Fig. 7.1) are often found in the tentacles and associated basal structural masses of sedentary polychaetes (particularly amongst the so-called (feather-duster' worms of the family Sabellidae). The results of chemical studies with *Eudistylia polymorpha* by PERSON and MATHEWS (1967), combined with additional histological observations, demonstrate that these tissues meet the criteria for identification of cartilage, i.e. they consist of cells suspended in a relatively rigid matrix (of varying abundance) rich in collagen and polyanionic glycan. The hydroxyproline content (based upon dry weight of tissue) is 3.2% as compared to 3.5% in embryonic (18-day) chick articular cartilage; the polyanionic glycan content is 2.5% compared to about 7% in the chick tissue (MATHEWS, 1967b). The polyanionic glycan content, however, exceeds that in lamprey cartilage, which is 1.5%.

The polyanionic glycan is similar to chondroitin 6-sulfate in composition except for the ester sulfate content, which is 1.88 moles per mole of galactosamine (Tables 7.1 and 7.2). Like its counterpart from

Table 7.1. Disaccharide periods of galactosaminoglycans identified by hydrolysis with chondroitinase according to SUZUKI et al. (1968) and SAITO, YAMAGATA, and SUZUKI (1968)

Symbol	Uronic acid type	Uronic acid sulfate position	Galactosamine sulfate position	Parent-type glycan[a]
O-S	glucuronic			ch
4-S	glucuronic		4	ch 4-S
6-S	glucuronic		6	ch 6-S
diS$_D$[b]	glucuronic	2 (or 3)	6	ch 6-S
diS$_E$	glucuronic		4, 6	ch 4-(or6-)S
4-S$_B$	iduronic		4	derm 4-S
6-S$_B$	iduronic		6	derm 6-S
diS$_B$	iduronic	2 (or 3)	4	derm 4-S
diS$_H$	iduronic		4, 6	derm 4-(or 6-)S
triS$_H$[c]	iduronic	2 (or 3)	4, 6	derm 4-(or 6-)S

[a] See Fig. 6.1
[b] A possible variant has sulfate on C-4 in place of C-6 of galactosamine.
[c] From hagfish (cyclostome) skin (see Chap. 9, Sect. 4a).

vertebrate tissues, this glycan still retains, after papain treatment, residual peptide, represented mainly by serine and glycine. However,

Table 7.2. Composition of polyanionic glycosaminoglycans of invertebrate cartilage (molar ratio to hexosamine = galactosamine + glucosamine)

Specimen	Galactosamine	Hexuronic acid	N	N-acetyl	SO_4	$[\alpha]_D^{25}$	$[\eta]$
L. pealii [a]	0.99	0.99	0.99	1.09	1.49	-56°	1.51
O. sloani pacificus [b]	1.00	0.99	1.04		1.50	-52	
A. dux [c]	0.99	1.17	1.00	1.00	1.22	-31	2.90
A. dux [c]	1.00	1.08	0.99		1.31		2.27
O. maculatus [c]	0.91	1.18	1.00		1.10		2.20
E. polymorpha [d]	0.97	1.07	1.20		1.88		0.32
L. polyphemus [a]	0.98	1.13	1.29	0.95	1.45	-58°	0.63
(chondroitin 4-sulfate) [e]	1.00	1.15	1.20	1.00	0.97	-33°	0.32

Notes: Values of $[\alpha]_D^{25}$ corrected for moisture; $[\eta]$ determined in 0.4 *M* NaCl with 0.01*M* phosphate buffer pH 7.0, in units of g/100 ml.
a MATHEWS, DUH, and PERSON (1962). b KAWAI, SENO, and ANNO (1966).
c Unpublished; forpreponderant fraction. d PERSON and MATHEWS (1967).
e Reference standard prep. (MATHEWS, CIFONELLI, and RODÉN, unpubl.).

its infrared spectrum differs slightly from that of chondroitin 6-sulfate, possibly due to the additional sulfate present. It is this extra sulfate content which appears to be responsible for the slow rate of hydrolysis by testicular hyaluronidase of the glycan compared to chondroitin 6-sulfate, since chemical desulfation of the *Eudistylia* glycan eliminates this difference in rate of hydrolysis.

Preliminary studies suggested the presence of a glycan of similar chemical composition in cartilage from *Sabellastarte* sp., a member of the same family as *Eudistylia* (PERSON and MATHEWS, unpubl.). The polysaccharide was not appreciably hydrolyzed by testicular hyaluronidase. It must differ significantly in structure from the highly sulfated glycosaminoglycans of vertebrates (Table 7.3), since bacterial chondroitinase ABC was also without hydrolytic effect (courtesy of Drs. SUZUKI and YAMAGATA). The contents of hydroxyproline (3.0%) and polyanionic glycan (2%), on a dry weight basis, were similar to values found for *Eudistylia* cartilage.

b) Mollusca

Cellular endoskeletal tissues resembling cartilage in appearance occur widely among molluscs. However, chemical studies of glycans are limited to tissues of gastropods and cephalopods.

i) Gastropod Odontophore Cartilage

PERSON and PHILPOTT (1969a) have briefly reviewed the histology (Fig. 7.1) and biochemistry of gastropod odontophore cartilage. These tissues, of mesodermal origin, function in support of the radulae of *Gastropoda*. They contain, as in *Busycon canaliculatum*, collagen-like fibrils and fibers with a banding periodicity of 600 to 650 (PERSON and PHILPOTT, 1963).

Table 7.3. Disaccharide period composition (%) of cartilage and notochord polyanionic glycans of Cyclostomata and Chondrichthyes

Specimen	Tissue	O-S	4-S	6-S	diS_D	diS_E	$4\text{-}S_B$	$6\text{-}S_B$	diS_B	diS_H	SKS[a]	S/H[a]
					Cyclostomata							
Hagfish *(E. stouti)*	cart.[b]		7	2		91						1.80
" *(E. stouti)*[c]	noto.		5	10	<5	5(?)	10		20	45		2.04
" *(M. glutinosa)*[c]	noto.				<5	<5	30		20	40		1.80
Lamprey *(P. marinus)*	cart.		5	95								1.00
" "	noto.		95	5								0.98
					Chondrichthyes							
Ratfish *(H. collei)*	cart.		10	80	10						<5	1.11
Hornshark *(H. franscisci)*	cart.		10	65	5						20	1.05
Tiger shark *(G. cuvier)*	cart.		10	75	10						5	1.13
Dogfish *(S. acanthias)*	cart.	5	30	55	10						<5	1.15
Lemon shark *(N. brevirostris)*	cart.		30	20	15						35	1.33
Bull shark *(C. leucas)*	cart.		20	35	15						30	1.22
Bull shark *(C. leucas)*[d]	cart.		20	30	20						30	1.25
Seven-gill shark *(N* maculatum)	cart.	15	5	75	5						<5	0.90
" " " "	noto.		20	75	5						<5	1.07
Sawfish *(P. perotteti)*[d]	cart.		35	25	10						30	1.12
Stingray *(Dasyatis* sp.)	cart.		5	75	20						<3	1.20
Freshwater ray *(P. circularis)*	cart.[e]	5	5	90	tr							0.95

a See Table 7.1 for definition of disaccharide periods; SKS = skeletal keratan sulfate. S/H = molar ratio of sulfate to galactosamine of uronic acid-containing glycans. b Cartilage is cranial except where otherwise indicated. c Incomplete enzyme hydrolysis; composition of recovered disaccharides (ANNO et al., 1971). d From fresh water of Lake Nicaragua. e Vertebral cartilage.

LASH and WHITEHOUSE (1960) isolated from *Busycon* odontophores a high-molecular-weight sulfated glycan, free of uronic acid and amino sugar, in yield of 27% of the dry weight. The only monosaccharide isolated upon hydrolysis was D-glucose; the ester sulfate content was approximately equimolar with that of glucose. Since β glycosidic linkages appeared likely, a structural relationship to cellulose was considered possible.

ii) Cephalopod Cartilage

Cartilage tissues composed of branching cells suspended in an abundant matrix are widely distributed among many cephalopod genera (PERSON and PHILPOTT, 1969a). They are found associated with the eye as scleral cartilage and the equatorial ring cartilage of the iris, in the head as cranial cartilage, and as a supporting structure in the neck of squid. PHILPOTT and PERSON (1970) have made a detailed light- and electron-microscope study of the head cartilage of the squid *Loligo pealii* and found a strong resemblance to many vertebrate hyaline cartilages (Fig. 7.1). Banding periodicities were not detectable in any of the fibrillar structures. However, the tissue gave the typical wide-angle X-ray pattern of collagen and had an amino acid composition that suggested a high collagen content.

Cranial and nuchal cartilages of *L. pealii* contain a polyanionic glycan (MATHEWS et al., 1962) that resembles chondroitin 4-sulfate in composition except for a molar ratio of ester sulfate to galactosamine of 1.49 (Table 7.2). Like the polyanionic glycan from vertebrate tissues, the squid preparation was rapidly and completely hydrolyzed by testicular hyaluronidase, suggesting similar glycosidic linkages. A very similar preparation (Table 7.2) was obtained from cartilage of the squid *Ommastrephes sloani pacificus* by ANNO et al. (1964) who isolated (KAWAI et al., 1966) by partial acid hydrolysis 2-acetamido 2-deoxy-3-O-(β-D-glucopyranosyluronic acid)-D-galactose (chondrosine) in 45% yield. Subsequently, KAWAI (1968) demonstrated the presence of diS_E periods (see Table 7.1). She estimated that the composition in disaccharide periods was 60% diS_E, 30% 4-S and 10 6-S.

Preparations of somewhat different sulfate contents were obtained from cranial cartilage of other cephalopods (MATHEWS, unpubl.). A preparation (0.7% of the wet tissue weight; $[\eta] = 1.00$) from *L. opalescens* was fractioned to give a minor component with a molar sulfate/hexosamine ratio of 1.35 and a major component with a sulfate/hexosamine ratio of 1.77. A preparation (8% of dry weight) from a specimen of giant squid *Architeuthis dux* (30 ft in length) was separated into fractions which ranged in sulfate/hexosamine ratio from 0.39 to 1.22. The bulk of the original preparation was represented by the fraction with a sulfate/hexosamine ratio of 1.22 (Table 7.2). This fraction had a weight-average molecular weight of 107,000, a value substantially greater than any for sulfated glycosaminoglycans of vertebrate cartilage. The major residual amino acid residues were serine and glycine. Another specimen of this species yielded a preparation (Table 7.2) representing 6% of the tissue dry weight.

Cartilage from a group of 10 specimens of *Octopus maculatus*, ranging in weight from 0.5 to 1.5 kg, yielded a preparation (3.0% of dry tissue weight; Table 7.2) that contained a relatively high proportion of glucosamine. The infrared spectrum was unusual and did not resemble that of the preparations from squid. The mixture was resolved by anion exchange chromatography into two principal fractions: one containing only galactosamine, and another with a glucosamine/galactosamine ratio of 1:4. Both fractions had similar ratios of uronic acid/hexosamine and retained serine and glycine as the major residual amino acid residues, but they differed greatly in susceptibility to hydrolysis by testicular hyaluronidase. *Octopus* cartilage thus differs in composition from homologous tissues of the other cephalopods.

It is particularly interesting that many cephalopods possess dermal scales with the histological and histochemical appearance of cartilage (PERSON, 1969). Such structures have their counterparts in the calcified cartilage of the dermal armor of ostracoderms (ØRVIG, 1968).

c) Arthropoda

Cartilage-like tissues have been detected histologically and histochemically in insects, scorpions, spiders, mites and in the horseshoe crab *Limulus polyphemus* (PERSON and PHILPOTT, 1969a). *L. polyphemus* gill cartilage, however, is the only tissue which has been extensively studied.

PERSON and PHILPOTT (1969b) have described details of the light- and electron-microscopic appearance of *Limulus* gill cartilage. The tissue strongly resembles the so-called "cellular" cartilage of lower vertebrates and early embryonic lim-bud cartilage of higher vertebrates. It also has many plant-like features. Amino acid residues, such as 3-hydroxyproline, 4-hydroxyproline, and hydroxylysine, characteristic for collagen, were found in amounts suggesting that collagen comprised only about 10-15% of the dry weight.

The only study to date of glycosaminoglycans of Limulus gill cartilage is that by MATHEWS et al. (1962). A preparation obtained by proteolytic hydrolysis of the tissue from animals 33 to 38 cm long represented 2.5% of the dry weight. Except for an additional peak at 800 cm^{-1} the infrared spectrum was similar to that of chondroitin 4-sulfate. Hydrolysis by testicular hyaluronidase was complete when determined by the turbidity method (MATHEWS et al., 1951). However, analyses by bacterial chondroitinase methods (SAITO et al., 1968) showed only 33% release of unsaturated disaccharide, almost entirely derived from periods of 4-S disaccharide (Table 7.1) in the glycan. A small amount of O-S disaccharide was also detected. The remaining 67% remained unidentified.

d) Evolution

Morphological criteria are generally adequate for identification of cartilage in the vertebrates, where anatomical homologies as well as developmental continuities are well established. Some difficulties, however, are encountered in recognizing as cartilage those tissues in which the relative proportion of intercellular matrix and cells is much reduced, for example, fibrous cartilage. But identification of cartilage solely on morphological and histological grounds becomes highly uncertain when our survey is extended to include invertebrate tissues.

In attempting to describe a cartilage-type tissue so as to encompass tissues of both invertebrates and vertebrates in a definition, we find it essential to utilize other criteria. Gross function is fundamental but too distant from molecular structure; purely morphological criteria, whether at the histological and histochemical levels or even at the molecular level, may be too restrictive. However, both adaptation for support function and morphology follow from the biosynthesis of specialized macromolecules. It is no surprise, then, that when these molecules are produced in certain proportions and amounts relative to the cell population, tissues of similar histological appearance of functional properties result.

The type of molecule common to all the invertebrate tissues just discussed is the collagen molecule. The second essential component is a polyanionic glycan, usually present as a proteoglycan. It is the organization of these two components, principally, that is the foundation of those properties of the intercellular matrix that are necessary to make a tissue cartilage-like, i.e. relatively rigid and resistant to a variety of such forces as compression, tension, and shearing. It is unlikely that these mechanical properties would be found in a tissue where the collagen content and the polyanionic glycan content are less than 10% and 1% of the dry tissue weight, respectively, or where the ratio of polyanionic glycan to collagen is less than 1:20. Apart from placing certain limits upon the contents and proportions of collagen and polyanionic glycan, our definition of cartilage also requires that the polyanionic molecule be a linear glycan of high linear charge density. It is likely that more than one anionic charge per disaccharide is necessary. A further requirement of proteoglycan-type structure is probable.

Although glycosaminoglycans in proteoglycan structures are universal components of cartilage in vertebrates and are present in cartilage of some invertebrates, the above definition is not limited by a requirement for hexosamine-containing glycans. For example, *Busycon* odontophore has the morphological, histological and cytological appearance of a typical cartilage tissue but contains a D-glucan sulfate in addition to collagen. Nevertheless, it may be considered cartilage, albeit a rather specialized type, by virtue of some unique characteristics (PERSON and PHILPOTT, 1967). A contrary view was expressed by LASH and WHITEHOUSE (1960), who preferred to emphasize the differences between the polyanionic glycan in this tissue and those present in vertebrate cartilage. However, a justification for regarding the glucan sulfate and chondroitin sulfate molecules as biologically equivalent may be given on physical chemical grounds. Both molecules are linear chains consisting of β-linked saccharides. The linear polyanionic charge density is similar for both, one containing two sulfate groups per disaccharide period, the other containing one sulfate group and one carboxyl group per disaccharide period. This essential similarity in polyelectrolyte structure, which indicates a similarity in interaction with collagen (MATHEWS and DECKER, 1968b) and with small ions (MATHEWS, 1964), suggests similar roles for both glycans in tissue organization and function (see also HUNT, 1970, p.29).

What, then, may be said of possible evolutionary relationships between the polyanionic macromolecules of invertebrate species and their apparent counterparts in vertebrates? It is evident from Fig. 2.2 that the invertebrate species considered here belong to phyla distant from the line of chordate evolution. Also, it is unlikely that cartilage was present in some remote common ancestor of both lines of the lower metazoa. On a purely morphological basis, one may presume that the appearance of cartilage tissue types among these very distantly related species was a result of biochemical convergence (see Sect. 1e). The basis for such biochemical convergence undoubtedly lay in the capacity of ancestral metazoa to synthesize both collagen-type molecules and some types of polyanionic glycans (see Chap. 8).

Some limited conclusions concerning macromolecules can be drawn. The collagen molecules evidently evolved along separate paths after divergence of the main lines of invertebrate evolution. It should be of particular interest to determine whether convergence at the morphological level was conditioned by convergence of the structure of collagen in vertebrate and invertebrate cartilage. The general structural similarities of the polyanionic glycans of invertebrate cartilages, particularly those of the glycosaminoglycan type, to the polyanionic glycosaminoglycans in vertebrate cartilages have already been noted.

Vertebrate cartilage more closely resembles squid cartilage than other invertebrate cartilage. Other structural similarities and convergences between vertebrates and cephalopods have been noted, including development of a complex nervous system, a brain protected by a cartilagenous case, and an image-perceiving eye. Perhaps more than any other invertebrate, the cephalopods have evolved along the same lines as the fast-moving predatory aquatic vertebrates. In this connection, the presence of diS_E glycan periods in squid cartilage and also in hagfish cartilage, alone among vertebrates, appears particularly striking (see Sect. 4a). Biochemical convergence often seems to be a necessary condition of morphological convergence. On this basis, we might have expected that the polyanionic glycans of cartilage of other invertebrates would show structural similarities to vertebrate glycosaminoglycans to a lesser degree. Although differences in glycan structure are evident between different classes of Mollusca and to a minor degree between different species within the same order of cephalopoda, some of the latter differences may relate only to the more variable secondary chemical features of structure, such as sulfation.

3. Lower Deuterostomia

A number of varied groups of animals have been combined with the vertebrates as the Deuterostomia (see Chap. 3, Sect. 7) in the widely accepted belief that they share a common ancestry with the vertebrates. The Deuterostomia are of interest here because many of these "invertebrates" possess endoskeletal structures resembling those of vertebrates. Although cartilagenous tissues have not been detected, structures homologous to the notochord of vertebrates are believed to be present in some groups, for example, in larval Echinodermata and Tunicata, and in adult Hemichordata and Cephalochordata. However, BARRINGTON (1965) cast considerable doubt on the implied homology of the 'notochord' of Hemichordata. Chemical studies of the connective tissues of these groups have not yet appeared in the literature.

The group of animals commonly called amphioxus is of especial interest to biologists. The amphioxus appears to possess several distinctive characteristics of the phylum Chordata and is considered to resemble some ancient ancester of this phylum (YOUNG, 1962). The amino acid composition of its collagen (Chap. 3, Sect. 8) accords with its taxonomic status.

The notochord of amphioxus is a highly elongated cylinder extending the entire length of the animal and situated dorsally just below the nerve chord. It is mainly cellular, surrounded by a thin sheath, and is embedded in a relatively acellular connective tissue of gelatinous consistency. Unlike the situation in vertebrates, the connective tissue surrounding the notochord is continuous, although the adjacent musculature is metameric. The main bulk of the connective tissue appears to be around the notochord and nerve chord, with a minor proportion distributed just below the epidermis and between muscle layers as septa. Upon staining with toluidine blue, the notochord sheath shows both metachromasia and dichroism (see also TENBAUM, 1955), suggesting the possible presence of sulfated glycans. The notochordal sheath is composed of two layers: an inner one of circular fibers, and an outer one of longitudinal filaments. Electron micrographs (EAKIN and WESTFALL, 1962) indicate that the fibers are collagenous, as suggested by TENBAUM (1955). Extensive descriptions have been given by DRACH (1948), HYMAN (1959) and PIETSCHMANN (1962a).

Amphioxus *Branchiostoma floridae* contains hyaluronate and highly sulfated glycosaminoglycans that are chemically unrelated to the sulfated glycosaminoglycans of vertebrates (MATHEWS, unpubl.). Histochemical observations suggest that these sulfated glycans are associated in part with the notochordal sheath. Further indication of the unusual character of amphioxus notochord was supplied recently by FLOOD et al. (1969) who showed that this structure is a uniquely specialized muscular organ of a type not found in any other chordate (or achordate). The demonstration of chitin (SANNASI and HERMANN, 1970) in gill bars of *B. floridae* adds additional weight to contentions that amphioxus is closer to invertebrates than to vertebrates.

4. Vertebrates

The living vertebrates are a relatively homogeneous taxonomic assembly. However, there is considerable uncertainty concerning the relations between the main groups of this subphylum, which appear already well established at the beginnings of the fossil record in the early Ordovician period (about 500 million years ago). Phylogenetic charts, such as that of Fig. 2.2, must be regarded as speculative in this respect (JARVIK, 1968).

a) Cyclostomata

Living Agnatha are represented today by two distinct branches, free of mineralized tissues: the Myxinoidea and the Petromyzontia (described by FONTAINE et al., 1958, and by PIETSCHMANN, 1962b). The first branch contains the marine hagfish, unique among vertebrates in many physiological features and resembling invertebrates in maintaining, largely by ionic means, a serum isotonic with seawater (BROWN and STEIN, 1960; URIST, 1963; ROBERTSON, 1963). A significant fraction of the intracellular osmotic pressure is furnished by near 200 m*M* each of trimethylamine oxide and an unknown acidic substance (BELLAMY and JONES, 1961). *Myxine* thus resembles both elasmobranchs and cephalopods in the quantity of trimethylamine oxide in the tissues, but is more like cephalopods with respect to the content of an acidic component. MORRIS (1960) and ROBERTSON (1963) consider the osmotic physiology of hagfish to be primitive and indicative of a derivation from marine chordate ancestors. Comparison of blood proteins led MANWELL (1963) to conclude that living vertebrates other than cyclostomes evolved from ancestors much closer to lampreys (Petromyzontia) than to the hagfish. This conclusion was reinforced by the findings of VAN PILSUM et al. (1967) that the hagfish *(Eptatretus)*, like those few invertebrates that also contain creatine, is lacking in transaminidinase and guanidinoacetic acid methylpherase, enzymes which are involved in creatine biosynthesis and are present in lamprey and higher vertebrates.

However, proposed relationships of hagfish to invertebrates remain highly speculative in view of the obscure phylogeny of cyclostomes

(HEINTZ, 1963). It is not clear whether recent cyclostomes are a monophyletic group which divided comparatively recently into myxinoidea and Petromyzontia, or, as argued by STENSIÖ (1968), a diphyletic group combining two independent branches that diverged in pre-Ordovician times. The absence of hagfish characters in a recently discovered fossil lamprey supports the view that the common ancestor of lampreys and hagfish lived prior to the Pennsylvanian period (BARDACK and ZANGERL, 1971). It is of interest to note that while the hagfish are restricted to a marine habitat, the lamprey enter freshwater streams to spawn and may adapt to permanent life in lakes (HARDISTY and POTTER, 1971).

i) Hagfish

The cartilage of hagfish (*Myxine glutinosa* and *Eptatretus stouti*, Table 7.3) contains at least two polyanionic glycans that are remarkable in possessing the unusual disaccharide period diS_E, which is also a major unit in squid cartilage. This disaccharide period has not been detected in cartilage of any other vertebrate, but is possibly present in small amounts in polyanionic glycans of coelacanth notochord (see later), hog intestinal mucosa, and fibrotic rat liver (ANNO et al., 1971). The polyanionic glycans of hagfish notochord are also very unusual in possessing a new disaccharide period, diS_H. To date, this type of period has been found elsewhere only in polyanionic glycans of coelacanth notochord (also possibly in hog intestinal mucosa and fibrotic rat liver).

ii) Lamprey

The polyanionic glycans of the cartilage and notochord of lamprey (*Petromyzon marinus*) have a relatively simple structure, being essentially the parent polyanionic glycosaminoglycans, chondroitin 6-sulfate and chondroitin 4-sulfate, respectively. Similar results were obtained with tissues of salt-water *P. marinus* that had just entered freshwater streams from Atlantic to spawn (MATHEWS, unpubl.). It is possible that the small amount of 6-S disaccharide periods detected in glycan preparations from the notochord arise from chondrocytes of the closely associated cartilagenous neural processes.

As will be established on the basis of the glycan type present in the two connective-tissue structures, the lamprey is closely associated with the non-cyclostome vertebrates while the hagfish is set apart. One possible interpretation of these findings is that the *parent*- type glycan components of cartilage and notochord that are shared by lampreys and non-cyclostome vertebrate embryos represent a primitive condition derived from the common ancestor of both groups. However, in their connective-tissue biochemistry as well as in many physiological characteristics, hagfish reflect an inheritance from a still more remote chordate ancestor, i.e. one shared by both acraniates and vertebrates. Alternatively, all the major lines of vertebrate evolution established in the Cambrian period or earlier may have possessed the full complement of glycosyl transferases and sulfotransferases needed to produce various polyanionic glycans. Subsequent evolution led to selective loss of enzymes in each line, resulting in either similarities or dissimilarities in biochemical composition between homologous tissues in different groups.

b) Chondrichthyes

The group of vertebrates classified as Chondrichthyes includes the sharks and their close relations, the skates and rays that form the subclass Elasmobranchii, and also the "ratfishes", or chimaeras, that form the subclass Holocephali. Although the last of the major groups of fishes to appear in the fossil record, the Chondrichthyes cannot be readily traced to their ancestral origins. Commonly, a derivation from some early placoderm form is accepted. However, the sharks and the chimaeras may have had separate origins in the now extinct class of Placodermi (ROMER, 1966). It is of particular interest that both groups maintain in their body fluids very high concentrations of urea (250-450 mmol/liter) and high concentrations of trimethylamine oxide (100-120 mmol/liter), apparently with osmoregulatory functions (SMITH, 1936), although the levels in Holocephali are somewhat lower than those in Elasmobranchii (READ, 1971).

The most distinctive feature of the class in implied in the name "cartilagenous fishes", i.e. an internal skeleton totally lacking in bone. Although bone as an internal skeletal structure was present in some early vertebrates long before the emergence of the chondrichthyans, there is no clear evidence to indicate whether the cartilagenous state of the endoskeleton in this class was a primitive condition or, as believed by ROMER (1966), derived from endochondral bone through a reductive process. However, it is probable that the dermal skeleton was strongly reduced from the ancestral state. It is of interest that, although elasmobranch cartilages may become heavily calcified, they never show characteristic bone structure.

The improbability of preservation of soft tissues in the fossil record has inhibited acquisition of knowledge of the axial skeleton of antecedent forms. It appears likely, however, that the wholly cartilagenous vertebral column and the closely associated notochord of Chondrichthyes were primitive characteristics. These skeletal structures may receive varied morphological expression: in some extant adult members of the group, the notochord is retained with limited growth of vertebra; in others this structure is greatly constricted by extensive vertebral development, with the sheath invaded by cartilage and replaced by centra that are calcified to varying degrees. Detailed descriptions have been given by BIGELOW and SCHROEDER (1948), DANIEL (1934) and GOODRICH (1958). Analytical data on the composition of cartilage and notochord glycosaminoglycans are given in Table 7.3.

The presence in chondrichthyan cartilage of a type of chondroitin sulfate containing more than 1 mole of sulfate per mole galactosamine has been known since 1940 (see FURUHASHI, 1961a). Most of the ester sulfate in the chondroitin sulfate is bound to C-6 of N-acetylgalactosamine with only a minor amount (0-30%) bound to C-4 (FURUHASHI, 1961a, b; FURUHASHI and UCHIDA, 1961; MATHEWS, 1962; LLOYD et al., 1963). SUZUKI (1960) presented evidence that the additional sulfate group in the glycan was bound to the uronic acid moiety of the repeating disaccharide (diS_D). Skeletal keratan sulfate of high sulfate/hexosamine ratio (1.7-2.0) may constitute up to 50% of the total polyanionic glycans.

i) Ratfish (Holocephali)

The ratfish, or chimaeras, of the subclass Holocephali are ordinarily denizens of the high seas. The cartilage from several 45-cm-long specimens of *Hydrolagus collei* yielded glycosaminoglycans containing mainly disaccharide of the 6-S type with a smaller contribution from the disulfated, diS_D type (MATHEWS, unpubl.).

The Holocephali possess a persistent and unconstricted notochord with a thick fibrous sheet (GOODRICH, 1958). Mesoblastic cells invade and spread throughout the sheath to form complete and sparsely calcified rings, rather than the centra typical of many chondrichthyans. Only very low concentrations of chondroitin sulfate-type glycans are present in the notochordal 'canal'. The glycosaminoglycans of the sheath are similar to those in cranial cartilage (Table 7.3).

ii) Sharks (Pleurotremata) of Marine Habitat

The cartilage of adult marine sharks (Table 7.3) contains glycosaminoglycans which have the characteristic periods of the diS_D and 6-S types and, in addition, variable proportions of SKS.

The hexanchid sharks are a primitive group that has remained relatively unchanged for about 150 × 10^6 years; centra remain uncalcified and the notochord is variably constricted (SCHAEFFER, 1967; ROMER, 1966). A member of this group, the seven-gilled shark *Notorhynchus maculatum* (DANIEL, 1934; BIGELOW and SCHROEDER, 1948) is unusual in respect to two features, namely, its lack of any calcification in spinal cartilage, although the jaws are well calcified (URIST, 1961), and the retention in the adult form of a large, only slightly constricted notochord. A female specimen 77 cm long showed about 4% of calcium salts in the jaw cartilage and no insoluble salts in cranial, gill, or vertebral cartilages. The cranial cartilage contained no diS_D, although these periods were present to a small extent in the other cartilages. Three larger specimens, from 160 cm to 180 cm in length, showed about 13% of calcium salts in jaw cartilage and about 3% of calcium salts in cranial cartilage but none in vertebral cartilage. The average glycosaminoglycan composition of cranial cartilage of these larger specimens is given in Table 7.3. An appreciable proportion (15%) of the amino sugars are unsulfated, a small amount (5%) of diS_D is present, and SKS is absent. Similar results were found for jaw and vertebral cartilage except for small increases in diS_D content. Analytical results for "notochordal tissue" show the absence of O-S disaccharide periods and an increase in 4-S disaccharide periods. It should be recalled, as mentioned earlier, that the "notochordal tissue" in sharks may reflect added contributions from skeletonogenic cells of mesodermal derivation. The content of 6-S periods may be due to such cells.

At first thought, both skeletal development and biochemistry may suggest a neotenous condition in *N. maculatum*. However, the more immediate, probable ancestral forms and the more remote placoderms had little or no ossification of the vertebral skeleton, as well as a persistent notochord (ARAMBOURG, 1958a; ARAMBOURG and BERTIN, 1958a; ROMER, 1966). Thus, the cartilagenous state of the vertebral elements is probably primitive. However, the cartilages of the cranium may be considered neotenous on the possible basis of a derivation from a common ancestor with placoderms accompanied by reduction of endoskeletal ossification (ØRVIG, 1951; ROMER, 1966).

iii) Rays and Skates (Hypotremata) of Marine Habitat

In Triassic times or earlier, the sharks divided into the Pleurotremata (considered above) and the Hypotremata, the group which includes the flattened, bottom-living rays and skates. Cartilage of a 90-kg stingray *(Dasyatis sp.)* captured in Cape Cod waters had a glycosaminoglycan composition (Table 7.3) characterized by a high content of 6-S and diS_D disaccharide periods.

A sample of sawfish cartilage from a 394-cm-long female *Pristis perotteti* taken from Lake Nicaragua had a glycosaminoglycan composition that was characterized by an elevated content of 4-S and SKS periods (Table 7.3). These sawfishes, like the bull sharks found in the same waters, are very likely immigrants from the Caribbean Sea.

iv) Marine Sharks of Fresh Water

Sharks of fresh water have been known for a long time (SMITH, 1936). However, until the recent studies by URIST (1962) and THORSON (reviewed in THORSON, 1967), the only published study of osmoregulation of freshwater elasmobranchs was that published in 1931 by SMITH (1931). This investigator studied the blood of sharks, rays and sawfish that had invaded the Perak River in the Federated Malay States and noted that the urea levels had dropped to near 30% of that of the corresponding marine species and that inorganic ion levels had declined to the approximate levels of teleosts.

Particular interest has attached to the sharks and sawfish of Lake Nicaragua because of the general consensus that these populations of fish were landlocked. BIGELOW and SCHROEDER (1948) had accepted classification of the lake shark as *Carcharhinus nicaraguensis*, on the basis of differences in several minor morphological characteristics from *C. leucas*, but later (BIGELOW and SCHROEDER, 1961) placed both groups in synonymy as *C. leucas*. Since only a small number of specimens had been measured, the question was not satisfactorily resolved until morphological comparisons of a large number of the freshwater sharks with marine *C. leucas* by THORSON, COWAN, and WATSON (1966a, b) left little doubt that the marine bull shark and the Lake Nicaragua shark were indistinguishable. Careful investigation showed also that there appeared to be no impassable obstacle along the 160-km length of the Rio San Juan for passage of Caribbean bull shark into Lake Nicaragua. Subsequently, a tagging program demonstrated unequivocally that bull sharks migrated freely between the lake and the sea (THORSON, 1971).

In 1963 I was able to obtain tissues from a freshly captured male specimen (180 cm length) of *C. leucas*, taken from Lake Nicaragua just north of the town of Granada. The cranial cartilage appeared to be very highly calcified; a section of the tissue contained about 15% of the wet weight as insoluble calcium salts, predominantly deposited as an outer layer on the chondrocranium. Proteoglycosaminoglycans obtained from this specimen were similar in composition and molecular parameters to those from cartilage of a 195-cm-long male bull shark of marine habitat. The glycosaminoglycan composition was also similar in both specimens. Cranial cartilage from three additional bull shark specimens, females 171 cm, 180 cm, and 205 cm in length, from Lake Nicaragua yielded similar glycosaminoglycan compositions, except that the SKS contents were somewhat higher, i.e. 35%, 30% and 42%, respectively. An immature male bull shark of 120 cm length obtained from Lake Izabel, Guatemala, had a similar cartilage glycosaminoglycan composition. That Caribbean sharks and sawfish have easy access to Lake Izabel seems unquestionable, and morphological measurements of specimens of sharks in Lake Izabel agree with those of *C. leucas* (THORSON et al., 1966b). It may be significant for the carcharinid group that all the bull sharks, including one immature specimen, and the lemon shark have cartilage with a high content of skeletal keratan sulfate.

In sum, no appreciable difference in cartilage glycosaminoglycans was evident between marine and freshwater forms of *C. leucas*.

v) Rays of Freshwater Habitat

Quite different from the migrant sharks and sawfishes of Lake Nicaragua are the South American stingrays of the family Potamotrygonidae, in particular those of the genus *Potamotrygon*, which inhabit and breed in the drainage waters of the Orinoco and Amazon rivers as much as 4500 km from the sea. Potamotrygon is placed in a separate family because of an unusual pelvic skeleton, but its characteristics agree in all other respects (other than fresh water habitat) with typical Dasyatidae (CASTEX, 1967; BIGELOW and SCHROEDER, 1953). THORSON et al. (1967) made the striking finding that urea was nearly completely absent (1 to 2 mmol/liter) from the serum and other body fluids of a number of specimens taken at locations approximately 3200 km from the mouth of the Amazon River. Although the history of the occurrence of Potamotrygon in these river sites is unknown, these authors suggested that the genus must have lived in fresh water for at least some thousands of years. In this respect, *Potamotrygon* differed greatly from the sharks of Lake Nicaragua and other fresh waters whose blood urea levels, although much reduced, still remained at levels of about 150 mmol/liter. Furthermore, it appeared likely that adaptation to very low urea levels in *Potamotrygon* would have required basic changes in many other facets of their physiology. *Potamotrygon* probably derived from marine ancestors that possessed mechanisms for retaining urea; however, the rather remote possibility exists that they descended from ancestors that never left fresh water, unlike the other members of the class Chondrichthyes.

Additional evidence supporting the suggestion of the permanent freshwater adaption of *Potamotrygon* was adduced by GOLDSTEIN and FORSTER (1971), who demonstrated that activity levels of ornithine-urea cycle enzymes of liver were much lower in *Potamotrygon* than in marine rays, that the renal tubules of the freshwater genus were relatively poor in ability to actively resorb urea, and that, in contrast to marine elasmobranches, a salt-secreting rectal gland was absent. Also freshwater stingrays are unable to increase their urea concentration appreciably when exposed to a saline environment, (THORSON, 1970) and cannot survive in seawater (GRIFFITH et al., 1973), suggesting that any ancestral mechanisms for osmotic regulation through elevated levels of blood urea had been lost. Apparently, neither urea retention nor special salt-secreting mechanisms are needed for a permanent freshwater existence.

Since both the urea-retention habitus and the presence of extra-sulfated disaccharide periods are common characteristics of the class of chondrichthyes, the question arose as to the degree to which these two adaptations were physiologically integrated. Accordingly, cartilages of several specimens of *Potamotrygon* were analysed. In one specimen taken from drainage waters of the Rio Orinoco approximately 1200 km from the sea, the glycosaminoglycan of cranial and of jaw cartilages consisted virtually entirely of disaccharide periods of 6-S with a small proportion of O-S; diS_D was not detectable. A second specimen from a side channel of the Amazon River about 3000 km from the ocean had been identified as *P. circularis* and was 56 cm long with a disc 33 cm long and 31.5 cm wide. Glycosaminoglycans of vertebral cartilage consisted (Table 7.3) nearly entirely of 6-S periods. In this case, however, a trace of diS_D periods was detected.

Two additional specimens of *P. circularis* that had been maintained in freshwater aquaria for 2 to 4 years were also examined. Their measurements for disk width, disk length, and overall length were 10.5, 11.0, and 18.5 cm, and 36.0, 36.2, and 60 cm, respectively. The cartilage glycosaminoglycans of both specimens had a virtually identical composition to that of the previous specimen (Table 7.3), including a trace of diS_D periods. The notochordal tissue was not examined since the centra were well calcified, resulting in a segmented notochord.

vi) Conclusions

Although the Chondrichthyes include two groups (the elasmobranchs and the holocephalans) whose divergent paths might entitle each to separate class status (SCHAEFFER, 1967), they are united by such common characteristics as the presence of an entirely cartilagenous endoskeleton and the urea-retention habitus. To these two class characteristics of the Chondrichthyes should now be added the presence of an active sulfotransferase with the specificity to synthesize diS_D disaccharide periods. However, the ability to retain urea in blood and tissues and to synthesize diS_D periods is not unique to this class among vertebrates. High concentrations of urea have been shown to occur in a few other vertebrate species, notably the crab-eating frog *Rana cancrivora*, the aestivating African lungfish *Protopterus aethiopius*, and the coelacanth *Latimeria chalumnae*. The latter is also capable of synthesizing cartilage glycosaminoglycans with diS_D periods; these are known to occur in significant amounts among other vertebrates only in codfish (acellular bone). However, very small or trace amounts of diS_D periods may be present in tetrapods (see below).

The functional significance of diS_D periods of glycosaminoglycans in the chondrichthyans is uncertain; further discussion is deferred. However, the absence, or very great reduction of diS_D in *Potamotrygon*, associated with loss of the urea-retaining habitus in a freshwater environment, suggests the possible contribution of extra sulfation to stabilization of tissue structure and/or to electrolyte physiology in marine chondrichthyans.

The tendency for cartilage to accumulate large proportions of skeletal keratan sulfate is perhaps surprising in animals which have a capacity for continuous growth (see Chap. 8). High contents of this glycan in cartilage may be characteristic of certain families, i.e. the carcharhinids, and may be associated with calcification of the tissue accompanied by 'degeneration' of chondrocytes. For example, vertebral cartilage of a 200-cm-long bull shark whose spine was very heavily calcified and scoliotic contained skeletal keratan sulfate in excess of 50% of the total glycosaminoglycans. However, there is no general correlation between skeletal keratan sulfate content and calcification among elasmobranchs since calcification is found also in vertebral and jaw cartilage of *Potamotrygon* specimens that do not contain this glycan.

Convincing evidence is still lacking on the primitive or neotenous nature of the cartilagenous endoskeleton and the primary or secondary character of the lack of osteocytes in the Chondrichthyes. However, there can be no question of the capacity of cartilage to function as a primary support tissue. It seems likely that some protovertebrate developed cartilagenous endoskeletal tissue by much the same evolutionary means as did the cephalopods and some other invertebrates. Subsequently, such cartilagenous tissues became associated with bone (perichondral and endochondral) in various early phylogenetic lines but may well have had a largely independent and continuous development in the line that led to the Chondrichthyes.

Calcified cartilage, although probably inferior to bone, is an adequate means of strengthening the skeleton. Such factors as mechanical stress, equilibrium, buoyancy, and protection have played a role in determining the distribution and density of calcified cartilage and bone in the adult fish skeleton (SCHAEFFER, 1961). Their complex interplay seems to have resulted in the preferential calcification of jaw and vertebral cartilages over cranial cartilages of most of the different species of Chondrichthyes examined. Similar considerations may account for the adaptive significance of the addition of cartilagenous vertebral elements to the axial skeleton in association with a primitive notochord and for their subsequent calcification to various degrees.

c) Osteichthyes

The Osteichthyes comprise three major subdivisions, the Actinopterygii, the Dipnoi and the Crossopterygii, whose origin and relationships have been disputed for some time. SCHAEFFER (1968) grants these groups equal rank while ROMER (1966, 1968) combines the last two in the category Sarcopterygii. Of ancient origin, the Osteichthyes appeared in the early Devonian period well before the Chondrichthyes. It is of special interest that the tetrapods and the Crossopterygii most likely derived from a common ancestral stock.

The Devonian forms of the Osteichthyes possessed endoskeletons that were commonly well ossified. Subsequent evolution led to reduction in ossification in many groups. In the "first" osteichthyan, ossification probably was extensive in the cranium but variable and thin in the postcranial skeleton; the notochord was persistent and unsegmented (SCHAEFFER, 1968).

i) Actinopterygii

The ray-finned fishes are commonly divided into three groups, the chondrosteans, the holosteans, and the teleosts, corresponding to three different levels of evolutionary development.

The chondrosteans, a relatively minor group, inhabit salt as well as fresh waters. The rock sturgeon *Acipenser fulvescens* and the hackleback or shovelnose sturgeon *Scaphirhynchus platorhynchus* are found in the Mississippi River drainage and are restricted to fresh water. Similar polyanionic glycans have been isolated from cranial cartilage and notochord of both species. As in the lamprey, cranial cartilage from hackleback sturgeon of 40 cm average length contains primarily chondroitin sulfate with 6-S disaccharide periods while the notochord glycan consists nearly exclusively of 4-S disaccharide periods. Cranial cartilage from a 60-cm-long specimen of the white sturgeon *Acipenser transmontanus*, known in the waters off the west coast of the United States, contained a higher proportion of 4-S disaccharide periods. All chondrosteans have substituted cartilage nearly completely for the ossifications of precedent forms. The state of the cranium thus appears to be neotenous, but the paleontological evidence (LEHMAN, 1958a) indicates that the postcranial axial skeleton is probably mainly primitive, with possible reduction of bone in some vertebral elements. The progressive loss of bone during the evolution of the Acipenseriformes may be related to a uniquely low concentration of calcium in plasma (URIST and VAN DE PUTTE, 1967).

The holosteans are archaic fish of which only two genera, *Lepisosteus* and *Amia*, survive today in fresh waters of North America (ARAMBOURG and BERTIN, 1958b). Much of the ancestral ossification has remained in these modern representatives, although reduction of bone is evident in some dermal and cranial elements. Vertebral ossification is incomplete and a notochord is still evident. The polyanionic glycans of cartilage from 120-cm-long specimens of alligator gar *L. spatula* contain nearly equal amounts of disaccharide periods of 4-S, 6-S, and SKS types. The skeletal keratan sulfate isolated contained a molar ratio of sulfate to hexosamine near 1.3. Similar results were obtained with a 75-cm-long specimen. The cartilage glycosaminoglycans of a specimen of longnose gar *L. osseus* 117 cm in length contained more 4-S than 6-S periods, and also a trace of diS_D periods. However, cranial cartilage of two specimens of the bowfin *Amia calva* , of 60-cm length, contained glycosaminoglycans with only a minor proportion of keratan sulfate and a preponderance of 6-S over 4-S periods; DiS_D periods were not detected.

The extensive experiments by the bony fish with many types and degrees of ossification reached their culmination in the well-calcified teleosts of today, whose centra are universally ossified, thus obliterating the notochord (BERTIN, 1958; LEHMAN, 1958a). Cartilage of adult specimens of the trout *Salmo gairdneri*, the carp *Cyprinus carpio*, and the cod *Gadus morhua* contains polyanionic glycans composed largely of 6-S disaccharide periods accompanied by very little, if any, SKS (Table 7.4).

ii) Dipnoi

The dipnoans, or lungfish, are represented today by very few species. While in the estivating state, they produce and store large amounts of urea. Their skeleton is largely cartilagenous with a persistent unconstricted notochord. Earlier forms generally show more ossification of the axial skeleton than do modern representatives (ARAMBOURG and GUIBE, 1958; LEHMAN, 1966; ROMER, 1966), with ossification of the vertebral column commonly incomplete and ossified centra rarely present. The condition of the internal skeleton has probably remained mainly primitive with possible neotenous reduction of bone in some elements (YOUNG, 1962). Data (Table 7.4) on fin cartilage of a specimen of African lungfish (*Protopterus* sp.) show 6-S disaccharide periods predominating; this specimen had been captured during the aquatic phase and was 3 to 4 years of age. Cranial cartilage from two specimens (60 cm in length) of *Protopterus* sp. contained glycosaminoglycans of similar composition. The notochordal glycosaminoglycans, however, contained a preponderance of 4-S periods (Table 7.4).

iii) Crossopterygii

This group includes the now extinct group of Rhipidistia and the Coelacanthini, whose fossil record does not extend beyond the Cretaceous period. The latter, in contrast to the rhipidistians, had passed into seawater at an early date and subsequently diminished very greatly in numbers before the Cretaceous period (ARAMBOURD, 1958b; ROMER, 1966). In 1938, the scientific world was astounded to learn of the capture of a live specimen of *Latimeria chalumnae* Smith, which resembled its last known fossil antecedant among the Coelacanthini to a remarkable degree. By 1973, about 70 more specimens had been captured. The unusual history and anatomy of the species have been reviewed by MILLOT and ANTHONY (1958a, b, 1965). THOMSON (1969) has extensively

Table 7.4. Disaccharide period composition (%) of cartilage and notochord polyanionic glycans of Osteichthyes

Specimen	Tissue	O-S	4-S	6-S	diS_D	diS_E	$4\text{-}S_B$	$6\text{-}S_B$	diS_B	diS_H	SKS[a]	S/H[a]
				Actinopterygii								
Chondrosteans												
Hackleback sturgeon	cart.[b]	5	10	85							< 2	0.95
(*S. platorhynchus*)	noto.		98	2								0.98
White sturgeon (*A. transmontanus*)	cart.	5	30	65							< 2	0.97
Holosteans												
Alligator gar (*L. spatula*)	cart.		30	40							30	1.00
Longnose gar (*L. osseus*)	cart.		35	20	tr						45	1.03
Bowfish (*A. calva*)	cart.		30	60							10	0.99
Teleosts												
Trout (*S. gairdneri*)			30	70							< 2	1.00
Carp (*C. carpio*)		5	30	65							< 2	0.95
Cod (*G. morhua*)			30	70							< 2	1.00
				Dipnoi								
Lungfish (*Protopterus* sp.)	cart.[c]		20	80							< 2	1.00
" "	noto.		60	40							< 2	1.02
				Crossopterygii								
Coelacanth (*L. chalumnae*)	cart.[c]		20	45	15						20	1.15
" "	noto.		5	5	5	5	45	15		20		1.30

[a] See Table 7.1 for definition of disaccharide periods; SKS = skeletal keratan sulfate; S/H = molar ratio of sulfate to galactosamine of uronic acid-containing glycans.
[b] Cartilage from skull unless otherwise indicated.
[c] Fin cartilage.

discussed the biology of *Latimeria* and of related groups, i.e. rhipidistians and dipnoans.

The internal skeleton of crossopterygians followed different courses in the two main branches. The rhipidistians retained an ossified braincase and an entirely bony postcranial skeleton with ossifications surrounding the notochord. However, the neurocranium of the related coelacanths became largely cartilagenous while the visceral skeleton remained mostly ossified. The fin skeleton has been sporadically ossified throughout coelacanth history; the neural and haemal spines only are perichondrally ossified (SCHAEFFER, 1952). A distinction between types of ossification of cartilage is important since Devonian coelacanths (as well as dipnoans) have extensive perichondral but little endochondral bone. The highly varied expression of the latter character among the earliest representatives makes it difficult to postulate an ancestral condition (SCHAEFFER, 1968).

Latimeria has many unusual anatomical and physiological features. Worth noting are a two-part, hinged skull (unique to crossopterygians) with the notochord extending far forward, a very simple heart compared to other fishes, a very small and simple brain, ventral, partially-fused kidneys, an intestine with a complex spiral valve, and a very large notochord with thickened sheath as the only longidutinal support (vertebral centra are absent). Of particular interest are the findings that the coelacanth uses high concentrations of urea (PICKFORD and GRANT, 1967) and trimethylamine oxide (LUTZ and ROBERTSON, 1971), comparable to those in Chondrichthyes, to maintain its serum near the osmolarity of sea water and that urea (BROWN, Jr. and BROWN, 1967) is synthesized in the liver by the ornithine-urea cycle, as in typical ureotelic vertebrates.

The composition of the glycosaminoglycans in cartilage (Table 7.4) of an adult coelacanth resembles that of Chondrichthyes, especially in including diS_D periods. Surprisingly, the notochord glycosaminoglycans of *Latimeria* most closely resemble those obtained from notochord of hagfish in containing a high proportion of disaccharide periods with L-iduronic acid residues as well as diS_H periods.

Any biological interpretation of these apparent biochemical relationships between elasmobranchs, hagfish, and crossopterygians is necessarily tentative at this time, for the three groups belong to entirely separate lineages whose ancestry is obscure. THOMSON (1969) has offered the following explanation of the striking similarity of urea metabolism between elasmobranchs and sarcopterygians (especially the coelacanths): Many of the enzymes of the ornithine pathway for urea production were possibly present in early gnathostome vertebrates inhabiting fresh waters (a direct phylogenetic relationship between elasmobranchiomorphs and Sarcopterygii cannot be ruled out). Also, Sarcopterygii and Elasmobranchii had a similar history in progressing from an early feshwater stage either to a marine environment or to environmental conditions in which water was scarce. Accordingly, the adaptation of urea retention that is common to elasmobranchs, coelacanths, and dipnoans was a similar response to the need to conserve water under dehydrating conditions, i.e. salt water or desiccation. Although the ornithine pathway for urea production was retained by the land-dwelling tetrapods, it was irretrievably lost by the Actinopterygii. Thus, the similarity in urea metabolism of the separate groups is a result of convergent evolution. Presumably, a similar interpretation can be given for the presence of highly sulfated chondroitin sulfate (diS_D periods) in the cartilage of both elasmobranchs and coelacanths. In this instance, however, the adaptation in glycosaminoglycans does not extend to the dipnoans and thus may be associated with the retention of very high levels of urea in serum and tissues.

Alternatively, the retention of high levels of urea is a primitive feature of jawed vertebrates, secondarily lost in actinopterygian fishes and most tetrapods upon the invasion of freshwater environments, but retained in an unmodified state in such groups as the elasmobranchs holocephalians and coelacanths, which remained in the sea (GRIFFITH et al., in press).

The singular occurrence of similar glycosaminoglycans of a unique structure in notochordal tissues of the coelacanth and the hagfish is not readily explained. The presence of L-iduronic acid, and in particular of diS_H periods, in glycosaminoglycans of notochord may represent a secondarily acquired characteristic, but more likely reflects the persistence of a very ancient condition. However, there is no direct evidence for a relationship between hagfish and coelacanths - all three groups of sarcopterygians were already extremely specialized when they first appeared in the fossil record (THOMSON, 1969).

d) Tetrapoda

The earliest amphibia appeared in the late Devonian period, having most likely derived from crossopterygian stock. These primitive forms gave rise in turn, not only to the various modern groups classed as amphibia, but also to the more fully terrestrial populations that do not breed in water - the Amniota, including reptiles, birds, and mammals (ROMER, 1966). The loss of buoyancy that took place upon leaving an aquatic medium necessitated a considerable strengthening of supporting tissues and led to extensive development of internal skeletal structures with a generally high degree of ossification. Among the many physiological and biochemical changes that accompanied this transition from aquatic to terrestrial life are those that occurred in the macromolecular constituents of cartilage matrices.

Available data on tetrapod cartilages show that their glycosaminoglycan compositions depend somewhat upon anatomic site, but may vary greatly with development and age, changing at a minimum rate during the young adult stage. Thus, comparisons between species that do not take these factors into account may be misleading. For practical reasons as well, the *species-characteristic composition* is defined as that of cartilage in a young adult. However, of possibly greater biological significance is the maximal ratio of 4-S to 6-S disaccharide periods in glycosaminoglycans attained during the life cycle. For tetrapods, such values are reached in postmetamorphic or postnatal growth periods and generally exceed unity (Table 7.5).

i) Amphibia

The glycosaminoglycans of cartilage of nearly maximum size bullfrogs *(Rana catesbiana)* contain 4-S and 6-S periods in a ratio of 2:1. However, the glycosaminoglycan composition of cartilage of the mature mudpuppy *Necturus maculosus* is atypical of tetrapods in respect to a ratio of 4-S to 6-S periods of only 1:3 and a high content of O-S periods. The presence of a preponderance of 4-S periods in glycosaminoglycans of notochord is generally consistent with the vertebrate pattern for this tissue when allowance is made for the probable invasion of the sheath by skeletonogenic cells. The apparent identity of glycosamino-

Table 7.5. Disaccharide period composition (%) of cartilage and notochord polyanionic glycans of Tetrapoda

Specimen	Tissue	O-S	4-S	6-S	SKS[a]	Remarks
			Amphibia			
Bullfrog *(R. catesbiana)*	skull cart.		70	30		adult, 12-15 cm length
Mudpuppy (*N. maculosus*)	gill cart.	20	20	60		adult, 30 cm length
" "	notochord	15	60	25		adult
" "	gill cart.	25	20	55		larva, 18 cm length
			Reptilia			
Crocodile *(C. latirostris)*	costal cart.		60	30	10	1 kg
			Aves			
Chicken	xiphoid cart.		65	30	5	4 mo.
"	" "	15	55	30		newborn
			Mammalia			
Rabbit	xiphoid cart.		65	35		7 mo.
"	" "		50	50		2 mo.
Hog	tracheal cart.		80	20		5 mo.
Ox	nasal cart.		70	30		calf, 3 mo.
"	" "		70	25	5	calf, 1-2 yr
Camel *(C.dromedarius)*	nasal cart.		60	20	20	adult, 15 yr
Rat	costal cart.		95	5		9 mo.
"	" "	5	70	20		newborn
Man	costal cart.		30	45	25	adult, 23 yr
"	" "		50	50		infant, 2 mo.

[a] See Table 7.1 for definition of disaccharide periods; SKS = skeletal keratan sulfate.

glycan compositions of cartilage in mature and larval stages (Table 7.5) is probably a reflection of the unusual development of this species of salamander. Unlike other salamanders, *Necturus* remains aquatic throughout its life history and does not undergo metamorphosis, retaining most of the enatomical features, including the external gills, of its larval stage (NOBLE, 1931). The mature mudpuppy is evidently neotenous in general anatomy and in many aspects of its biochemistry (COHEN, 1966).

ii) Reptilia

Cartilage of the crocodile *Caiman latirostris* contains glycosaminoglycans with a characteristic preponderance of 4-S periods (MATHEWS, unpubl.).

iii) Aves

A major content of 4-S periods in the glycosaminoglycans of cartilage is found in the newborn as well as in the adult chicken.

iv) Mammalia

Cartilage of the mature rabbit, hog, ox and camel *(Camelus dromedarius)* contains glycosaminoglycans in which 4-S periods are major components. Elastic ear cartilage of the only species examined, rabbit and ox, contains glycosaminoglycans that do not differ significantly in composition from those of the other cartilages (Table 7.5) in the respective species. Glycosaminoglycans of elastic epiglottis cartilage from rabbits and from a 2-year-old steer contain about 90% of chondroitin sulfates and 3-4% of unsulfated anionic glycan (GILLARD and WUSTEMAN, 1970). However, glycosaminglycans of the homologous tissue from humans (30-60 years old) consist of equal amounts of chondroitin 6-sulfate and hyaluronate (SOLHEIM, 1968a). Fibrous cartilage, i.e. the semilunar cartilage (meniscus) of the human knee joint is unusual in containing hybrid glycan polymers with 6-S and $4\text{-}S_B$ periods (HABUCHI et al., 1973). The usually small amounts of hyaluronate that are present in some cartilage tissues (HARDINGHAM and MUIR, 1973b) have generally been disregarded.

For the adult rat, 6-S periods are present in very small proportion compared to 4-S periods. For man (MATHEWS and GLAGOV, 1966a), the maximum proportion of 4-S periods appears at a very early age. These findings and additional data on the variation of glycosaminoglycan compositions with age are discussed in Chapter 8. Small amounts of diS_D and diS_E, of the order of 1% of the total disaccharide periods, were detected in human costal cartilage from individuals over 20 years of age by IWATA (1969). Hyaluronate contents also increased with age.

It is interesting that the whale *(Baelenopter physalus)*, though adapted to a wholly aquatic life, contains cartilage with glycosaminoglycans consisting predominantly of 4-S periods. Keratan sulfate, and a small amount of hyaluronate are also present (LLOYD et al., 1963; SENO and ANNO, 1961). The keratan sulfate fraction isolated from nasal cartilage had a high content of sialic acid (>11%), which was identified as N-acetylneuraminic acid (TODA and SENO, 1970).

5. Calcified Cartilage, Bone and Teeth

Polyanionic glycosaminoglycans, as well as glycoproteins (see Chap. 5), have been detected in bones and teeth of mammals (BALAZS and ROGERS, 1965). However, neither their 'native' state (HERRING, 1968a, b, 1969) nor their role is clear (HERRING, 1968a, b, 1972). The major polyanionic glycosaminoglycan in human enamel and dentine-cementum appears to be chondroitin sulfate. ENGFELDT and HJERPE (1972) found chondroitin sulfate to be the sole glycosaminoglycan (over 98% of the total present) in predentine and dentine with the ratio of 6-S to 4-S periods significantly greater in the chondroitin sulfate of predentine than in that of dentine. Bovine cortical bone (HERRING, 1968a), cortical bone (SOLHEIM, 1965) and fracture callus (ANTONOPOULOS et al., 1965) of rabbits, and pathological calcification in myositis ossificans of man (SOLHEIM, 1968b) contain chondroitin 4-sulfate as the major glycosaminoglycan. Similar observations have been made for normal bone matrix of rat, man, rabbit, and dog by IWATA and URIST (1972). The glycosaminoglycans of epiphyseal cartilage of young dogs differ from those of compact bone in containing a significant component of 6-S periods and in having a lower molecular weight (HJERTQUIST and VEJLENS, 1968).

The composition of the polyanionic glycosaminoglycans of skull cartilage, partially calcified vertebral cartilage, and heavily calcified centrum or bone in the same organism is compared in Table 7.6. In the

Table 7.6. Disaccharide period composition (%) of cartilage, calcified cartilage, and bone polyanionic glycans[a]

Specimen	Skull (or other) cartilage % 4-S	Vertebral cartilage % 4-S	Centra or bone % 4-S	Remarks
	Chondrichthyes			
Spiny dogfish	25 (5 SKS)	20 (5)	75	calcified centra
Seven-gill shark	5	15		centra absent
Horn shark	10 (20 SKS)	20 (20 SKS)	50 (10 SKS)	
	Osteichthyes			
Hackleback sturgeon	10	30		centra absent
White sturgeon	30		80	dermal bone of skull
Buffalo *(Letiobus)*	50		75	dermal bone of skull
Carp	30		75	bone, endochondral
Cod	30		70[b]	bone, acellular
Alligator gar	30 (30 SKS)		83	dermal bone of skull
	Tetrapoda			
Chicken	65 (5 SKS)		100	sternum; femur cortex
Rabbit	65		85	sternum; tibia cortex
Calf	70 (5 SKS)		90	femur cortex
Man (male, 58 yr.)	0 (40 SKS)		83	articular cart., femur

[a] See Table 7.1 for definitions of disaccharide periods; SKS = skeletal keratan sulfate; % 6-S + % diS_D = 100 minus values for 4-S and SKS in table. [b] Acellular bone includes 15% 6-S and 15% diS_D.

sharks, a small increase in percentage of 4-S periods may be seen for vertebral cartilage over skull cartilage; a large increase is noted for calcified centra. In bony fish, the percentage of 4-S periods of glycosaminoglycans is strikingly greater in bone than in cartilage.

The presence of a high percentage of diS_D periods, a characteristic that is associated with chondrichthyan cartilage, in cod *(Gadus morhua)* bone is surprising. Acellular bone, so-called because it is totally lacking in enclosed osteocytes, occurs only in "higher" orders of Teleost fish (MOSS, 1963) but was present in some early agnatha of the Silurian period. The adaptive significance of acellularity of bone and of the presence of diS_D periods in the polyanionic glycan component as well as the reappearance in higher-order teleosts of a 'primitive' type of bone is not clear (MOSS, 1963, 1964, 1968b). It is of incidental interest that chondroitin sulfates (type not determined) are also present in carp and cod scales, which contain collagen and a related protein, ichthylepidine (MOSS, 1963).

Although 4-S periods predominate in the glycosaminoglycans of cartilage of tetrapods, their proportion in bone glycosaminoglycans rises still higher, approaching 100%. Thus, there appears to be a highly significant association of the chondroitin 4-sulfate type of glycan with calcified tissues of vertebrates. It is interesting that this difference in glycans between cartilage and bone parallels a difference in collagen type associated with each kind of tissue.

6. Discussion

The capacity of invertebrates as well as vertebrates to produce the macromolecular building blocks of cartilagenous tissues indicates the probable evolutionary antiquity of the basic biosynthetic mechanisms. A considerable amount of chemical latitude is possible with respect to glycan backbone structure, provided only that the molecules are linear chains and bear a large number of anionic charges in solution. Evolutionary convergence probably accounts for those instances of close similarity of glycosaminoglycan structure in invertebrates and vertebrates.

Clear directions of evolution cannot be deduced from consideration only of an indirectly homologous character such as polyanionic glycan structure. Nevertheless, it is likely that the enzymes responsible for biosynthesis of the polyanionic glycans in vertebrates, with the possible exception of hagfish, are closely homologous. The hagfish are clearly set aside from the remaining vertebrates, but there remain curious similarities of their notochordal glycans with those of the coelacanth, another evolutionary oddity. Is the latter some phylogenetic relative of the Chondrichthyes or has it obtained its glycosaminoglycan with high diS_D content by convergence? It would be of interest to compare its unusual notochordal fibrillar glycoprotein (Chap. 5) with those of other vertebrates.

The high content of diS_D periods in cartilage glycosaminoglycans of Chondrichthyes and of the coelacanth is associated with the urea-retention habitus, high osmolarity of body fluids, stabilization of interactions with collagen and other proteins, and possibly with calcium binding in cartilage. The sulfotransferase activity responsible for uronic acid sulfation in formation of diS_D periods is repressed, partially or totally, in freshwater rays and in species of other vertebrates classes, including mammals (Chap. 8).

The isomeric 4-S and 6-S disaccharide repeating periods may have a tissue-specific distribution in notochord and cartilage of some lower vertebrates or may appear mixed in cartilage of other species. It is possible that this specificity extends to the polypeptide moieties of the proteoglycans (Chap. 6) and to the collagen of these two kinds of tissue as well. Chondroitin 4-sulfate may be present in high proportion in cartilage of adult tetrapods because a high degree of ossification is necessary to terrestrial animals, whereas chondroitin 6-sulfate and its variants are retained in the cartilage of lower vertebrates that have undergone phyletic reduction of bone. Alternatively, 6-S periods may be associated with glycosaminoglycans of noncalcifying cartilage as a primitive condition.

Chapter 8

Polyanionic Glycans in Development and Aging of Vertebrate Cartilage

"In biology one is never far from history; and each living organism carries fragments of this history with it, as vestiges and recapitulations, anatomical and, as we shall see, biochemical." WALD (1960).

"The concept of physiological aging as a process advancing relentlessly with time under ordinary environmental conditions has two components so far as cartilage is concerned: 1. loss of some genetically governed, self-replicating capacity of the chondrocyte; and 2. physicochemical changes in the intercellular matrix of the tissue." SOKOLOFF (1969)

1. General Considerations

The principle of paleogenesis - that descendent ontogenies tend to recapitulate ancestral ontogenies - is clearly revealed in the vertebrates, where developmental sequences tend to be conservative (SMITH, 1960). If the assumption of a molecular counterpart to this morphogenetic principle is valid, as suggested by WALD (1960) and by COHEN (1966), then developmental data may indicate the nature of ancestral biochemistry.

At first thought, it seems possible that analysis of the biochemical changes that precede or accompany the transformations in anatomy and ecology during vertebrate development and metamorphosis (WALD, 1960) could lead to identification of "primitive" macromolecules. However, as pointed out by ZUCKERKANDL (1968), this hope is illusory. In the only instance where complete data are available at the level of informational macromolecules, i.e. in the case of the fetal and adult polypeptide chains of hemoglobin, the findings do not accord with the theory of recapitulation. ZUCKERKANDL (1968) suggests that there may be, to some extent, a dissociation in organisms between evolution of form and evolution of function. Recapitulation may then be largely a matter of temporary reproduction in the embryo of ancestral ratios of rates of synthesis of macromolecules. And "...evolution of form may depend more directly on the influences of the structural changes in DNA on the regulation of synthesis of polypeptide chains, whereas evolution of function may result more directly from their influences on the physicochemical and steric properties of the polypeptide chains."

"Aging" is an elusive concept, which still lacks a commonly accepted definition (COMFORT, 1964). The process is evident as a progressive deterioration of function with time. At all biological levels, structural correlates of functional changes are illuminating; in many instances, it has been possible to demonstrate relationships at the molecular level.

At the tissue level, the focus is upon an organized group of cells in interaction with its organic environment. For connective tissues in general and for cartilage in particular, the intercellular matrix exceeds the cellular elements, thus creating a special situation. Intercellular distances are much greater than those present in parenchymatous tissues and mechanical and physicochemical activities are localized to a large degree in the intercellular matrix. As a consequence, the matrix macromolecules tend to assume added importance with regard to tissue functions.

However, this division of a tissue into cells and matrix should not deflect attention from the vital relationships that exist between cells and matrix: the cells produce the matrix components and in turn are supported and nourished by their surroundings. The phenomena of tissue development, growth and aging reflect this constant interplay between cells and immediate environment (WEISS, 1961, 1966) as well as much more complex relationships within the entire organism (NEEDHAM, 1964).

The methodologies employed in obtaining information on glycosaminoglycans of cartilage impose certain limitations upon the possible interpretation of such data. Reliable identification and quantitation of mixtures generally requires the use of samples greater than 0.1 g of wet tissue. Since sample dimensions are very large compared to the diameter of a chondrocyte and to the average distance between cells, the data obtained represent average compositions only. Even a relatively simple cartilage like that of nasal septum shows variations in composition of tissue sections taken at increasing distances from the perichondrium. However, variability may be small for some components. Although the proportion of proteins varies greatly in the sections, little change is seen except for the chain length of the glycosaminoglycans (SZIRMAI et al., 1967). The uneven distribution of total dry mass concentration with distance from the cartilage cell surface appears to be due mainly to collagen and other proteins and not to glycosaminoglycans. The highest protein concentration occurs as an envelope surrounding the cell at the periphery of the lacunar region (SZIRMAI, 1969).

Metabolic, molecular structural, and topological heterogeneity of proteoglycans has been shown for growing cartilage. Readily extractable fractions from rat cartilage are more metabolically active than insoluble fractions (GROSS et al., 1960). Chondroitin sulfate is apparently synthesized at a higher rate in the vicinity of the osteochondral junction than in other portions of mouse costal cartilage (HERBAI and LINDAHL, 1970). Proteoglycans of calf-rib cartilage may be separated into several fractions differing in chemical and physical parameters as well as in metabolic activity (KLEINE et al., 1971).

Tissue organization varies not only in various topological regions within a single anatomical structure but also between cartilage structures in different anatomical sites of the same organism. Furthermore, it is not surprising to find that the biochemistry and structure of cartilage matrices vary with the life cycle of the organism. Finally, it should be noted from the paradigm of complementarity (mutual dependence) of cells and matrix that alterations in matrix imply alterations in cellular behavior.

2. Lower Vertebrates

Complete studies of developmental changes in cartilage biochemistry of lower vertebrates have yet to be carried out. The glycosaminoglycans of cranial cartilage of a 20-cm-long embryo of spiny dogfish *S. acanthias* consisted of 80% of 6-S periods, 10% of diS_D periods, and 10% of O-S periods. Both 4-S and SKS periods were absent. Comparison of these results with data from a 75-cm-long adult specimen (Table 7.3) shows that maturation results in an increase in the degree of sulfation accompanied by a replacement of 6-S periods by 4-S periods and in the appearance of SKS. It is of interest that the embryonic glycosaminoglycans already contain the diS_D period that is characteristic of the class chondrichthyes. The size of the embryo indicated that it was probably approaching term; gestation is unusually long in this species, being about 2 years.

DOYLE (1969) analysed pectoral fin cartilage of 43 specimens of *S. acanthias*, ranging in length from 28 to 98 cm. Contents of uronic acid, glucosamine and galactosamine underwent no change in relative amount with aging, although the mineral content increased to about 50% of the dry weight. He estimated that keratan sulfate constituted 10% of the total glycosaminoglycans. However, this inference, which is based only upon constituent sugar analysis of whole tissue, cannot be entirely reliable (MATHEWS and GLAGOV, 1966a).

3. Amphibia

Anuran metamorphosis represents a postembryonic developmental process by which the tadpole, adapted to aquatic life, is transformed into the land-dwelling frog. The dramatic, adaptive changes that occur in morphology, cytology, and biochemistry have been extensively investigated. Of particular value to the experimenter is the fact that the process can be rapidly induced by administration of thyroid hormones (FRIEDEN, 1967). Some of the biochemical changes appear to be preadaptive, occurring ahead of gross morphological changes (COHEN, 1970).

Data on glycosaminoglycan composition of cranial cartilage in the giant bullfrog *Rana catesbiana* are given in Table 8.1 (MATHEWS and HINDS, 1963). Prior to hormone treatment and emergence of hind legs, O-S periods are prominent; 4-S periods constitute less than 50% of the total periods but are proportionally greater in number than are 6-S periods. Further maturation prior to metamorphosis, indicated by the emergence of hind legs and their growth to a minor degree (L/O = 0.11), results in a considerable reduction in the proportion of O-S periods and a corresponding increase in 6-S periods. Subsequent metamorphosis and maturation to the adult terrestrial form leads to a fully sulfated chondroitin sulfate composed mainly of 4-S periods.

The stimulatory effect of thyroxine treatment on the biosynthesis of glycosaminoglycans in cartilage was demonstrated by observations of *in vivo* and *in vitro* incorporation of labelled acetate and sulfate ions. The normal tadpole glycosaminoglycans consist of a mixture of chondroitin sulfate molecules separable into fractions having a molar ratio of sulfate to galactosamine ranging from 0.25 to 1.00. The low sulfated fractions have a high metabolic activity (turnover) and the highly sulfated fractions show a preferential incorporation of sulfate as 4-S over 6-S. Although thyroxine induction stimulates incorporation

Table 8.1. Disaccharide period composition (%) of cranial cartilage glycosaminoglycans of the frog *Rana catesbiana* [a]

Specimen	Maturity index [b]	0-S	4-S	6-S
Tadpole	L/O = 0.00	40	40	20
Tadpole	L/O = 0.00	35	40	25
Tadpole	L/O = 0.11	10	45	45
Tadpole [c]	L/O = 0.20	10	45	45
Adult	15 cm length		70	30

[a] See Table 7.1 for definition of disaccharide periods.
[b] L/O = average ratio of length of hind leg to over-all length.
[c] After 7 days treatment of tadpoles of maturity index = 0.11 thyroxine.

of precursors into all fractions, it increases the degree of preferential biosynthesis of 4-S over 6-S periods and the degree of sulfation so that the bulk of the newly synthesized chondroitin sulfate now has equimolar proportions of sulfate and galactosamine residues. Thus, the shift in biosynthesis is toward the production of a glycosaminoglycan composition characteristic of the adult form (for data on skin, see Chap. 9, Sect. 4a).

These changes of the biosynthetic pathway are absent from the cartilage of amphibia, such as the mudpuppy, which do not undergo metamorphosis. The glycosaminoglycan composition of the larval form, which resembles that of the premetamorphic tadpole, is identical with that of the sexually mature adult (Table 7.5).

4. Birds

Interest in problems of differentiation and development has stimulated considerable work with readily accessible avian models. Realization of the utility of characteristic molecular markers, such as proteoglycans and collagen, has directed attention to connective tissues generally and to cartilage in particular (see review by THORP and DORFMAN, 1967). However, in most cases, paucity of material has restricted investigators to identification of the glycan moiety of proteoglycans and to detection of hydroxyproline residues of presumptive collagen molecules.

Early studies indicated the presence of a sulfated glycan (presumably a chondroitin sulfate) in prospective structures (stages 11-23) long before cartilage tissue could be demonstrated histologically (FRANCO-BROWDER et al., 1963). Chondroitin sulfate was identified in the periphery of the notochord, but not in somites, of 1.5 day (stage 9) chick embryos (T. YAMAGATA, priv. comm.). MEDOFF (1967) found that chondroitin sulfate and an unidentified sulfated glycan insensitive to testicular hyaluronidase were synthesized by the prospective limb regions of 3-day (stage 15-16) chick embryos and stage 17-18 limb buds more than two days before the appearance of metachromatic staining material. Sulfotransferase and chondroitin sulfate proteoglycan polymerase activities were detectable by stage 19; the hyaluronidase-insensitive glycan was no longer present by then. Limb-bud mesoderm from stage 19 onward, when cultured *in vitro*, produced sulfated glycans but no hydroxyproline until 5-day (stage 26) cells were used (MOTTET, 1967). However, LEVITT and DORFMAN (1972) observed synthesis of both chondroitin sulfate and hydroxyproline by stage 24 limb-bud cells. The microenvironment of the mesenchymal cells appeared to have an important influence on biosynthesis (see also Chap. 10, Sect. 5). In later stages, tissue concentrations of the only detectable glycosaminoglycan, chondroitin sulfate, vary topologically in tibial cartilage, with higher concentrations found in the hypertrophic zone, especially at a time of accelerated bone formation (HUFFER, 1970).

The proportions of glycans of the axial region of the developing chick embryo during early somite differentiation (stages 17-28; 2.5 to 5.5 days of age) undergo extensive changes (KVIST and FINNEGAN, 1970a, b). In the region that includes notochord, dermatome, myotome, scleratome and neural tube, the hyaluronate concentration was 2.5 times that of chondroitin sulfate at stage 17; the ratio was reversed by stage 28. The authors inferred the presence of traces of dermatan sulfate from the presence of iduronic acid. However, the glycan identification should be regarded as uncertain. The observations on the proportions and distribution of hyaluronate and chondroitin sulfate are curiously reminiscent of the situation in amphioxus (Chap. 7, Sect. 3a).

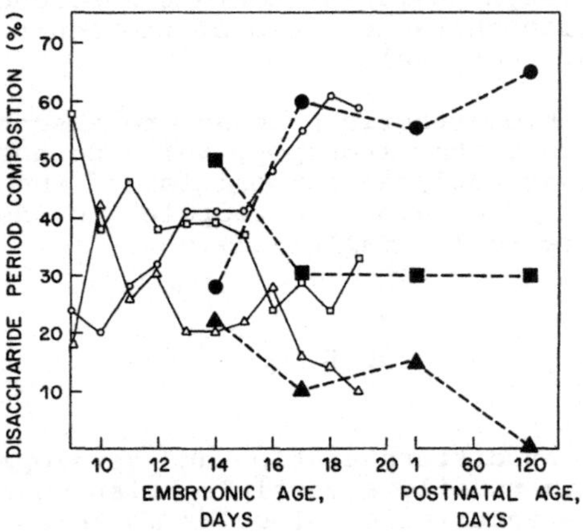

Fig. 8.1. Biosynthesis of disaccharide periods of chondroitin sulfates of chick embryos at various ages (from data of ROBINSON and DORFMAN, 1969, on incorporation of acetate-^{14}C into minces of condyles of tibias and femurs): △, 0-S periods; ○, 4-S periods; □, 6-S periods. Also included are data (MATHEWS, 1965a, 1967a, b) for the disaccharide composition of chondroitin sulfates in embryo condyles and in postnatal xiphoid (includes 5% SKS at 120 days): ▲, 0-S periods; ●, 4-S periods; ■, 6-S periods. For definitions of disaccharide repeating periods, see Table 7.1

Data on the composition of the glycosaminoglycans of older chick embryos and of chickens are given in Fig. 8.1. The glycans of 14-day chick embryo epiphyseal cartilage consist predominantly of 6-S and 0-S periods, which during further development *in ovo* are replaced by 4-S periods. Postnatal growth results in complete loss of 0-S periods and the appearance at 120 days of a small percentage of SKS.

With respect to the relative proportions of the three periods, the glycans that are newly synthesized by embryonic cartilage cells at successive stages change with increasing age in a manner corresponding generally to that shown by the total glycans (Fig. 8.1). Such similarity in direction of change is not surprising. Thus, if the proportion of each type of period were a monotonic function of age, then the newly synthesized composition may be expected to correspond to the total composition at a more advanced age. This situation was indeed found by THORP and DORFMAN (1963), who compared intracellular (i.e. newly synthesized) glycosaminoglycans with extracellular glycosaminoglycans. The former contained a lower proportion of O-S periods and a higher proportion of 4-S periods than did the latter.

The relationships between intra- and extracellular glycan become more complex when studied at the proteoglycan level. KIMATA et al. (1971) separated nascent chondroitin sulfate-protein from the Golgi apparatus of chondrocytes in 12-day-old embryonic chick epiphyseal cartilage into two fractions by density gradient centrifugation. The $^{35}SO_4$ activity ratio of 4-S to 6-S was different for the two fractions. Although the high-density fraction was readily hydrolyzed by pronase, the low density fraction was not. It was suggested that the latter fraction contained an unusual polypeptide core.

In contrast to epiphyseal cartilage (Fig. 8.1), vertebral cartilage of 10-day-old chick embryos contains chondroitin sulfate with mainly 4-S periods (SHULMAN and MEYER, 1968). A similar glycan composition is produced by the vertebral chondrocytes in culture (SHULMAN and MEYER, 1970). The discrepancy may be due to intrinsic differences between the tissues or to the low resolution of infrared spectra for analysis of sulfated glycan structure.

5. Mammals

a) Rat

Previously cited data (Fig. 7.5) show the change in glycosaminoglycan composition of rat (Sprague-Dawley) costal cartilage from 5% of O-S, 70% of 4-S and 20% of 6-S in the newborn [small amounts of Hy and SKS may be present (HANDLEY and PHELPS, 1972)] to 95% of 4-S and 5% of 6-S in the 9-month-old animal. Very little, if any, further change in either male or female 14-month-old animals, although the glycan content of the cartilage declines greatly (WEIL et al., 1969). The extractable proteoglycans in cartilage of rat at 2 years of age contain galactosamine and glucosamine in a ratio of 20:1 (KAO et al., 1970). However, a portion of the glucosamine could be due to glycoprotein contaminants.

The retention of 4-S periods and the exclusion of skeletal keratan sulfate from cartilage of advanced age is an unusual condition (see Sect. 5) that may be related to a growth pattern peculiar to the rat among mammals. Although a majority of the epiphyses close early in adult life (80-120 days), many remain open for more than 3 years. As a consequence, the rat is capable of slow but continual growth after sexual maturity (BECKS and EVANS, 1953). In this connection it is noteworthy that, unlike the cow and man, the rat continues to lay down collagen in articular cartilage well into adult life (BAILEY and SHIMOKOKOMAKI, 1971). However, costal cartilage may differ in this respect. The metabolic and topological heterogeneity of the tissues were noted by GROSS et al. (1960), ANDERSON and ODELL (1960), SZIRMAI (1969), and HOPWOOD (1972).

b) Hog

Metaphyseal cartilage of newborn pig contains chondroitin sulfate consisting approximately of 55% of 4-S periods and 45% of 6-S periods (CASTELLANI et al., 1962). SKS, virtually absent at birth, may rise to near 5% at 9 months (BRANDT and MUIR, 1969). Cartilage from older animals contained relatively more of lower-molecular-weight protein-polysaccharides. The bulk of the changes in articular cartilage occurred during the first few weeks of postnatal life (SIMUNEK and MUIR, 1972).

c) Rabbit

Rabbit embryo costal cartilage contains glycosaminoglycans consisting mainly of 6-S and O-S periods, which are replaced during postnatal growth by 4-S periods (Fig. 8.2). The glycans in young growing animals turn over rapidly but seem to be virtually inert in 12-month-old rabbit cartilage (DAVIDSON et al., 1961). Protein synthesis is rapid in articular cartilage of 2-month-old rabbits but drops precipitously to very low values within the next four months and remains constant up to 2 years of age (MANKIN and BARON, 1965).

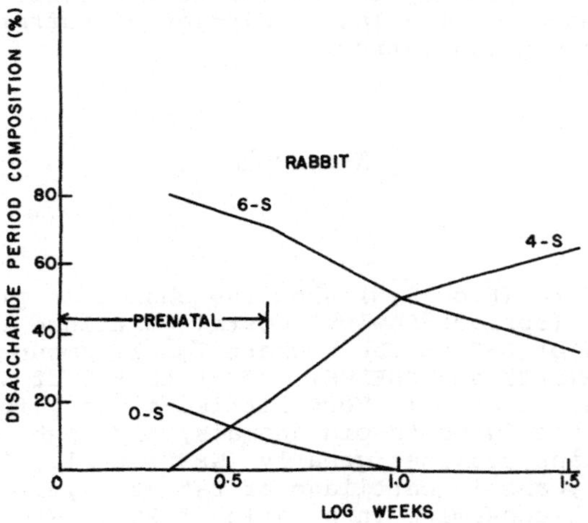

Fig. 8.2. Disaccharide period composition of glycosaminoglycans of rabbit costal cartilage at various ages (MATHEWS and GLAGOV, 1966b). For definitions of disaccharide repeating periods, see Table 7.1

d) Ox

Data on the composition of glycosaminoglycans of costal and nasal cartilage of ox embryo and calf and of a 12-year-old cow are given in Fig. 8.3. Changes in the embryonic glycans with development are toward increased sulfation with a rising ratio of 4-S to 6-S periods. At 7 months of embryonic development the glycan composition of costal and nasal cartilage is indistinguishable. At an advanced adult age, a drop

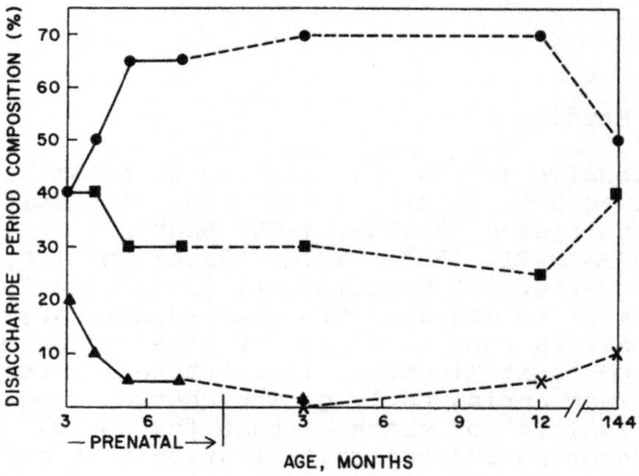

Fig. 8.3. Disaccharide period composition of glycosaminoglycans of bovine cartilage at various ages: ▲, O-S periods; ■, 6-S periods; ●, 4-S periods; X, SKS periods; ——, articular cartilage; -----, nasal cartilage (MATHEWS and GLAGOV, 1966b). Data on crown - rump length of embryos were correlated with age according to ROBERTS (1956). For definitions of disaccharide repeating periods, see Table 7.1

in the ratio of 4-S to 6-S periods is noticeable. Skeletal keratan sulfate is absent from nasal cartilage as well as from costal and tracheal cartilages of fetuses (KAPLAN and MEYER, 1959). First noted in appreciable proportion (5%) in the 12-month-old calf cartilage, it increases little with age. A somewhat higher proportion (20%) of skeletal keratan sulfate was found in tracheal cartilage (age unspecified) by GREILING et al. (1964). Small amounts of hyaluronate are present (HOPWOOD, 1972).

The metabolic and possible topological heterogeneity of glycosaminoglycans in 12-year-old cartilage is indicated by the observation that alcohol fractionation separates these glycans into fractions with ratios of 4-S to 6-S from 3 to 0.7. However, the isolation of fractions with different relative proportions of 4-S and 6-S periods cannot be accomplished by application of the same procedure to glycosaminoglycans of one-year-old cartilage. The average chain length of chondroitin 4-(6-) sulfate declines while that of keratan sulfate increases with age (HOPWOOD, 1972).

GOH and LOWTHER (1966) isolated 'proteoglycans' from bovine nasal cartilage of embryos (3 months), 2-year-old calves and 14-year-old steers. From analyses for hexose, glycosamine, galactosamine, and uronic acid they concluded that the ratio of skeletal keratan sulfate to chondroitin sulfate varied from 0.1 to 0.4. It is likely, however, that these authors were including in their assays associated glycoproteins, which may increase in concentration with age (see Chap. 5, Sect. 3d).

e) Man

i) Histochemical Studies

The early and extensive microscopic studies of human hyaline, fibrous and elastic cartilages demonstrated the great complexity of these types of connective tissue (HANSEN, 1905; BENNINGHOFF, 1925; SCHAFFER, 1930; AMPRINO and BAIRATI, 1933, 1934a, 1934b; AMPRINO, 1938). The tissues undergo a variety of histological transformations from early embryonic development to old age. Of interest here are the differences in chronology of alterations noted for hyaline and other cartilage types dependent upon anatomic site, i.e. tracheal cartilage lags behind costal cartilage during early growth changes. Degenerative changes also occur later in life for tracheal than for costal cartilages. The asbestos-form alteration and the calcification that occurs by the 4th decade is a constant feature of costal cartilage but is less frequent in tracheal cartilage. AMPRINO and BAIRATI (1933) also observed the constant appearance of secondary cartilage growth in tracheal cartilage over age 40. They noted further the appearance of young cartilage within zones of asbestos-form degeneration in costal cartilage.

Our knowledge of the changes of glycosaminoglycans with age was advanced by development of differential staining methods. In costal cartilage, trachea, and primary bronchi, the distribution, localization, and concentration of the different glycans varies according to age and with type of hyaline cartilage, area of a section, and distance from the surface of the chondrocyte cell (QUINTARELLI and DELLOVA, 1966). Areas of appositional growth contain mainly chondroitin sulfate; territorial and interterritorial areas of cells in the central zone, beginning with the second decade of life, contain both this glycan and keratan sulfate. With advancing age, keratan sulfate gradually disappears from the interterritorial matrix and is confined primarily to the matrix adjacent to the chondrocyte in costal but not primary bronchial cartilage (MASON, 1971). The regressive changes that are initiated in costal cartilage at 12 to 15 years and advance with time are evident to a lesser degree in trachea and primary bronchi. Generally similar observations have also been made by STOCKWELL and SCOTT (1965, 1967). In addition, for articular cartilage, the territorial predominance of skeletal keratan sulfate was noted in the deeper parts next to the calcified layer from about the beginning of the fourth decade.

ii) Chemical Studies

Present histochemical techniques are incapable of precise chemical identification of polyanionic glycans in tissue sections. Yet the current methodology that permits reliable chemical characterization of polyanionic glycans is much more restricted in topological resolving power than are light- and electron-microscopic procedures. Thus, it appears essential to utilize information from both microscopic and chemical studies in order to arrive at a complete understanding of the macromolecular organization of connective tissue.

Most of the chemical studies of the macromolecular composition of cartilage have been based upon analyses for hexoses, uronic acid, hexosamines, and sulfate residues of whole tissue, or of multicomponent preparations from tissues. The procedures have numerous qualitative and quantitative inadequacies (MATHEWS and GLAGOV, 1966a). As a consequence, many of the inferences drawn by early workers (KAPLAN and MEYER, 1959) require modification in the light of the more recent work to be described.

Costal, cricoid, thyroid, and tracheal cartilages of the same individual were analyzed by MATHEWS and GLAGOV (1966a). The individual glycans were isolated and characterized by chemical and physical methods; data for embryonic and postnatal ages appear in Fig. 8.4. During the last two trimesters of embryonic development 4-S and 6-S periods are present in near equal proportions, while the proportion of O-S periods

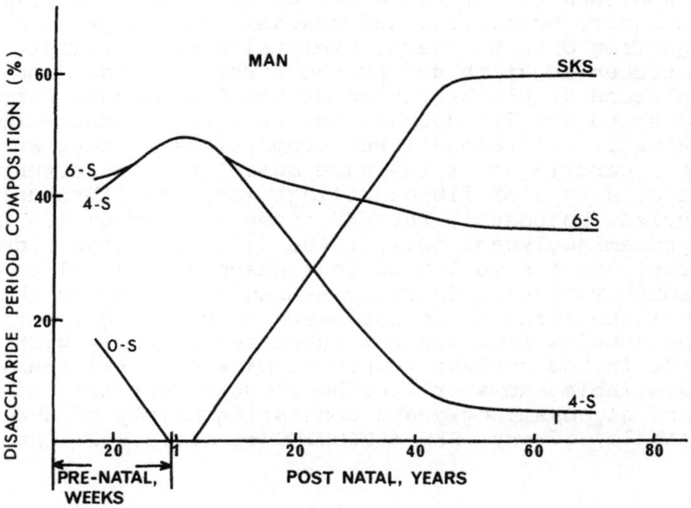

Fig. 8.4. Disaccharide period compositions of glycosaminoglycans of human costal cartilage at various embryonic and postnatal ages (smoothed data from MATHEWS and GLAGOV, 1966a). For definitions of disaccharide repeating periods, see Table 7.1

steadily declines [human bone matrix also contains glycans with O-S periods in fetal but not adult stages (IWATA and URIST, 1972)]. During postnatal growth, maturation, and subsequent old age, 6-S periods decline to a small extent while 4-S periods fall almost to vanishing point. SKS, first detected toward the end of the first decade, rises rapidly until the end of the fourth decade and thereafter remains approximately constant into old age.

Some individual variability was noted. It was also evident that the changes of glycosaminoglycan composition in tracheal, cricoid, and thyroid cartilages occur less rapidly with advancing age than do the comparable changes in costal cartilage, in agreement with the histological observations on these tissues (AMPRINO and BAIRATI, 1933; QUINTARELLI and DELLOVA, 1966; STOCKWELL and SCOTT, 1965, 1967).

Although less extensive data are available for cartilages other than those cited above, it is likely that they undergo similar age-dependent changes. The glycosaminoglycans of articular cartilage of a 58-year-old femur contained 40% of SKS and 60% of 6-S periods (MATHEWS and GLAGOV, 1966). The proportion of SKS periods was 39% at 20 years of age, and rose to 53% at 52 years of age, while the total glycosaminoglycan content remained essentially constant (GREILING et al., 1964). A similar trend was indicated by the tissue content of glucosamine and galactosamine (BENMAMAN et al., 1969) and of uronic acid (BOLLET and NANCE, 1966; GREILING and STUHLSATZ, 1969; MAROUDAS et al., 1969). Progressive increase with age of the proportion of keratan sulfate in the total glycosaminoglycans of articular cartilage from femoral condyles (HJERTQUIST and LEMPERG, 1972) and from lower femoral epiphyses (HJERTQUIST and WASTESON, 1972) was also noted. The average molecular weight of the chondroitin 4-(6-) sulfate declined with age.

Results generally similar to those indicated in Fig. 8.4 were obtained by MASON and WUSTEMAN (1970) for a series of samples of primary bronchial, secondary bronchial, and tracheal cartilage from individuals ranging in age from 0 to 65 years. Skeletal keratan sulfate increased rapidly from traces at birth during the first 2 decades and thereafter slowly to a plateau at the beginning of the 5th decade. Further change was not noted up to the 7th decade. However, the tracheo-bronchial cartilage generally maintained lower proportions of keratan sulfate than did costal cartilages of the same age. For intervertebral discs, which consist of a type of fibrocartilage (annulus fibrosus) surrounding a gel (nucleus pulposus), the SKS rises from about 10% to 50% of the total glycosaminoglycans between the first and eighth decades, whereas the ratio of 4-S to 6-S periods changes from 2:1 to 1:2 (BUDDEKE and SZIEGOLEIT, 1964). However, HALLÉN (1962) found that the glucosamine/galactosamine ratio increased only during the first 3 decades in the annulus fibrosus but increased linearly with age throughout life in the nucleus pulposus. Data on fetal annulus fibrosus are not available. However, the homologous cartilage of fetal whales contains glycosaminoglycans consisting mainly of 4-S and 0-S periods and little, if any, SKS (LUDOWIEG et al., 1972; Chap. 9, Sect. 4e).

The semilunar cartilage (meniscus) of the human knee joint is unusual in that the glycosaminoglycans are copolymers of 6-S and derm 4-S periods. The glycan composition of this fibrous cartilage does not change with age (HABUCHI et al., 1973).

The suggested presence of small amounts of "oversulfated" chondroitin sulfate in human cartilage (BARKER et al., 1965) was confirmed by use of the newer enzymatic methods. By use of chondroitinase, IWATA (1969) showed the presence of both diS_D and diS_E periods (see Table 7.1) in costal cartilage of persons over 20 years of age. The proportion of each of these types of periods in the total glycosaminoglycans varied in the cartilage of different individuals and averaged near 1% of the total periods.

The extensive data (43 cases) of IWATA (1969) on the relative proportions of 0-S, 4-S, and 6-S periods present at different ages agreed with the pattern indicated in Fig. 8.4. The glycosaminoglycans of 14-week fetal cartilage averaged 15% of 0-S, 35% of 4-S and 50% of 6-S periods. By 32 to 36 weeks, the proportions in the glycans of fetal cartilage had altered to 5% of 0-S, 53% of 4-S and 42% of 6-S periods. For one-year-old infants, the proportion was 2% of 0-S, 49% of 4-S and 49% of 6-S periods. By the eighth decade the average proportion of 4-S to 6-S periods had fallen to 1:4. A very variable, but significant proportion (1-10%) of the total glycosaminoglycans of mature cartilage appeared to be hyaluronate, possibly related to secondary alterations (TOOLE and GROSS, 1971).

Studies of solubilized proteoglycans of aging cartilage yield data, interpretation of which is limited by the heterogeneity of the extracts obtained and also by the retention of major proportions of these components in the insoluble residue (KRÖZ and BUDDECKE, 1967). Most of the keratan sulfate of adult tracheal cartilage remains in the insoluble residue (MASON, 1970).

6. Discussion

A thorough interpretation of the data presented in this chapter not only depends upon an understanding of the evolutionary factors involved in vertebrate phylogenesis and ontogenesis, but also requires a knowledge of the functional roles of proteoglycans. The lack of information relative to the latter sharply limits the scope of molecular interpretations. As a basis for discussion, only a brief listing is presented here; an extensive analysis of proteoglycan properties and functions appears in Chapter 10.

a) Functional Roles of Proteoglycans

Proteoglycans have a major role in the structural organization of the intercellular matrix. These polyanionic macromolecules are involved in electrostatic and steric interactions with other macromolecules of the matrix, including collagen and elastin. In particular, they may control initiation and growth in diameter of collagen fibrils and link covalently to collagen (KOBAYASHI and PEDRINI, 1973). Within the interfibrillar spaces, the proteoglycans contribute importantly, especially in cartilagenous tissues, to osmotic and water-retaining properties, and to ion-binding and transport functions of the matrix. In addition, they appear to be involved in calcification and ossification processes.

The high degree of structural heterogeneity within a single class of proteoglycan molecules permits a wide range of functional flexibility. Macromolecular structure may vary greatly (Table 6.5). The resultant heterogeneities have temporal as well as topological dimensions.

b) Embryonic Development and Maturation

The changes in composition of the glycosaminoglycans of vertebrate cartilage during ontogeny proceed broadly from a low density of sulfation of chondroitin sulfate towards a high density of sulfation and from a high to a low ratio of 6-S/4-S periods. Thus, the corresponding temporal order of involvement of component activities of the biosynthetic mechanism is apparently at first: the enzyme complex for synthesis of proteoglycan, followed by the 6-sulfotransferase and the 4-sulfotransferase. Whether these enzyme activities relate directly to enzyme levels or to other aspects of control of biosynthesis is unclear. Unfortunately, too little is known of the sulfotransferases involved in introduction of a second sulfate group into a disaccharide period to suggest any conclusions relative to changes in their activities during development.

Two alternate phylogenetic interpretations may be offered. The first, based upon the concept of biochemical paleogenesis, leads to the inference that the formation of a polyanionic proteoglycan containing little or no sulfate group was an essential step in the histogenesis of cartilage in the earliest vertebrates. Sulfation of the glycan at the C-6 position of the amino sugar was subsequently favored and increased the linear density of anionic charge. At this stage of evolution the degree of sulfation of the glycan at the C-4 position was nil, or very small. The introduction of a major proportion of 4-S disaccharide periods, a probable preponderant component of the polyanionic glycosaminoglycans of primitive notochord, occurred at a later period, possibly in association with the calcification process (see Chap. 7, Sect. 5).

The alternate, and more plausible interpretation is based on the hypothesis that the genetic capacity for producing the complete repertoire of enzymes necessary for synthesis of all types of proteoglycans was present in the earliest vertebrates. Subsequent evolution involved primarily the elaboration of mechanisms for control of biosynthetic patterns. The presence of a wide range of sulfated disaccharide periods in glycosaminoglycans of cyclostomes (Chap. 7, Sect. 4a) supports the basic assumption. This viewpoint is consonant with that of MOSS (1964), who considers that the entire group of vertebrate skeletal tissues forms a series connected by common embryonic inductive mechanisms under the influence of extrinsic factors of mechanical function. It is also consistent with the suggestion of ZUCKERKANDL (1968) that recapitulation may be largely a temporary reproduction in the embryo of ancestral ratios of rates of synthesis of macromolecules (see Sect. 1 above).

Within the broad limits of the pattern set for tetrapods, the temporal changes during development in the composition of glycosaminoglycans of cartilage differ appreciably among the vertebrate classes (Figs. 8.1 to 8.4). Detailed comparisons, however, are unjustified since the data do not apply uniformly to homologous cartilage structures. However, it is probable that at any given embryonic stage the composition and organization of the macromolecular components reflect the specific, local functional requirements for skeletal development. For example, it is probable that the proteoglycan with 4-S periods has a specific role in calcification in view of its preponderance in calcified tissues over the related proteoglycan with 6-S periods (Chap. 7, Sect. 5). A comparable change in collagen type also occurs from $[\alpha 1\ (II)]_3$ of cartilage to $[\alpha 1\ (I)]_2\alpha 2$ of endochondral bone. The significance of this correlated shift in sulfation and collagen type for the macromolecular organization of the tissues remains to be explored (Chap. 10).

The considerable biochemical differences among the species are not related in a simple fashion to general features of development. For example, both the domestic rat and domestic rabbit *(Sylvilagus)*, with gestation periods of 22 days and 26-30 days, respectively, are born helpless, hairless and with closed eyes. In contrast, the domestic newborn calf, which has undergone a gestation period of 280 days, has a hairy coat and is ready to walk. However, with respect to glycosaminoglycan composition, the resemblance is close between rat and ox, whereas the newborn rabbit alone retains a high ratio of 6-S/4-S periods.

Unlike domestic rabbits, newborn hares, which belong to the same family (Leparidae) as rabbits, are well-furred, open-eyed and capable of moving about shortly after birth. Also, the various species of *Lepus* have longer gestation periods, in the range of 36 to 47 days (WALKER, 1968). Developmental similarities may be noted. For example, the ribs of the wild cottontail rabbit at birth consist of uncalcified cartilage along about 50% of their length. In this respect, as well as in the average disaccharide composition of cartilage glycosaminoglycans, the wild rabbit newborn are indistinguishable from newborn of the domestic genus. However, the ribs of adult cottontails at 2 to 3 months of age are fully calcified whereas uncalcified cartilage still constitutes near 40% of the rib length of adult domestic rabbits at 4 months (MATHEWS, unpubl.). Apparently, the developmental pattern of glycosaminoglycans is conserved within the family but postnatal ossification of ribs is greatly delayed in the domestic rabbit. It would be of interest to determine whether the prolonged retention of costal cartilage is due principally to an intrinsic genetic pattern or to extrinsic influences.

Extrinsic influences undoubtedly influence the chemical composition and structures of connective tissue matrices (Chap. 10). As suggested earlier (Chap. 7), vertebrates probably differ little in their inherent capacity to synthesize glycosaminoglycans, as indicated by the presence in species of all classes of a wide range of homologous glycosyltransferases and sulfotransferases. Thus, the major differences between different cartilage tissues may possibly be explained by differences in extrinsic physiological and physical factors (MATHEWS, 1973).

c) Aging

The composition of the glycosaminoglycans of cartilage continues to change during the period of postnatal growth and maturation and through senescence. The characteristic changes, which are evidently species-dependent, are clearly seen in man, for whom the data are most definitive. Within a few years after birth, cartilage tissues in many anatomical sites begin to show a progressive increase in the ratio of 6-S/4-S periods and in the proportion of skeletal keratan sulfate periods to reach ultimately plateau values in the 5th decade and remain approximately constant thereafter. The direction of change in the ratio of 6-S/4-S periods is the reverse of that seen during tetrapod embryogeny. However, sulfation of periods is complete, i.e. 0-S periods do not appear. The presence of keratan sulfate as a predominant component of aged cartilage indicates a major physiological transformation from the embryonic state. Minor components, such as diS_D and diS_E periods, which are first noted in costal cartilage at 18 years of age (IWATA, 1969), similarly signal a shift in metabolic activities of chondrocytes with increasing age.

Histochemical studies generally agree with the biochemical findings and add an important dimensional refinement. STOCKWELL and SCOTT (1965) detected negligible staining for keratan sulfate in infant cartilage. From the second decade onwards, keratan sulfate is localized in the interterritorial matrix. However, in the central region of mature costal cartilage, the most intense staining for keratan sulfate is in the territorial matrix. A similar territorial predominance is seen in the deeper parts of articular cartilage next to the calcified layer, from about the beginnings of the fourth decade. Similar conclusions concerning a fundamental difference in cell differentiation and maturation between different areas of hyaline cartilage, dependent upon depth from the perichondrial surface, were drawn by QUINTARELLI and DELLOVA (1966). Thus, the biosynthesis of keratan sulfate appears to be the expression of a chondrocyte with an altered biosynthetic activity, if not a reduced viability. Since cartilage cells are dependent for their nutrition upon metabolite diffusion through the matrix, it seems likely that the distance of the cell from its nutrient source can have a profound influence upon its metabolism, thus affecting relative rates of synthesis of different glycan chains (STOCKWELL and SCOTT, 1965; see also MASON, 1971 and Chap. 6, Sect. 4d) and, possibly, relative activities of different sulfotransferases. The striking changes with aging in the distribution of keratan sulfate with distance from the cell surface may result from the steric barriers confronting macromolecules, such as proteoglycans and glycoproteins that are secreted into a matrix containing a network of collagen fibrils.

Although the proportion of keratan sulfate increases with age in human cartilage, the relationship of the glycan to the observed 'degenerative' changes is obscure. It is generally accepted that the biosynthetic activity of human costal cartilage decreases greatly after childhood whereas that of articular cartilage is retained into old age (SOKOLOFF, 1969). However, at age 58, articular cartilage has about the same glycosaminoglycan composition as costal cartilage (MATHEWS and GLAGOV, 1966; see also BALAZS, 1969). Similarly, in the rat, costal cartilage shows a rapid decline in capacity to incorporate radioactive sulfate into glycosaminoglycans, accompanied by considerable signs of "degeneration" by 5 months of age, although articular cartilage appears in vigorous health (WEIL et al., 1969). Yet keratan sulfate is not found in rat costal cartilage, even up to 14 months. Thus, the histological changes that are seen in aging costal cartilage are not necessarily associated with a shift in types of glycan synthesized.

The major questions relative to the aging process center about the influences thereon of two classes of factors: 1. physicochemical changes in the intercellular matrix and 2. genetic factors (SOKOLOFF, 1969; JUNGE-HÜLSING and WAGNER, 1968). With respect to the first class, it is evident that much more needs to be learned of the chemistry and structure of intercellular matrices and of the effect of microenvironments upon cellular processes before significant insights can be gained (see Chap. 10, Sect. 5). The main chemical findings thus far are the decline with age in the ratio of 4-S to 6-S periods and the relative increase in concentration of keratan sulfate. Such changes cannot be ascribed solely to altered biosynthetic activity. It may be noted that the degradative systems involved in turnover of proteoglycans have received very little attention to date (BOSTRÖM, 1969; GROSS, 1969; WASTESON et al., 1972) and that the concentration of any component of the matrix is the result of a balance between synthetic and degradative processes.

The role of genetic factors in aging proves equally difficult to comprehend. The changes in the direction of a glycan composition characteristic of human aged costal cartilage (Fig. 8.4) are initiated in childhood and are significantly advanced by the end of the period of maturation and growth. According to COMFORT (1964), the growth rate of man is greatest from 0 to 4 yrs, slower from 4 to 12 yrs, and much slower therafter to about 20 yrs. The observation that similar changes are not seen in the rat, which is capable of continuous growth throughout life, suggests that the changes in man may be associated broadly with cessation of development and growth and possibly with a low metabolic turnover of matrix components.

One might expect similar conclusions to apply to cartilage structures in other warm-blooded vertebrates that also have a fixed adult size (COMFORT, 1964). Also, these changes should be absent from fish, whose growth rate is commonly considered to be continuous although declining with increasing age and size. However, the data do not appear to meet this expectation. For example, cartilage glycans of a number of species of bony fish and chondrichthyans have a high content of skeletal keratan sulfate (Tables 7.3 and 7.4). Although the relationship between glycosaminoglycan composition and growth is a very intriguing one, further analysis will be fruitless until more reliable data on skeletal growth of fish species become available. There is good reason to question the apparently unfounded idea that sharks continue to grow during their entire adult life (Dr. E. CLARK, priv. commun.; CLARK and VON SCHMIDT, 1965). It is interesting, however, that the cartilage of those adult bony fish and chondrichthyans that possess a neotenous (or primitive) skeletal structure are free of keratan sulfate.

The wide distribution of skeletal keratan sulfate among vertebrates and its association with adult or "old" cartilage structures raises the question of the adaptive significance of this cartilage component in evolution. Questions of this type probe the fundamental nature of the aging process itself, which so far has largely eluded satisfactory analysis (COMFORT, 1970). The prevailing impression is that the characteristic deterioration that arises with time appears to proceed as part of a genetically determined clock-like process that is influenced by the accumulation of mechanistic defects.

MEDAWAR (1946) proposed that senescence is a genetically determined program that arose by natural selection. He pointed out that selection will operate maximally against deleterious genetic manifestations at the time of maximal reproductive value, i.e. at puberty, and with progressively less pressure at greater ages. Also, there will be a tendency for genetic expression to stop at the age when the individual becomes reproductively redundant, when senescence might be expected to accelerate considerably. Indeed, there could be an advantage in most species, somewhat modified in man by the value of the experience and death after the age of reproductive redundancy. Thus, the significance of senescence lies with reproduction rather than growth.

WEISS (1968) has emphasized the accumulation and compounding of errors and defects as life proceeds. The net effect is a gradually increasing disharmony of mutual relations among various component activities, which manifests itself as a decline in plasticity, adaptability, and efficiency with age. Related views have been expressed by MEDVEDEV (1966) and extended to connective tissues, in particular, by JUNGE-HÜLSING and WAGNER (1969) and VELICAN and VELICAN (1967).

The changes in metabolic activity and matrix structure that result in increased contents of keratan sulfate, diS_D and diS_E periods with advancing age may be the expression of a broadly programmed system in adaptive response to intrinsic and extrinsic conditions. Although the newly introduced matrix constituents have roles which as yet are unknown to us, we may presume that they tend to compensate for the disturbances of aging. That they do not entirely succeed in doing so may be due to inherent inadequacies and imperfections of the homeostatic mechanism or to an inexorable genetic determination.

Chapter 9

Polyanionic Glycans of Other Tissues

1. Introduction

Polyanionic glycans are ubiquitously distributed in the animal organism, appearing not only in supporting tissues, but also in a very great variety of glandular secretions and specialized tissues and organs. Their study in secretions and in tissues other than cartilage, notochord, and bone (Chap. 7) has been comparatively unsystematic, frequently confined to a single species of a class, and meager with respect to chemical structural details. Many presumptive identifications of specific types of polyanionic glycan structures depend solely upon histochemical observations. These must necessarily be considered uncertain and will be given little weight or attention here. Others are far from adequate with respect to extent of chemical investigation and to criteria applied to characterization of chemical groups and deserve little better consideration. However, a considerable body of sound information has become available in recent years and is worthy of full discussion. It is hoped that future investigators will be stimulated to extend it further.

2. Invertebrates

HUNT (1970) has prepared an excellent comprehensive review of the literature on polyanionic acid glycans in invertebrates. Accordingly, the more significant and better established findings that are treated by HUNT are presented below in a broad, summary fashion only, with extended descriptions reserved for the more recent findings. Further details of many topics, including histochemical observations, are given in his monograph. However, a word of caution to the reader: there has been an unfortunate tendency of many investigators to relate their observations in invertebrates to the commonly better understood situation in vertebrates. As a consequence, many "identifications" of specific vertebrate polyanionic glycans, such as hyaluronate and chondroitin sulfates, in invertebrate tissues have been founded upon little more than similarity in staining properties or the detection of common residues such as uronic acid, hexosamine, and sulfate residues.

a) Protista

The Protista are included here although their relationship to the plant and animal kingdoms is still a debated question. Of particular interest is a group, the Coccolithinea, which produce intracellular structures of calcium carbonate for subsequent transport extracellularly. The complex, species-specific calcified forms are known as coccoliths (WILBUR and WATABE, 1967). ISENBERG et al. (1965) demonstrated that the organic phase of the coccoliths of the phytoflagellate *Hymenomonas* consisted of protein and a polyanionic glycan. The collagen-like protein fraction was characterized by the presence of hydroxyproline and a paucity of sulfhydryl and aromatic amino acids. The glycan, judged to be sulfated from its staining reactivity, contained glucose, hexuronic acid, two pentoses, a methylpentose and possibly a dideoxy sugar.

b) Porifera

A carbohydrate component is tightly held in two types of collagenous structures, termed spongin A and spongin B, of the common bath sponge *Spongia graminea* (GROSS et al., 1956; see Chap. 3, Sect. 6). After alkali extraction, spongin B contained glucosamine and galactosamine in equal amounts, as well as uronic acid, glucose, galactose, mannose, and arabinose. Spongin A contained the same sugars as spongin B but in higher concentrations and, in addition, fucose. Subsequently, the arabinose of spongin A from the sheepswool sponge, *Hippospongia gossypina* (order Keratosa, class Demospongiae) was conclusively identified by KATZMAN et al. (1970) as D-arabinose, a sugar of extremely rare occurrence in the animal kingdom.

These workers further noted the presence of sulfate residues in their spongin A preparation. They also isolated from a trypsin digest of the sponge a partially purified glycan fraction of the composition shown in Table 9.1. Glycans of generally similar composition can be

Table 9.1. Composition (%) of polyanionic glycans of sponges

Component	*H. gossypina*[a]	*C. nucula*[b]	
		Fr. B	Fr. F
Arabinose	11.2	4.3	16.2
Fucose	16.4	23.4	21.0
Galactose	10.9	20.6	1.2
Mannose	1.5	0.7	8.1
Xylose	0.8	1.3	0.0
Glucose	c	1.3	0.0
Galactosamine	8.6 (total	2.0	9.7
Glucosamine	hexosamine)	2.1	19.3
Hexuronic acid	16.4	10.8	1.0
Sulfate residues	30.0	30.9	12.4
Amino acid residues	4.1	2.7	11.9
Hydroxyproline	c	<0.05	0.30

[a] Normalized from KATZMAN, LISOWSKA, and JEANLOZ (1970).
[b] MATHEWS (unpubl). [c] Not analyzed.

obtained from other species of the class Demospongiae. The order Carnosa has been of particular interest because its members have greatly reduced mineralization, or are entirely devoid of hard skeleton, deriving their support mainly from a thick, firm mesoglea containing fine collagenous fibrils but none of the thick, branching spongin fibers that are characteristic of the Keratosa (HYMAN, 1940). Papain digestion of *Chondrilla mucula* (SCHMIDT) taken from the area near Port Royal, Jamaica, yields a glycan preparation in amounts near 2% of the original dry weight (MATHEWS, unpubl.). Only about 50% of the glycan is sufficiently anionic and of large enough molecular size to be precipitated as the cetyl pyridinium salt. This fraction is still very heterogeneous in molecular size and composition (Table 9.1). Similar preparations have been obtained from *Chondrilla* sp. found in waters near La Parguera, Puerto Rico, and also from *Chondrosea* sp. (order Carnosa). The presence of hydroxyproline in Fraction F suggests the possibility that a collagen-derived peptide is covalently linked to a glycan component of the complex heteropolysaccharide mixture. Linkage of a glycoprotein to collagen has been shown in coelenterates (Chap. 3, Sect. 6).

STEMPIEN (1966) has briefly described his histochemical and chemical studies of glycans in six orders of Demospongiae. In both the Keratosa and the Carnosa, polyanionic glycans are confined to the mesoglea, with large concentrations of sulfated glycans in the dermal area and linings of canals; in the Haplosclerina, Hadromerina, and Poecilosclerina, the polyanionic glycans are confined to cells, choanocytes, and "amebocytes"; in the Epipolasida, both cells and mesoglea contain polyanionic glycans. *Haliclona viridis* (order Haplosclerina) contains only nonsulfated polyanionic glycans whereas *Microciona prolifera* (order Poecilosclera) contains a mixture of sulfated and nonsulfated glycans. Preparations from these two species, both of which possess spongin fibers, contained similar monosaccharides, viz. uronic acid, hexose, pentose, glucosamine, and fucose. The structure of the uronic acid is unknown but it differs from glucuronic, galacturonic, mannuronic, and iduronic acids.

The sponge cell-aggregation factor resembles vertebrate polyanionic proteoglycans in chemical composition and in some molecular parameters (HENKART et al., 1973; CAULDWELL et al., 1973).

c) Coelenterata

Physalia float as well as the derived gelatin, contains glucose, galactose, arabinose, uronic acid, glucosamine, galactosamine, and possibly ribose, fucose, and rhamnose, whereas *Metridium* corium contains none of these sugars (GROSS et al., 1958). KATZMAN and JEANLOZ (1970 a-c) isolated the total carbohydrate from the body wall of the sea anemone *Metridium dianthus* and found principally galactose and glucose, minor amounts of fucose, mannose and glucosamine, and a trace of ribose. No anionic residues, i.e. sialic acid, hexuronic acid, sulfate, or phosphate, could be detected. BOCQUET et al. (1972) found glucose, galactose, mannose, arabinose, and a small amount of glucosamine in the mesoglea of the jellyfish *Rhizostoma pulmo* L. but no hexuronic acid. Analyses for other acid groups were not performed. KATZMAN and JEANLOZ (1970d) remarked upon this unique occurrence of collagen in the complete absence of polyanionic glycans.

However, small amounts of polyanionic glycans appear to be localized elsewhere at specific sites in coelenterates. Both uronic acid and hexosamine are present in near equimolar amounts in nematocysts of *Metridium senile* (PHILLIPS, 1956). Slight metachromasia with threonine staining indicates the presence of polyanionic glycans in discharged nematocysts of the sea anemone *Aiptasia pallida*. The nematocysts from this species, as well as from other sea anemones, are particularly rich in hydroxyproline 2-8.5%), suggesting the presence of a collagen-like protein (BLANQUET and LENHOFF, 1966). A protein from nematocysts of *Hydra* contained 22% of hydroxyproline (LENHOFF et al., 1957). Polyanionic glycans in the secreted cuticle of Zoanthus are presumably responsible for a strong metachromatic reaction with toluidine blue (TRENCH, 1970).

d) Annelida

Many of the polychaete worms secrete a protective tube in which they live. The structure of a D-glucan phosphate polymer, onuphic acid, in tubes of the marine annelid *Hyalinoecia tubicola* (formerly *Onuphis tibicola*) has been extensively investigated (see review by HUNT, 1970). The glucosidic links may be (1 → 3) with an ester phosphate at the 4 or 6 position; a small amount of main chain phosphodiester links are found and some L-fucose is present. A detailed review of the structure and composition of tubes of *H. tubicola* and of other tube-forming polychaetes has been given by DEFRETIN (1971). Uronic acids and hexosamines are detected in tubes of some species; hyaluronate is thought to be present.

RAJULU and KRISHNAN (1967) suggested that the cuticle of the earthworm *Megascolex mauritii* may contain hyaluronate. Polyanionic glycans have been detected by histochemical methods in many tissues, including cuticles of other worms (see HUNT, 1970). The botryoid granules of the horseleech *(Haemopis sanguisuga)* contain a sulfated proteoglycan consisting of glucose and galactosamine and including phosphate residues but no uronic acid (FISCHER and KLUJBER, 1970).

e) Arthropoda

Connective tissues are sparsely distributed in insects. Only since about 1950, with the advent of the newer techniques of electron microscopy and chromatography, has it been possible to establish the presence of typical collagenous fibrils and neutral and polyanionic glycans. ASHURST (1968) has reviewed the recent literature which relies primarily upon histochemical methods for identification of polyanionic glycans. In most cases, the characterizations must be regarded as presumptive only. Chemical evidence supports the presence of hyaluronate and the absence of sulfated glycans in the midgut of the wax moth (ESTES and FAUST, 1964). Recently, very careful histochemical studies have suggested the presence and inhomogeneous distribution of glycans identical with, or very similar to, vertebrate hyaluronate, chondroitin sulfate, dermatan sulfate, and keratan sulfate in the connective-tissue layer around the ejaculatory duct of the locust and in various neural tissues of the cockroach, locust, wax moth, and stick insect (ASHURST and COSTIN, 1971a, b, c), and in the peritrophic membranes of *Melipona quadrifasciata* (MELLO and VIDAL, 1972).

Polyanionic glycans were isolated from the digestive gland and the cuticle of the crab *Hemigrapsus nudus*. Both tissues yield preparations of similar composition containing fucose, galactose, and glucose, with the last-named sugar more abundant than the first two. The sulfur content (presumably as ester sulfate) was in the range of 2.0 to 2.4 mg per 100 mg glucose (MEENAKSHI and SCHEER, 1959, 1961).

f) Mollusca

i) Mucins

Sulfated glycans are widely distributed in secretions of invertebrates. The many studies of gastropod mucins have been discussed at length by HUNT (1970). Only a brief summary will be given here.

The mucins of the hypobranchial gland of gastropods probably serve a largely hygienic function and protect organs present in the pallial cavity. Generally, the component glycans are heterogeneous in chemical constitution and difficult to characterize. Two different types of glycans, termed charonin sulfuric acids, are present in mucin from the large whelk *Charonia lampas*. These are: 1. a highly sulfated linear D-glucan containing β-(1 → 4)glycosidic linkages similar to those in cellulose, with a near equimolar ester sulfate and glucose, and 2. a low sulfated, branched-chain D-glucan probably containing both α-(1 → 4) and β-(1 → 4) glucosidic bonds.

The mucin from *Buccinum undatum*, a large prosobranch marine gastropod closely related to *C. lampas*, contains a mixture of a glycoprotein and a D-glucan sulfate similar to the highly sulfated fraction in *C. lampas* mucin. The glucan sulfate is linked to a polypeptide (2.2%) in a macromolecule of molecular weight 1.7×10^5. After extensive proteolysis, the major amino acids remaining are approximately one mole each of serine, glycine, and glutamic acid per mole of glycan (HUNT and JEVONS, 1965, 1966). These residual amino acids are the same as those that remain after proteolysis of vertebrate chondroitin sulfate proteoglycan (see Chap. 6). However, the sugar sequence linking glycan to peptide cannot be the same because xylose and galactose are absent from the invertebrate glycan. Since the linkage of glycan to peptide is alkali stable, it was also thought not to be an O-glycosyl link to serine (but see MATHEWS, 1971). Also, unlike the vertebrate proteoglycan, the native glycan peptide is apparently a single-chain macromolecule.

A similar glucan sulfate is present in mantle mucin of the lamellibranch mollusc *Pecten maximus*, and possibly also in the odontophore cartilage of *Busycon canaliculatum* (see Chap. 7, Sect. 2). The latter, however, has an unusual hypobranchial mucin which contains a highly acidic protein noncovalently linked via Ca^{++} bridges to a sulfated glycosaminoglycan. The glycan is composed of alternating, β-(1 → 4) linked D-glucosamine and D-galactosamine residues, each containing a sulfate group, probably present as *N*-sulfate.

Foot mucin of Japanese snails (unspecified species) contains a glycan, termed limacoitin sulfate, consisting of ester sulfate, L-fucose, galactose, mannose, galacturonic acid, and D-glucosamine (see HUNT, 1970). PANCAKE and KARNOVSKY (1971) characterized a glycan in mucus secreted by the North African land snail *Otella lactea*. This substance consists of glucosamine, L-iduronic acid, acetyl and sulfate residues in molar ratio of 1:1.5:1:1 and has a molecular weight of about 8×10^5. The liver (digestive gland) of *C. lampas* contains a family of glycan-peptides, containing, in various proportions, D-glucosamine, D-galactosamine, sialic acid, ester sulfate, and hexoses, including fucose, mannose, galactose, and glucose. This material might be of exogenous origin.

ii) Glycans of Shell and Mantle

Many workers have suggested that sulfated glycans in shells and in the closely associated mantle tissues are involved in the mechanisms of calcification (WILBUR and SIMKISS, 1968). HUNT (1970) has briefly reviewed various studies of preparations isolated from mantle tissues and from the organic matrix of shells. However, generally only estimates of contents of protein, hexosamine, and ester sulfate are cited. Recently, WADA and FURUHASHI (1970) isolated a polyanionic glycan-peptide in a yield of 1 to 2% of the dry weight of the organic matrix from shells of the bivalve *Hydriopsis schlegeli*. Analyses showed 29% of protein, 2.8% of hexosamine (i.e. glucosamine plus galactosamine), 3.3% of ester sulfate and 0.1% of hexose. The preparation stained metachromatically and had an electrophoretic mobility on paper near that of chondroitin sulfate. Similar substances were isolated from mantle mucus of the oyster *Pictada fucata* and the clam *Chlamys nobilis* (WADA and FURUHASHI, 1971). Previously, WADA (1964a, b) had shown that tracer calcium was localized in the epithelium, subepithelium and in secreted mucin of the mantle and also selectively localized in the metachromatic substance in the process of calcification. WADA (1966) suggested further that the conchiolin (protein) of mineralized layers is probably a protein complex of a glycan-protein containing ester sulfate.

iii) Mactin and Spisulan

The initial demonstration of a heparin-line (anticoagulant) factor in clams by THOMAS (1951) spurred a series of further studies. Anticoagulant activity, as well as metachromatic intensity, was concentrated in mantle, gills, and labial palps of the common surf clam *Spisula (Mactra) solidissima* but was also present in other tissues. High concentrations were noted in certain regions of the mantle-edge connective tissues that also concentrate calcium (THOMAS, 1954). Heparin-like glycan preparations with high anticoagulant activity were isolated from *S. solidissima* and *Cyprina (Artica) islandica* (FROMMHAGEN et al., 1953; LOVE and FROMMHAGEN, 1953) and designated Mactin A and Mactin B, respectively. A chemical resemblance to mammalian heparin was noted. Large-scale preparations from the two species of clams were further purified by BURSON et al. (1956). Both mactin A and mactin B were reported to contain only glucosamine and glucuronic acid as carbohydrate constituents and thus appeared similar to heparin, except for their optical rotations (higher than heparin) and molecular weights (about 25,000 compared to about 14,000 for heparin).

Further study was indicated when mactin B was subsequently shown to contain appreciable amounts of galactose, galactosamine and threonine, and mactin A to contain a sulfated galactan component (CIFONELLI and MATHEWS, 1966). Both mactin preparations contained at least three separate components; the major component, designated simply mactin or clam heparin, had a composition and physical chemical parameters within the range of mammalian heparins (CIFONELLI and MATHEWS, 1972a). Mactin was unusual in that it contained D-glucuronic acid rather than L-iduronic acid as the major uronic acid (TAYLOR et al., 1973), which may explain why its sulfate content is somewhat lower than that of mammalian heparin (see Chap. 6). Molecular weights, estimated at near 50,000, are in the high range of values for mammalian heparins. The biosynthesis of mactin, like that of mammalian heparin, involves the participation of phosphoadenosine-phosphosulfate as sulfate-group donor (DeMEIO et al., 1967).

In addition to mactin and sulfated galactan, clams contain a third component, a sulfated glycan consisting of galactosamine, galactose, fucose, threonine, proline, and ester sulfate. Preparations by CIFONELLI and MATHEWS (1972) of this substance, termed spisulan, from *C. islandica* and from the common quahog *Venus merceneria* L., have been tentatively assigned the structure shown in Fig. 9.1. The preparation had a molecular size greater than that of mactin and about the same electrophoretic mobility.

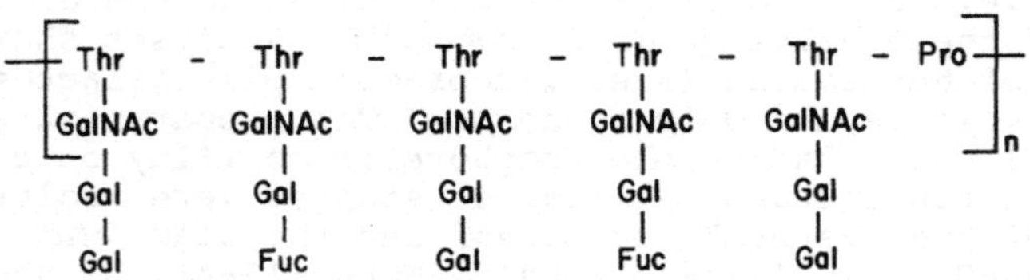

Fig. 9.1. A possible structural arrangement for spisulan. The placement of fucose is arbitrary and merely reflects its terminal position and average proportion to other residues. Ester sulfate residues are contained in molar equivalence to the sum of all the sugar residues but have been omitted here because of uncertainty regarding their location. Appreviations: GalNAc, N-acetylgalactosamine; Gal, galactose; Fuc, fucose; Thr, threonine; Pro, proline

Since all three polyanionic glycans, isolated after extensive proteolytic treatments, still retain significant amounts of bound peptide, it is probable that they are fragments of larger molecular entities. They are nearly equivalent in charge density and likely to bind basic metachromatic dyes in an equivalent manner and hence are not readily distinguishable by histochemical techniques alone. However, it may be possible to utilize the strong anticoagulant activity of the heparin-like mactin to determine its histological location, since spisulan is devoid of anticoagulant activity and the sulfated galactan component, although of undetermined anticoagulant activity, is a very minor component. The presence of these glycans in several species of clams suggests that the polysaccharides are widely distributed in molluscs and possess important common functions, possibly in digestion, hygiene, tissue structural organization, mineralization, etc. Mactin also has the unusual property of completely and reversibly inhibiting cell division of cultures of L-M cells (LIPPMAN et al., 1971).

iv) Glycans of Cephalopod Eye and Skin

The squid cornea, unlike the analogous tissue in the vertebrate eye, is a direct continuation of the skin. Its glycosaminoglycans are quite unlike that of vertebrate cornea (see Sect. 3 below) in possessing more galactosamine-containing glycan than glucosamine-containing glycan (ANSETH, 1961a; MOCZAR and MOCZAR, 1973). Also, a relatively higher proportion of the galactosamine appears to be present in glycans of low sulfate content. Unfortunately, identification of the glucosamine and galactosamine with keratan sulfate and chondroitin sulfate, respectively, is uncertain. The vitreous body, unlike that of most vertebrates, is a liquid rather than a gel and contains a high concentration of hexuronic acid and hexosamine, presumed to be present as hyaluronate (BALAZS et al., 1959).

The polyanionic glycans that can be isolated from the skin of the squid *Ommastrephes sloani pacificus* in a yield of about 2.4% of the dry weight have been demonstrated to consist predominantly (85%) of "chondroitin", a glycan with the same polysaccharide structure as chondroitin sulfate of vertebrates but lacking sulfate (ANNO et al., 1964; KAWAI, 1968). Small amounts of glycans containing 4-S and diS_D periods (see Table 7.1) are also present (ISOBE and SENO, 1971). Low-temperature aqueous extraction of skin released over 75% of the glycans which consisted nearly exclusively of chondroitin bound to peptide. Most of the sulfated glycans remained in the residue. The purified chondroitin-peptide was reasonably homogeneous by various criteria. However, gel electrophoresis (MATHEWS and DECKER, 1971) showed the preparation to have a molecular weight (about 28,000) much lower than expected and only about 10% greater than that of single glycan chains obtained after proteolytic hydrolysis (MATHEWS, unpubl.). The difference in size appeared to be due to the presence of a small peptide. The low molecular weight could be a reflection of either 'native' state or inadvertent proteolysis attendant upon isolation.

Table 9.2. Amino acid and neutral sugar residues (moles per glycan chain of 25,000 daltons) of chondroitin-peptide from squid skin (values recalculated from ISOBE and SENO, 1971)

Amino acid	Moles	Neutral sugar	Moles
Serine	3.2	xylose	0.7
Glycine	6.5	galactose	2.2
Glutamic acid	4.9	mannose	0.7
Aspartic acid	3.3	fucose	1.6
Proline	2.1	glucose	4.8
Threonine	1.8		
Alanine	2.9		
Valine	1.6		
Isoleucine	1.3		
Leucine	1.7		
Total	∿ 29		∿ 12

Analytical values obtained by ISOBE and SENO (1971) on the chondroitin-peptide have been recalculated and appear in Table 9.2; the molar ratio of glucuronic acid to galactosamine was 1.07. Perhaps most unexpected is the finding of a large amount of glucose, which was also noted in a glycopeptide derived from the polysulfated chondroitin sulfate of squid *(Loligo opalescens)* skin (RADHAKRISHNAMURTHY et al., 1970). Glucose is not a linkage-region component of vertebrate chondroitin sulfate proteoglycans (Chap. 6, Sect. 2). It has been detected only in very minor proportion in relation to xylose and galactose in the chondroitin-peptide fraction but not the chondroitin 4-sulfate fraction of bovine corneal stroma (GREILING et al., 1970).

The polyanionic glycans in skin of another species of squid, *L. opalescens*, consist predominantly of polysulfated chondroitin sulfate with only 25% chondroitin and no dermatan sulfate (SRINIVASAN et al., 1969). RADHAKRISNAMURTHY et al. (1970) suggested that the protein moiety of the polysulfated chondroitin sulfate proteoglycan contains, in addition to the chondroitin sulfate side-chains linked through a xylosyl-serine residue, short oligosaccharide chains consisting of glucose, galactose, and glucosamine linked by an alkali-resistant bond. However, the isolation by HABUCHI et al. (1971) of a fragment of squid cartilage chondroitin polysulfate that contains glucose leaves little doubt that a fraction of the glucose in the proteoglycan is a part of the glycan-peptide linkage region of preparations from cartilage.

The differences in relative amounts of chondroitin and chondroitin sulfate in skin between the two species of squid just discussed may be related to a general variation in occurrence among cephalopod species of cartilagenous dermal scales. Such epidermal structures, which resemble vertebrate hyaline cartilage in histology and stain metachromatically with toluidine blue, probably contain a high concentration of polysulfated chondroitin sulfate. They have been described in the squid *Cranchia scabra* by PERSON (1969) and in the common squid *Loligo pealii* by MIROW and RHODIN (quoted by PERSON, 1969). It would be of interest to compare the skins of *O. sloani pacificus* and *L. opalescens* with respect to this morphologic feature.

g) Brachiopoda

Members of the class Inarticulata among the Brachiopoda have been intensively studied (JOPE, 1971) both because their fossil record is excellent from recent times to the Cambrian period and because of the association of collagen with phosphatic calcification of their shells. (see Chap. 3, Sect. 6h). Of special interest are *Lingula* and related genera, which have remained virtually unchanged since the Ordovician period. This group had developed a contractile pedicle, a long, muscular cylinder that permits the animal to burrow into sand when disturbed (RUDWICK, 1970).

A preliminary study (MATHEWS, unpubl.) has been carried out on polyanionic glycans of *Lingula unguis*. The composition given in Table 9.3 suggests a complex glycan-peptide or a mixture of glycan-peptides of similar charge density. The sulfated glycan-peptide represented about 5% of the dry weight of the cuticle; also present were about 2% of collagen (estimated from hydroxyproline content) and about 20% of chitin.

Table 9.3. Sulfated glycan-peptide from *Lingula* cuticle (moles/mole hexosamine)

Residue	Moles
Glucosamine	0.40
Galactosamine	0.60
Mannose	0.27
Fucose (or arabinose)	0.24
Xylose	0.11
Glucose	0.02
Hexuronic acid	0.0
Amino acids	1.6
Ester sulfate	1.33

h) Echinodermata

The echinoderms produce a large variety of polyanionic glycans, in most cases of the sulfated type and probably linked to protein, with highly diverse and commonly uncertain functions. Epithelial secretions of the alimentary tract probably serve a lubricatory function in the gut. Secretions of epithelial cells of appendages are often adhesive and probably serve to trap food particles or assist in locomotion. The jelly coats of eggs contain glycans that play an important role in the fertilization process (VASSEUR, 1954; MONROY, 1965). Polyanionic glycans are also found in supporting structures and presumably have structural functions. In most cases only histochemical procedures have been utilized for detection purposes; only a small number of glycans have been chemically investigated. HUNT (1970) and DOEZEMA (1969) have furnished reviews.

The egg jelly of the sea urchin *Echinus esculentus* contains a sulfated galactan-peptide which consists of (1 → 3) glycosidically linked L-galactose residues sulfated at position 4. *Echinocardium cordatum* egg jelly contains a sulfated glycan which contains only fucose; several structures have been proposed, one of which is a (1 → 4) linked chain with branch points at position 3 and ester sulfate in each residue at position 2. Jelly coats of eggs of other Echinoidea, such as *Strongylocentrotus droebachiensis* and *Paracentrotus lividus*, contain sulfated glycans consisting primarily of fucose, with glucose, xylose, galactose, and hexosamines as minor components. The egg jelly of the sand dollar *Echinarachnius parma* contains only fructose; that of star fish *Asterias amurensis* (class Asteroidea) apparently contains a sulfated heteroglycan consisting of near equimolar amounts of fucose and galactose with about 15 mol % of hexosamine. It is generally assumed that only one type of molecule is present.

MOTOHIRO (1960a, b, c) extracted the body walls of sea cucumbers with 0.85% NaCl and isolated glycan-proteins containing about 60% protein and 40% protein from *Stichopus japanicus* and *Cucumaria japonicus*, respectively. It was suggested that *S. japonicus* contained a chondroitin sulfate-type glycan and a neutral glycan consisting of hexosamine and neutral sugar in equimolar amounts. The product from *C. japonicus* was deproteinized by "mild" procedures and shown to contain two components. One glycan contained L-fucose as the sole carbohydrate component with ester sulfate in equimolar proportion; about 15% of protein remained. The second glycan (containing about 6% of peptide) consisted of equimolar proportions of galactosamine, glucuronic acid, and ester sulfate; a structure similar to that of chondroitin sulfate was suggested (see also TANAKA et al., 1972).

Related glycans were isolated from the body wall of the sea cucumber *Thyone briareus* by KATZMAN and JEANLOZ (1969, 1973) by proteolytic digestion and purified by alcohol fractionation. The residues of the L-fucan sulfate component ($[\alpha]_D^{21} = -150^o$) were linked by $\alpha(1 \rightarrow 2)$ glycosidic bonds and were sulfated at position 4. The amino sugar-containing component ($[\alpha]_D^{21} = -42^o$) contained D-galactosamine, D-glucuronic acid, and ester sulfate in a molar ratio of 1.00:1.07:1.6. A portion of the sulfate was located in position 6 of the amino sugar residue. Apparently, the degree of sulfation of glycans in sea cucumbers is highly variable. The identity of the galactosamine-containing glycan with a vertebrate glycan remains to be established.

3. Chordata Acrania

Although this group of animals is of great potential importance in providing clues as to the possible invertebrate origins of the vertebrates, they have been subjected to comparatively little biochemical investigation.

a) Urochordata (Tunicata)

The tunicates have received a modicum of attention because of the presence in their tunics (outer layer of the body wall,) of a celluloseprotein complex, tunicin. The tunic in many ascidians gives a metachromatic color with toluidine blue, a reaction which has been interpreted as being due to a polyanionic glycan. Since the staining was sensitive to testicular hyaluronidase, it was thought to indicate the presence of chondroitin sulfate (HUNT, 1970). This glycan was also identified in the cytoplasm of testaceous cells of *Ciona intestinalis* L. (DeVINCENTIIS and MATERAZZI, 1964). Further work is necessary since examination of the polyanionic glycans extracted from the body walls, or tests, of four different species of ascidians (representing both orders of the class Ascidiacea) reveals a mixture of sulfated glycans containing mainly neutral sugars, little hexosamine and no hexuronic acid. It appears unlikely that chondroitin sulfate is present as more than a very minor component of the sulfated glycans in ascidian test (MATHEWS, unpubl.).

b) Cephalochordata

Hyaluronate and an unknown sulfated glycan have been found in amphioxus (see discussion in Chap. 7, Sect. 3).

4. Chordata Craniata (Vertebrata)

The connective tissues are ubiquitous in the vertebrate body. They appear in a great variety of morphological forms and are closely associated with virtually every organ. Since morphologic homologies are relatively well established in the subphylum (ROMER, 1970; SMITH, 1960), it is advantageous to organize comparative biochemical data on an anatomical basis.

Cartilage, calcified tissues, and notochord have been covered in previous chapters. Other connective-tissue structures for which reliable comparative biochemical data exist in adequate amount will be accorded major consideration here. Additional tissues and organs will receive a briefer treatment and are included because of their inherent importance in a comparative survey. However, no attempt has been made to achieve complete coverage of the literature for all tissues. Further information may be found in various reviews (CHVAPIL, 1967; BRIMACOMBE and WEBER, 1964; STACEY and BAKER, 1962; JEANLOZ and BALAZS, 1965; SCHILLER, 1966; MUIR, 1964). Qualifications relating to interpretation of data were discussed in Sect. 1.

a) Skin

The vertebrate integument - the skin with its accessory structures - is an organ system with varied and important functions including protection against mechanical injury (see Chap. 10, Sect. 4a), regulation of water content and temperature of the body, metabolite exchange, etc. The skin is not a single structural entity, but consists of two parts, epidermis and dermis, closely united but differing greatly in structure and origin. The epidermis is superficial in position and is essentially a cellular epithelium; the dermis is a mesenchymal derivative and contains the connective tissue cells and their biosynthetic products.

In the course of evolution of the anamniotes, the integument has for the most part been characterized by (1) a decrease in thickness and (2) a decrease in degree of ossification. These trends were advantageous in permitting greater mobility and, in amphibians, in permitting greater reliance upon the skin as a respiratory organ. In the amniotes, however, this evolutionary trend is reversed in part and the skin has become progressively thicker, especially in the dermal layers, primarily to prevent loss of water and body heat. The need for a thick skin is reduced only for those amniotes provided with such skin accessories as feathers or a thick coat of hair. The biochemistry of glycans in skin has been reviewed by PEARCE (1965).

i) Cyclostomes

The dermis itself, as well as numerous kinds of bony elements, may have derived phylogenetically from the primary dermal armor of the earliest vertebrates (HALSTEAD, 1968; SMITH, 1960). In the cyclostome derivatives of the ostracoderms, reduction of armor has proceeded further than in any other group of vertebrates leaving no hard vestiges, such as teeth or scales. Phylogenetic relationships of the two orders of extant cyclostomes are unclear (see Chap. 7, Sect. 4).

The skins of the hagfish *Paramyxine atami* and *Eptatretus burgeri* possess a significant amount of hyaluronate, and in addition, a highly sulfated fraction (about 25% of the total sulfated glycans) containing a novel disaccharide repeating unit. SENO et al. (1972) showed that this disaccharide unit had a structure which can be considered to derive from that of dermatan 4-sulfate (Fig. 6.1) by addition of an ester sulfate group to position 6 of the amino sugar moiety and another ester sulfate group to positions 2 or 3 of the uronic acid moiety. It differs from derm-H (diS_H) by the presence of an additional sulfate group in the uronic acid moiety and has therefore been designated $triS_H$ in Table 7.1.

Table 9.4. Disaccharide period composition (%) of highly sulfated fraction of polyanionic glycans from hagfish skin (SENO, AKIYAMA, and ANNO, 1972)

	4-S	6-S	diS_E	diS_B	$triS_H$
P. atami	22.8	8.2	4.1	38.2	26.7
E. burgeri	21.8	7.1	9.1	36.4	25.6

The disaccharide composition of the highly sulfated fraction of the polyanionic glycans from hagfish skin is given in Table 9.4. It is possible that the monosulfated periods 4-S and 6-S occur in combination with one or more of the polysulfated periods in the same molecule. Comparison with polyanionic glycans in notochord and cartilage of hagfish (Table 7.3) shows that triS_H is present only in skin and diS_H only in notochord; small amounts of diS_E are present in skin while diS_E is predominant in cartilage (and also in cartilage of squid (Chap. 7, Sect. 2b); hyaluronate is detectable only in skin. Is is evident that each tissue has a characteristic composition of glycosaminoglycan disaccharide periods. However, it should be noted that hagfish skin also contains unidentified sulfated glycans, consisting of glucosamine, galactose, xylose and other sugars, which require further investigation. Hyaluronate is a very minor component compared to sulfated glycans (MATHEWS, unpubl.).

Lamprey (species not identified) skin, which contains hyaluronate and dermatan sulfate, has a polyanionic glycan chemistry comparable to that of mammalian skin, according to SENO and MEYER (1963). A similar conclusion may be drawn from a study of *P. marinus* skin (MATHEWS, unpubl.) with the qualification that hyaluronate is very low or absent; the sulfated glycans consist of dermatan sulfate and only traces of unidentified components. It appears that the two cyclostome orders are biochemically set apart with respect to skin as well as to cartilage and notochord; a phylogenetic interpretation of the biochemical differences for the latter tissues has already been given (Chap. 7, Sect. 4a).

Sulfated heteropolysaccharides of hagfish skin mucin are a complex mixture of several glycans differing in sulfate, glucosamine, galactosamine, galactose, fucose, and mannose contents (LEHTONEN et al., 1966).

ii) Chondrichthyes

SENO and MEYER (1963) isolated polyanionic glycans in yields of 1.4% and 0.1% of the dry weight from skins of the blue shark and the Sandbar shark, respectively. The preparations contained 70% of hyaluronate, 5% of dermatan polysulfate, 20% of chondoitin 6-sulfate in the case of the blue shark, and 30% of hyaluronate, 60% of dermatan polysulfate, and 10% of chondroitin 6-sulfate in the case of the Sandbar shark. In both species, the dermatan polysulfate had a molar ratio of sulfate to galactosamine of about 1.6. A preparation comparable to the one from the Sandbar shark was obtained in a yield of 0.2% dry weight from skin of *H. franscisci* (MATHEWS, unpubl.; Chap. 7, Sect. 4b). The sulfated glycan fraction had the following disaccharide composition: 2% of 0-S, 2% of 4-S, 7% of 6-S, 2% of diS_D, 8% of diS_E, 37% of 4-S_B, 42% of diS_B and no diS_H. It may be noted that skins of both species that had the lower total content of polyanionic glycans contained predominantly dermatan polysulfate, whereas the skin of the third species that was relatively richer in these glycans contained predominantly hyaluronate.

iii) Osteichthyes

The skin of the electric eel (species unknown) contains hyaluronate and dermatan monosulfate (SENO and MEYER, 1963).

iv) Amphibia

The layers normally found in the anuran skin, starting at the surface are: a thin stratum corneum of keratinized epithelium, the epithelium of 3-15 cells in thickness, a basal membrane, the stratum spongiosum which contains various glands in a network of collagen and elastic fibers, smooth muscle fibers, nerves, capillary blood vessels and pigment cells, the stratum compactum, consisting of a few or of many layers of thick collagen fibers, and the tela sucutanea which forms the outer lining of the subcutaneous lymphatic space. An additional layer, interposed between the stratum spongiosum and the stratum compactum, stains strongly with basic dyes and exhibits a metachromasia with toluidine blue that is abolished by testicular hyaluronidase. It has been termed the ground-substance layer (ELKAN, 1968). This layer apparently contains glycosaminoglycans and has, in addition, a high concentration of calcium salts.

The adaptation of amphibia, which have skins freely permeable to water, to various habitats ranging from fresh water to arid desert has occupied many biologists (BENTLEY, 1966). ELKAN (1968) has suggested that the ground-substance layer of the anuran skin has an important role in controlling the transport of water and electrolytes. Published data on glycosaminoglycan composition are meager (Table 9.5).

Table 9.5. Composition of glycosaminoglycans of adult anuran skin

	R. catesbiana [a]	*Bufo marinus* [b]	*Bufo bufo* [b]
Hyaluronate	30	0	10
Dermatan sulfate	50	15	45
Chondroitin 4-sulfate	20	0	0
Chondroitin 6-sulfate		85	45

[a] Calculated from data of LIPSON and SILBERT (1968) with correction for an apparent content of dermatan (nonsulfated) probably due to contaminating carbohydrates (LIPSON, CERSKUS, and SILBERT, 1971).
[b] Data of MATHEWS and ELKAN (unpubl.).
N.B.: all three species are terrestrial and demonstrate the presence of a ground-substance layer in the skin; skin of *B. bufo* also contained 4.3% of the dry weight as water-insoluble inorganic salts.

Profound alterations in biochemical properties have been noted to occur in amphibian tissues during the developmental stages preceding the transition from an aquatic to a terrestrial mode of life. The changes in composition and metabolism of cartilage were described earlier (Chap. 8, Sect. 2). Comparable data, for skin of *R. catesbiana* are shown in Fig. 9.2. Hyaluronate, which appears as virtually the sole glycosaminoglycan at the tadpole stage, declines rapidly during development in both absolute amount in skin and in proportion of total glycosaminoglycan content. Chondroitin sulfate and dermatan sulfate become important components at an early stage with dermatan sulfate becoming predominant. The relative amounts of the various glycans that are established by the end of metamorphosis are maintained in the adult frog skin, although the total amount in adult skin is greatly reduced from the amount present in skin during metamorphosis and at the juvenile stage. Thus, preparation for adult life on land not only involves changes in relative proportions of the different glycosamino-

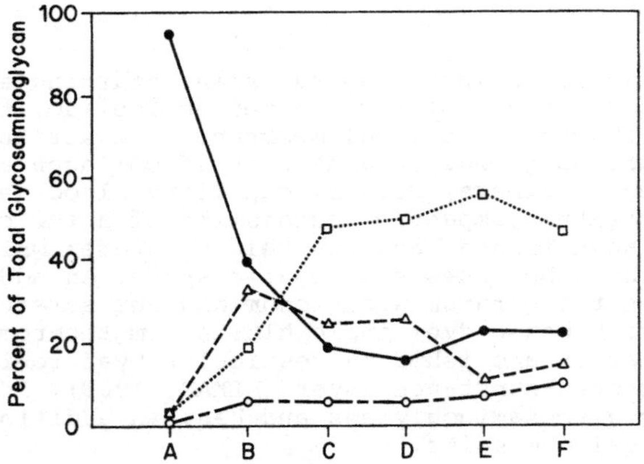

Fig. 9.2. Glycosaminoglycans of *R. catesbiana* back skin at various developmental stages (redrawn from LIPSON et al., 1971). Indicated are hyaluronate ●, chondroitin 4-(or 6-)sulfate △, dermatan sulfate □, and "dermatan" ○ (the latter glycan has a questionable identity, see text). The developmental stages are characterized as: A, tadpole - no visible appendages; B, similar to end of metamorphic stage 20 as described for *R. pipiens* by TAYLOR and KOLLROSS (1946); C, similar to metamorphic stage 22 of *R. pipiens*; D, similar to metamorphic stage 23 of *R. pipiens*; E, juvenile frog similar to metamorphic stage 25 of *R. pipiens* - tail and operculum completely resorbed, length 3 to 4 cm; F, adult bullfrog, body length 14 to 17 cm

glycans but also decreases the absolute amounts in skin. Dermatan sulfate is concentrated in the basal lamella of larval skin where collagen is localized (HATA and NAGAI, 1973).

At this point one may raise two related questions; are the changes observed in glycosaminoglycans during development functionally related to the change in habitat from aquatic to terrestrial? And, if so, do similar relationships hold for adults of different species in different habitats? Structural, functional, and topological data relating to glycosaminoglycans of skin are too scanty to warrant further discussion of the first question. An answer to the second question was sought by ELKAN (1968) on the hypothesis that the ground-substance layer of anuran skin (see above) functions as a protection against dessication. Histological examination of skins from anurans representing 105 species and 17 families revealed a significant correlation between the presence of a ground-substance layer and habitat. Of the aquatic species examined, 15 (of a total of 22) were totally or nearly completely devoid of the ground-substance layer; 69 (of a total of 83) of the amphibious (including predominantly terrestrial, fossorial and deserticolous species) species were positive. However, some striking exceptions were found: three predominantly terrestrial species were completely negative while two almost completely aquatic species were strongly positive.

Preliminary data obtained from an investigation of skin glycosaminoglycans of a limited number of species (MATHEWS and ELKAN, unpubl.) are presented in Table 9.6. A lack of correlation of glycosaminoglycan hexosamine with staining of the ground-substance layer or with habitat is evident, particularly when the three species at the bottom of the table are compared. Several possible explanations can be given: a) skin layers other than the ground-substance layer also contain glycos-

Table 9.6. Contents of glycosaminoglycan (hexosamine) and collagen (hydroxyproline) of anuran skin (MATHEWS and ELKAN, unpubl.)

Species	Habitat[a]	Ground-subst. layer[b]	Hexosamine, % dry wt.	Hydroxyproline, % dry wt.	Hexosamine/hydroxyproline	Dermatan sulfate[c]
Bufo regularis	ter	+	0.35	5.0	0.071	10
B. spinulosus	ter	+	0.31	5.1	0.060	
B. viridis	ter	trace	0.53	4.4	0.120	
Ceratophrys ornatus	ter	+	0.35	6.7	0.038	45
C. stolzmanni	ter	+scanty	0.20	6.7	0.030	
Discoglossus pictus	amph	+	0.27	4.6	0.059	
Odontophrys americanus	ter	+	0.48	4.0	0.120	
Rana esculenta	ter	+	0.33	5.1	0.064	50
R. grylio	ter	+	0.23	4.7	0.048	45
Xenopus laevis	aq.	trace	0.58	5.8	0.100	75
B. marinus[d]	ter	+abundant	0.45	4.6	0.097	15
B. bufo[d]	ter	+	0.07	6.4	0.011	45

[a] Abbreviations: ter = terrestrial, amph = amphibious, aq = aquatic. [b] + = readily stained with Alcian Blue (ELKAN, 1968). [c] % of total glycosaminoglycans. [d] See Table 9.5.

aminoglycans; b) variation occurs with age of the animal, with seasonal changes and other factors; and c) anurans utilize alternate mechanisms of adaptation to desiccating environments. Also, anuran species may differ greatly in the types of organization of the various skin layers and in possession of specialized elements (ELKAN, 1968). For example, skin of the toad *B. bufo* is heavily calcified and also contains poison glands, particularly numerous in areas near the ears. Astonishingly, the venom from these parothoid glands contains nearly 15% of the dry weight as chondroitin 6-sulfate (see Sect 4j below). The skin of *R. catesbiana*, however, is free from poison glands.

v) Aves

Data on glycosaminoglycans of chicken skin at various ages are compiled in Table 9.7. Rapid changes in collagen and glycan composition occur with increasing age. Hyaluronate declines but remains the dominant glycosaminoglycan in embryonic life and well into advanced age. Dermatan sulfate rises steadily with age to equal prominence. The smaller proportions of chondroitin sulfates in embryonic skin are reduced to a very minor proportion at birth. As these changes occur, the collagen content of skin rises rapidly during embryonic stages and early postnatal growth while the total content of glycosaminoglycans declines. Interestingly, KONDO et al. (1971) detected in their glycosaminoglycan digests a very small content of diS_B periods, which

Table 9.7. Glycosaminoglycans of chicken skin

Age	Hypro[a] %	HexN[b] %	Hy % of HexN	Derm 4-S % of HexN	C 4- (or 6-) S % of HexN
17 days (embryo)[c]	2.1	0.16	55	15	30
1 day[d]	3.4	0.15	70	∿ 25	∿ 5
4 wks[d]	5.4	0.11	60	∿ 35	∿ 5
23 mo.[d]	10.3	0.10	45	∿ 50	∿ 5

[a] Hydroxyproline content of dried skin. [b] HexN = total hexosamine of glycosaminoglycans of dried skin. [c] Calculated from data of MATHEWS (1967b) for special strain of breeding flocks from Agricultural Experimental Station of the University of Connecticut. [d] Calculated from data of KONDO, SENO, and ANNO (1971) for the white Leghorn chicken.

increased in amount with age. Heparan sulfate represents about 3% of the total glycosaminoglycans of adult skin (KONDO-OTSUKA et al., 1972).

vi) Mammalia

The principal glycosaminoglycans of normal adult mammalian skin appear to be hyaluronate and dermatan sulfate. Hyaluronate is readily extracted while a portion of the dermatan sulfate is tightly held to the collagenous residue (TOOLE and LOWTHER, 1966; FLEISCHMAJER, PERLISH, and BASHEY, 1972). In addition, a number of other sulfated glycosaminoglycans have been isolated in small amounts or detected as traces, including chondroitin 4- (or 6-) sulfate. Heparin and heparan sulfate are clearly identified components of skin in rat (SCHILLER and DORFMAN, 1960; SCHILLER, 1966b) and in hog (CIFONELLI and RODÉN, 1968). The heparin content of rat skin is high (apparently associated with the presence of numerous mast cells in skin of this species) while no heparin can be detected in guinea-pig skin (FABIANEK et al., 1965). Traces of keratan sulfate have been reported (BARKER et al., 1965) to occur along with the other glycosaminoglycans in human skin (BARKER, KENNEDY, and SOMERS, 1969; BARKER, KENNEDY, and CRUIKSHANK, 1969).

The concentration and metabolism of glycosaminoglycans varies with physiological state, hormonal influences, and age (SCHILLER, 1963, 1966a). Data which demonstrate that the concentration of total glycosaminoglycans in skin of many mammals declined from birth to old age by 50-70% are expressed on a dry weight basis. This method of expression of concentration may be misleading. MIER and WOOD (1969) point out that the observed decline is due mainly to an increase in collagen content of skin and that, at least for the guine-pig, no major variations in concentration with age are seen if values are expressed on a wet weight basis.

Unfortunately, quantitative analysis of the complex mixture of glycosaminoglycans of skin is a formidable technical problem, especially when only small quantities of tissue are available. As a consequence, much of the published data has been confined to the determination of hyaluronate and frequently of heparin, with the remaining glycans grouped in a third fraction. SCHILLER and DORFMAN (1960) showed that hyaluronate and heparin respectively comprise about 50 to 60% and 15% of the glycosaminoglycans in rat skin at 21 to 44 days of age. The proportion of hyaluronate remained relatively constant to over 30 months of age, while the heparin concentration fell sharply to about 5 to 10% at 2 months and remained at that level thereafter. PRODI (1967), however, found much lower values of heparin, in the range of 1 to 3%, throughout adult life of the rat; hyaluronate values declined from 80% at 1 to 5 days to 50% at 30 to 36 months. Neonatal rat skin glycosaminoglycans consisted of 56% of hyaluronate, 16% of dermatan sulfate, 9% of chondroitin 6-sulfate and smaller amounts of chondroitin 4-sulfate, heparan sulfate, heparin, and unidentified sulfated glycans (HARDINGHAM and PHELPS, 1970).

Earlier analyses by LOEWI (1961) of human skin from various anatomical sites at different ages were limited to determination of hyaluronate and dermatan sulfate only. In general, the proportion of hyaluronate declines from about 60 to 40% from birth to 70 years of age while that of dermatan sulfate increases from about 35 to 50% during the same span of time. At all ages, the thin skin at the back of the hand shows a lower ratio of dermatan sulfate to hyaluronate than does skin from the abdomen or leg regions. Also, analyses of different horizontal layers of skin showed no correlation between the amounts of collagen and dermatan sulfate, although a strong association between these two components was indicated for bovine (HOFFMAN et al., 1957; TOOLE and LOWTHER, 1966), rabbit (PRODI et al, 1971), and hog (LOEWI and MEYER, 1958) skin. BREEN et al. (1972) found that the glycosaminoglycan concentration of fetal skin fell tenfold from 3 months gestation to term, but was still twice that of adult skin at 41 years of age. Skin from a 3-month-old fetus contained only hyaluronate and chondroitin 4-(or 6-) sulfate. With fetal development from 3 months to term the proportion of hyaluronate declines from 70% to 40% of the total glycosaminoglycans, the remainder being largely dermatan sulfate (BREEN et al., 1972).

The content of dermatan sulfate (plus heparan sulfate) but not of hyaluronate declines with age (VAN LIS et al., 1973).

The glycosaminoglycans of hog-embryo skins taken from specimens 9-20 cm in length consisted of about 70% of hyaluronate, about 15% of dermatan sulfate, and about 15% of chondroitin sulfate. LOEWI and MEYER (1958) suggested that the last was chondroitin 6-sulfate, although the optical rotation of the preparation was closer to the value expected for chondroitin 4-sulfate. This composition was quite different from that of the glycosaminoglycans of adult hog skin, which consisted of 30% hyaluronate and 70% of dermatan sulfate.

BARKER et al. (1969) had suggested that human skin contains "oversulfated chondroitin sulfate". Analysis of a dermatan sulfate preparation from hog skin showed that about 3% of its disaccharide repeating periods possessed two ester sulfate groups as diS_B (MATHEWS, unpubl.).

b) Cornea

The biochemistry of the cornea (and other tissues of the eye) has been thoroughly reviewed by BALAZS (1965), MAURICE and RILEY (1970), and DISCHE (1970b); ultrastructural organization, macromolecular interactions, swelling, and other properties of corneal stroma are discussed in Chap. 10, Sect. 4b. The cornea of the squid, which is strictly a transparent part of the skin, was discussed earlier.

i) Chondrichthyes

The corneal glycosaminoglycans (hexosamine = 1.6% of the dry wt.) of the spiny dogfish were separated by PRAUS and GOLDMAN (1970a) into a minor glucosamine-containing fraction with a molar ratio of sulfate to glucosamine of 1.27 and a major galactosamine-containing fraction with a molar ratio of sulfate to galactosamine of 1.40. The minor and major fractions were identified on the basis of their elution behavior on cellulose-cetylpyridinium chloride columns as keratan sulfate and chondroitin sulfate, respectively, The relative proportions of these two components differed greatly from those in corneal stroma of animals of other vertebrate classes (PRAUS and GOLDMAN, 1970b; BALAZS, 1965).

Keratan sulfate preparations purified from corneal stromas of several species of shark *(G. cuvier, C. leucas, H. franscisci)* showed molar ratios of sulfate to glucosamine from 1.40 to 1.96 (MATHEWS and CIFONELLI, 1965). The chondroitin sulfate fraction from *C. leucas* cornea had a molar ratio of sulfate to galactosamine of 1.33 and the disaccharide composition: 50% of 4-S periods, 15% of 6-S periods, and 35% of diS_D periods (see Table 7.1).

ii) Osteichthyes

The glycosaminoglycan hexosamine of carp and tuna corneas were 1.1% and 1.4% of the dry-tissue weight, respectively, with glucosamine to galactosamine ratios of 2.1 and 3.4, respectively (ANSETH, 1961a). Keratan sulfate from whitefish *(Coregonus)* cornea had a molar ratio of sulfate to glucosamine of 1.24 (MATHEWS and CIFONELLI, 1965); the chondroitin sulfate fraction consisted entirely of polysaccharide composed of 4-S disaccharide repeating periods (MATHEWS, unpubl.). PRAUS and GOLDMAN (1970b), however, obtained a different result for carp corneas.

iii) Reptilia

The mixture of glycosaminoglycans from turtle cornea, representing 1.2% of the dry-tissue weight, had a glucosamine to galactosamine ratio of 3.5 (ANSETH, 1961a).

iv) Aves

The corneas of the adult owl, turkey, and chicken contain about 1.2% of the dry-tissue weight as glycosaminoglycans with a ratio of glucosamine to galactosamine of 1.8 (ANSETH, 1961a). In these respects, birds are similar to mammals. The keratan sulfate of adult chicken cornea is polydisperse and fractions with different molar ratios of sulfate to glucosamine are present. A major fraction from chicken cornea had a molar ratio of sulfate to glucosamine of 1.31 (MATHEWS and CIFONELLI, 1965). The galactosamine-containing fraction was virtually exclusively chondroitin 4-sulfate (MATHEWS, unpubl.). Much higher values of glycosaminoglycan (hexosamine = 1.55% of dry-tissue wt.) content of chicken cornea were obtained by PRAUS and GOLDMAN (1970b); the molar ratio of sulfate to galactosamine in the chondroitin sulfate fraction was also low (0.78).

Early studies (ANSETH, 1961b; ANSETH and LAURENT, 1962) indicated that the glycosaminoglycans of embryonic corneas were quite different in degree of sulfation, chain length, and proportion of "unsulfated glucosaminoglycans" from those of adult corneas. The sequence of biosynthetic events in embryonic development may be summarized as follows. Collagen and presumably hexosamine-containing macromolecules appear early (day 3, stage 17 1/2) in the presumptive cornea (HAY and REVEL, 1969). Sulfate incorporation was detected as early as day 5 (stage 26) by CONRAD (1970a). Rates of biosynthesis of sulfated glycosaminoglycans increase rapidly during subsequent embryonic development until by day 14 intense staining of sulfated glycans is evident. CONRAD (1970b) suggested that the "unsulfated glycosaminoglycan" was a low sulfated chondroitin sulfate. However, characterization of this (or a related) fraction as a glucosaminoglycan composed of equimolar uronic acid and glucosamine indicated that it was probably hyaluronate (PRAUS and BRETTSCHNEIDER, 1970, 1971). The concentration of the presumptive hyaluronate component was relatively constant from day 10 to day 15, reduced by about two thirds at day 20, and virtually absent from the 2-day-old postnatal chick cornea. TOOLE and TRELSTAD (1971) detected a rapid hyaluronate synthesis as early as day 4 (stage 24), which predominated over biosynthesis of sulfated glycosaminoglycans until day 8 (stage 34). At this time, hyaluronidase activity was noted in the stromal tissue (see Chap. 10, Sect. 4b).

v) Mammalia

The corneal stromas of adult man, owl monkey, rabbit, sheep, cattle, and whale all possess similar concentrations of glycosaminoglycan hexosamine, in the range 1.1 to 1.5% of dry-tissue weight, with glucosamine to galactosamine ratios about 1.7 (ANSETH, 1961a; but see PRAUS and GOLDMAN, 1970b).

The glycosaminoglycans of human fetal cornea consist predominantly of low sulfated chondroitin sulfate. With development the proportion of keratan sulfate increases to become the dominant component in the adult cornea (BREEN et al., 1972).

The keratan sulfate portion of bovine cornea is highly polydisperse in molecular weight, ranging from 4,200 to 19,100, however, the chondroitin sulfate portion is fairly homogeneous with a molecular weight near 40,000 but separable into fractions varying in degree of sulfation (LAURENT and ANSETH, 1961). GREILING and STUHLSATZ (1966) separated the keratan sulfate of bovine cornea into fractions with molecular weights varying over a similar range and also varying in molar ratio of sulfate to glucosamine from 0.96 to 1.19. A much wider range of variation was found by SALITERNIK-GIVANT and BERMAN (1970) for fractions obtained by chromatography of preparations obtained under mild conditions and without use of proteolytic enzymes. The chondroitin sulfate-peptide fractions had a molar ratio of sulfate to hexosamine varying from 0.43 to 0.89. The keratan sulfate-peptide fractions had a molar ratio of sulfate to glucosamine varying from 0.92 to 1.70 and also a varying molar ratio of galactose to glucosamine. The average degree of sulfation of keratan sulfate in bovine cornea increased with age (MATHEWS and CIFONELLI, 1965).

MEYER and ANDERSON (1965) suggested that, because of a content of 12% of glucose, the chondroitin sulfate of bovine cornea is distinguished from chondroitin sulfate of other mammalian tissues. GREILING et al. (1970) detected glucose in similar preparations but the amounts (not quantitated) were apparently much smaller. Purified keratan sulfate preparations from cornea of calf, whitefish and chick contained negligible amounts of glucose (MATHEWS, unpubl.). It is possible that glucose in some chondroitin sulfate fractions, as well as the extra galactose found by SALITERNIK-GIVANT and BERMAN (1970) in keratan sulfate-peptide fractions, may be components of structural glycoproteins that are difficult to separate from the glycosaminoglycans.

c) Vitreous and Other Tissues of the Eye

i) Vitreous

The biochemistry of various tissues of the eye with emphasis on the vitreous has been reviewed by BALAZS (1965), BERMAN and VOADEN (1970), DISCHE (1970a, b) and PIRIE (1969).

The vitreous contains few cells and is essentially an extracellular matrix. Hyaluronate, collagen (vitrosin, see Chap. 3, Sect. 10) and sialoglycoprotein are apparently the principal components responsible for the structural and mechanical properties of the system (Chap. 10, Sect. 4c). In a study of the vitreous of squid, frog, tortoise, chicken, turkey, pigeon, owl, rabbit, guinea pig, sheep, cattle, and owl monkey (for data on tuna, barracuda and carp, see BALAZS, 1965), BALAZS et al. (1959) observed that these macromolecular constituents were unevenly distributed throughout the tissue. Total concentrations of hyaluronate, collagen, and sialic acid were species-variable; structure varied from a liquid of low viscosity in the squid to a viscous fluid in the owl monkey, to part liquid and part gel in birds, to a gel state in most mammals. Data on additional species are given by BERMAN and VOADEN (1970). For example, the vitreous of the steer (1-3 years old) contains the following substances per 100 ml of gel: hyaluronate, about 70 mg; collagen, 14 mg in the anterior portion and 6.7 mg in the posterior portion; sialic acid 3.4 mg (corresponding to an unknown amount of sialoglycoprotein).

The value of 6.6 mg hyaluronate in the vitreous of trout (BERMAN and VOADEN, 1970) is much lower than values given for other teleosts by BALAZS (1965). HOFFERT and FROMM (1969) suggested that the latter values were distorted because of a high concentration of protein in teleost vitreous.

The chemical development of the vitreous in mammals is a process that continues not only during the pre-natal period but also for a prolonged period after birth. Species-specific temporal patterns are evident for the concentration of hyaluronate while hexosamine concentrations may follow a separate course. Adult levels of hyaluronate are usually reached at 1 to 3 years for cattle and near the beginning of the second decade for man. The concentrations of collagen and hyaluronate in the vitreous do not appear to change with age in man; however, starting at approximately 45 years, a liquid pocket begins to form in the central part of the vitreous gel. The mechanism of formation of the liquid vitreous is unknown (BALAZS, 1969).

ii) Other Tissues of the Eye

BALAZS (1965) and DISCHE (1970a, b) have reviewed the biochemistry of various tissues of the eye with particular reference to glycosaminoglycans and glycoproteins. Aqueous humor contains very low concentrations of hyaluronate. Both dermatan sulfate and chondroitin 4-sulfate have been demonstrated in sclera. The lens incorporates $^{35}SO_4$, possibly into sialoglycoproteins since known vertebrate glycosaminoglucurans are very low or absent. A glycosaminoglycan very similar to, if not identical with heparan sulfate is present in lens capsule. The interstitial matrix surrounding the photoreceptor cells of cattle matrix contains a partially sulfated chondroitin sulfate and an unsulfated sialoglycoprotein (BERMAN and BACH, 1968; BERMAN, 1969). Retrobulbar connective tissue in man contains hyaluronate and dermatan sulfate (PRAME et al., 1966) and, in the carp *Carassius carassius* L., hyaluronate, dermatan sulfate, and chondroitin 4-sulfate (PRAME, 1967) in addition to glycoproteins.

d) Blood Vessels

The biochemistry of glycosaminoglycans and glycoproteins in arteries has been investigated thoroughly because of the involvement of these vessels in arteriosclerosis (see reviews by MUIR, 1965, and by BERENSON et al., 1971). Consideration of both gross anatomic and histologic features is essential. Three layers can be distinguished in arteries: a) the inner coat or intima, b) the intermediate layer of media, and c) the external layer or adventitia. The boundary between the intima and media is formed by the internal elastic membrane. The adventitia consists of loose connective tissue that passes into the surrounding connective tissue without sharp boundaries and is commonly removed from the vessel prior to chemical analysis. Its major glycosaminoglycan is dermatan sulfate (GARDAIS et al., 1973). The intima may be separated from the media for analytical purposes. However, in arteries of large caliber such as the adult human aorta, the intima is poorly delimited from the media and is only about one tenth the thickness of the media (MAXIMOW and BLOOM, 1948). Although it may contribute relatively little to the gross strength and elasticity of the vessel the intima is involved in important pathologic alterations. Elastic membranes, collagenous fibers, and smooth muscle cells are major histological features of arterial walls.

The principal glycosaminoglycans of arteries are hyaluronate, chondroitin 4-sulfate, chondroitin 6-sulfate, dermatan sulfate, and heparan sulfate, commonly present in total amounts of 1 to 2% of the dry-tissue weight. The quantitative analysis of such a mixture is a formidable task. Additional difficulties stem from heterogeneities in both chemical and molecular size of the glycans (THUNELL et al., 1967). For these and other reasons, presently available data should be regarded as significant in a semiquantitative fashion only. It is known that the concentrations of each type of glycosaminoglycan vary with species and with anatomical location. They vary also during embryonic development and with postnatal age; differentiation of a large artery such as the aorta may not be complete until well into adult age. For example, the aorta in man develops continuously until about 25 years of age. The early onset of pathologic changes introduces further uncertainties in aging patterns (reviewed by MILCH, 1965) because of simultaneous occurrence of both degenerative and regenerative processes.

It is likely that all of the glycosaminoglycans, with the possible exception of hyaluronate, are present in the aorta as proteoglycans (Chap. 6). A proteoglycan containing chondroitin 6-sulfate as the principal glycan moiety was isolated by mild methods from pig aorta (FRANEK and DUNSTONE, 1968). Chondroitin 4-sulfate was also present in the preparation, as was a small amount of glucosamine and galactose, suggestive evidence for keratan sulfate. The molecular weight was low, being only 160,000. Data on composition are given in Table 9.8.

Table 9.8. Composition (%) of glycosaminoglycans of aorta

Source	Hyaluronate	Chondroitin 4- (6-)sulfate	Dermatan sulfate	Heparan sulfate	Remarks and references
Man	present	present	absent	absent	newborn; SCHMIDT and DMOCHOWSKI (1964)
Man	20	65	10	5	3rd decade; KAPLAN and MEYER (1960)
Man	-	65	15	-	6th decade; also 10% of O-S and 10% of diS_B; MURATA, HARADA, and OKUBO (1968)
Man	10	55	15	20	7th decade; KAPLAN and MEYER (1960)
Dog	5	70	5	20	unknown age; SIREK, SCHILLER, and DORFMAN (1964)
Pig	8	65	4	15	5 months; DUNSTONE (1967)
Cow	23	45	16	16	12-18 months; MEYER et al. (1956)
Chicken	12	88	present	trace	6 weeks; see text for references
Chicken	8	68	present	24	12-18 months; see text for references

i) Man

Evidence for differences in chemical composition with site and layer was obtained by DALFERES et al. (1971) from analyses of thoracic and abdominal aortas with minimal pathologic changes from males aged 12 to 14 years. The total concentration of glycosaminoglycans was about 40% higher in the abdominal portion than in the thoracic segment. The total concentration, and the concentrations of chondroitin 6-sulfate, chondroitin 4-sulfate, and heparan sulfate, were higher in the intima than in the media, while the reverse was true for hyaluronate and dermatan sulfate (see also HODARA et al., 1973).

ii) Other Vertebrates

The aortic glycosaminoglycans of shark, frog, turtle, chicken, pigeon, cat, rabbit, rat, mouse, pig, cow, monkey, dog, and hamster were analyzed by MULLINGER and MANLEY (1969) and by STEFANOVICH and AKIYAMA (1970), who also analyzed thoracic and abdominal segments separately. The relative amounts of each glycan were estimated from the intensity of staining of bands partially separated by electrophoresis; chondroitin 4- and 6-sulfate were not resolved. Dermatan sulfate appeared to be absent from aortas of dusky shark *Carcharhinus obscurus* and of smooth dogfish *Mustelus canis*. Frog and turtle aortas had the highest relative concentration of heparan sulfate. The highest concentrations of total glycosaminoglycans were observed in birds, including chicken, pheasant, and pigeon, and in dogfish.

e) Intervertebral Disc

The intervertebral disc comprises a central body of gel-like consistency in young animals, the nucleus pulposus, surrounded by a thick, laminated ring, the annulus fibrosus, a type of fibrous cartilage. The discs lie between the cartilaginous plates of the vertebral bodies. The glycosaminoglycans, primarily skeletal keratan sulfate and chondroitin 4-(and 6-) sulfate, occur as proteoglycans (see SCHUBERT and HAMERMAN, 1968; LUDOWIEG et al., 1972, and in particular, ZIRMAI, 1971). Data on composition are given in Table 9.9..

f) Eggs and their Matrices

Most animal eggs are surrounded by coatings of various types which not only serve as protection but also have important roles in fertilization and development (MONROY, 1965). Yet comparatively little is known of their polysaccharide or polysaccharide-proteins. More is known of the structural chemistry of egg coatings of invertebrates, in particular of echinoderms (Sect. 2), than of vertebrates.

i) Teleosts

The eggs of teleostean fish contain vacuoles with a polysaccharide content. In some species, cortical granules with sulfated polysaccharide are present (see reviews by ROTHSCHILD, 1958; YAMAMOTO, 1961).

Table 9.9. Composition (%) of glycosaminoglycans of intervertebral disc

Species	Tissue	Chondroitin 4- (6-)sulfate	Keratan sulfate	Remarks and references[a]
Man	disc	90	10	1st decade; 60% of 4-S, 30% of 6-S; BUDDECKE and SZIEGOLEIT (1964)
"	"	50	50	8th decade; 30% of 4-S, 60% of 6-S; above reference; see HALLÉN (1962)
Cow	nucleus	100	0	27-cm embryo; CIFONELLI, SAUNDERS, and GROSS (1967)
"	"	80	20	term embryo; above reference
"	"	40	60	1 year; above reference
Rabbit	"	65	35	1 month; DAVIDSON and SMALL (1963)
"	"	25	75	18 months; above reference
Finwhale	"	80	20	embryo; 40% of 6-S, 20% of 4-S, and 20% of O-S; LUDOWIEG et al. (1972)
"	"	70	30	23 years; 55% of 6-S and 15% of 4-S; above reference

[a] For abbreviations, see Table 7.1.

ii) Amphibia

In most amphibians, the egg is surrounded by a thick jelly envelope, which until recently was thought to be a glycoprotein similar in structure to glycoproteins of echinoderm eggs but differing by a high content of hexosamine and the absence of sulfate (MONROY, 1965). More recently, HUMPRIES (1970) showed that $^{35}SO_4$ injected into female newts *(Notophthalmus viridescens)* was incorporated into the oviduct and into the egg jelly layer nearest the oocyte, presumably bound as a sulfated glycan. Similar observations were made for the frog *Rana pipiens* (PEREDA, 1970). Polyanionic glycan-peptides isolated from egg jelly of *R. pipiens* and of the salamander *Ambystoma maculatum*, contain principally neutral carbohydrates, a small proportion of hexosamines, ester sulfate, but no uronic acid (MATHEWS and SALTHE, unpubl.). Sulfated glycans were detected in granules of toad eggs (OSANAI, 1960).

iii) Birds

Considerable interest is attached to the nature of the organic matrix of the avian eggshell and its role in calcification (see review by WILBUR and SIMKISS, 1968). About 11% of the organic material is of carbohydrate nature consisting of galactose, mannose, fucose, galactosamine, glucosamine, iduronic acid, glucuronic acid, and sialic acid. Chondroitin 4-sulfate and small amounts of dermatan sulfate comprise about one third of the carbohydrate. The concentration in matrix of chondroitin 4-sulfate and dermatan sulfate (?) is constant in different layers of the shell, while that of sialic acid decreases from the inside to the outside (COOKE and BALCH, 1970). Hyaluronate may be present also (ROBINSON and HEANEY, 1973).

Eggshell membranes appear to contain only hyaluronate while the isthmus region of hen oviduct, which secretes this tissue, contains chondroitin 4-sulfate, dermatan sulfate, heparan sulfate and hyaluronate (OSUOJI, 1971).

iv) Mammals

Histochemical studies indicate that hyaluronate is present in the cumulus oophorus of rat and rabbit eggs; ester sulfate may be present in the zona pellucida (SECHACHAR and BAGGA, 1963). Bovine follicular fluid contains 0.2 to 0.3% of a mixture of chondroitin 4- (6-) sulfate and hyaluronate (JENSEN and ZACHARIAE, 1958a, b).

g) Neural Tissue

Chemical data acquired within the last two decades have demonstrated that neural tissues contain glycosaminoglycans and glycoproteins (including sulfated glycoproteins; see Chap. 5) in addition to the earlier known glycolipids (see reviews by SVENNERHOLM, 1965; MARGOLIS, 1969). Most of the known vertebrate glycosaminoglycans are present in vertebrate neural tissues; only heparin and keratan sulfate appear to be absent. Difficulties inherent in the fractionation, characterization and quantitative analysis of complex mixtures, which are present in only small concentrations in tissue (0.05 to 1.0% of dry weight), and occasional neglect of developmental aspects, have led to frequent disagreements among investigators.

Compared to the brain, other neural tissues have received little attention. Glycosaminoglycans have been shown to be present in the peripheral nerve of frog and monkey (MARGOLIS, 1969), in ganglia of insects (ASHURST and COSTIN, 1971a, b, c), retina (BERMAN and BACH, 1968; BERMAN, 1969), and in electroreceptor and electric organs of fish (see below).

Neither the precise anatomic locations nor the biological functions of the glycosaminoglycans in neural tissue are known (MARGOLIS, 1969). The extremely low content of collagen in brain suggests that the glycosaminoglycans may have unique and unusual physiological roles in neural tissues (WOLMAN, 1957; CUSTOD and YOUNG, 1968; LANGLEY, 1970; and ELAM et al., 1970). Attempts at localization on the cellular level have depended entirely upon histochemical techniques of questionable specificity. Although specific chemical identifications are not possible, it is probable that high concentrations of polyanionic glycans and/or sulfated glycoproteins are present in the extracellular sites of brain tissues. At the level of gross anatomy, some differences in distribution of glycosaminoglycans have been noted for different areas of the brain (see MARGOLIS, 1969). Glycosaminoglycans appear to be absent from bovine myelin (MARGOLIS, 1967).

The question has naturally arisen whether glycosaminoglycans in the nervous system can be distinguished in some chemical and structural manner from their namesakes in other vertebrate tissues. Differences may be sought in macromolecular structure of the proteoglycans, in peptide sequences, in structure of the linkage regions, and in structure of the glycan moieties. However, information is not yet available on peptide compositions or sequences of proteoglycans. Also, due to the general use of proteolysis as a means of isolating glycan-containing components, data are limited to residual amino acid residues.

MARGOLIS (1967) purified hyaluronate from bovine brain and found the major residual amino acids to be glutamic acid, alanine, aspartic acid, serine, and threonine, in order of decreasing content. Since about 5% of the total hexosamine was galactosamine, there may have been minor contamination of the preparation. However, some of the residual amino acids may be covalently bound to the glycan. For example, residual amino acids in hyaluronate from human umbilical chord were aspartic acid, serine, and glycine, each present as approximately one residue per glycan molecule (RODÉN and MATHEWS, 1968). Approximately one mole arabinose per hyaluronate molecule from brain was reported by MARGOLIS (1967), who suggested that this sugar might participate in the linkage-region sequence (Chap. 6, Sect. 2b). Arabinose (and glucose; see WARDI et al., 1969) had been previously detected in hyaluronate of brain but has not been identified in hyaluronate of other tissues (LINDAHL and RODÉN, 1972). However, the presence of arabinose, even at trace levels, in hyaluronate is doubtful in view of the study by KATZMAN (1971).

The residual amino acids in chondroitin 4-(6-)sulfate preparations from bovine brain (MARGOLIS, 1967; SAIGO and EGAMI, 1970) are similar to those in the namesake glycan-peptide from bovine cartilage (MATHEWS, 1971). Additional data of MARGOLIS, MARGOLIS, and ATHERTON (1972) and SAIGO and EGAMI (1970) indicate that the glycan chains are linked primarily to serine by glycosidic bonds; this makes it likely that the linkage regions of chondroitin 4-(6-) sulfate proteoglycans of brain and of skeletal connective tissue (Chap. 6, Sect. 2a) are identical. There is a small proportion of disulfated disaccharide repeating periods (diS_D and diS_E) in the glycan (SAIGO and EGAMI, 1970).

Heparan sulfate of rat and rabbit brain also appears to be linked to serine (MARGOLIS, MARGOLIS, and ATHERTON, 1972; see heparan sulfate from other vertebrate tissues, Chap. 6, Sect. 2b). Two fractions of heparan sulfate from rat brain, although differing in total sulfate content, contain equal proportions of N-sulfate and O-sulfate. Most of the heparan sulfate chain appears to consist of large, completely N-acetylated segments that also contain most of the O-sulfate groups. These segments also carry the galactose, xylose, and residual amino acids of the linkage-region peptide (MARGOLIS and ATHERTON, 1972). It is in these quantitative aspects of distribution of acetyl and sulfate groups only that the brain glycan differs from the namesake glycan of other tissues (CIFONELLI, 1968; CIFONELLI and KING, 1970).

Any attempt to summarize published data on glycosaminoglycans in brain tissue is hindered by widespread conflicts between analytical values obtained by different investigators. Since a marked improvement in analytical methodology has occurred since 1968, one is inclined to place greater confidence in more recent results than in the older data (see MARGOLIS, 1969). However, conflicting results are still appearing in the current literature. A summary assessment of recent data of SINGH et al. (1969), SAIGO and EGAMI (1970), and MARGOLIS and ATHERTON (1972) indicates that the glycosaminoglycans in adult brain of several mammals, including man, and of the chicken, consist of 30 to 50% of hyaluronate, 20 to 40% of chondroitin 4-(6-)sulfate and 10 to 20% of heparan sulfate. A small percent, 1 to 5%, may be dermatan sulfate; nonsulfated disaccharide periods of chondroitin sulfate may be present. Disaccharide periods of chondroitin 4-sulfate greatly predominate over disaccharide periods of chondroitin 6-sulfate, except in chicken-brain glycosaminoglycans, which also have an unusually high proportion of heparan sulfate (SINGH et al., 1969). The disaccharide composition of the chondroitin sulfate fraction from bovine brain glycosaminoglycans is about 75% of 4-S, 12% of 6-S, 7% of O-S, 4% of diS_D, and 2% of diS_E (see Table 7.1; SAIGO and EGAMI, 1970). Rat and pig brains contain primarily hyaluronate and heparan sulfate (SRINIVAS et al., 1972).

Very little information is available as yet on the metabolism of glycosaminoglycans of brain (MARGOLIS, 1969). These glycans, as well as sulfated glycoproteins, are distributed among several metabolic pools of different turnover times ($t_{1/2}$) (MARGOLIS and MARGOLIS, 1973). For sulfated glycoproteins of adult rat brain, $t_{1/2}$ varies from 2.5 days to 14 days. The average $t_{1/2}$ of chondroitin 4-(6-)sulfate is about one week and that of hyaluronate much longer, perhaps 30 days (MARGOLIS, MARGOLIS, SANTELLA, and ATHERTON, 1972). Lysosomal hyaluronidase plays a role in the turnover of the two glycans. DORFMAN and HO (1970) have clearly demonstrated the synthesis of hyaluronate, chondroitin 4-(6-)-sulfate, and heparan sulfate by glial cells from a rat glial tumor in clonal culture.

h) Electroreceptors and Electric Organs

A number of groups of fish have developed electroreceptor organs specialized for the detection of electric fields and electric organs capable of producing electric discharges sufficiently strong to serve as weapons, either defensively against predators or offensively in securing food. Comparable organs appear to be absent from invertebrates and amphibia. Electric organs have probably evolved from primitive electroreceptors. All known electroreceptors are modifications of mechanoreceptor organs of the lateral-line system (BENNETT, 1971a).

i) Electroreceptors

The characteristic electroreceptors of elasmobranchs, the ampullae of Lorenzini, are located mostly in the head region. Each unit tends to lie deep beneath the skin and is associated with a jelly-filled tube which may be many centimeters long and which opens at the surface. The detailed anatomy and electrophysiology of this organ have been discussed by WALTMAN (1966), MURRAY (1965), and BENNETT (1971a).

Of particular interest here is the chemistry of the polyanionic glycans of the jelly. The first preparation characterized was one isolated by DOYLE (1967) from the spiny dogfish *Squalus acanthias*. The glycan consisted of N-acetylated hexosamine (ratio of glucosamine to galactosamine of 4:1), galactose, and ester sulfate in approximately equimolar proportions; uronic acid was absent. It was named "lorenzan sulfate". Studies were extended to fifteen other species (DOYLE, 1968). All glycan preparations contained equimolar proportions of hexosamine and galactose (DOYLE, priv. comm.). Ester sulfate was very low or absent only in the glycan from the large shark *Galeorhinus galeus*. The glucosamine to galactosamine ratio in glycans from fourteen species of sharks and rays varied greatly.

A variability in chemical composition and considerable structural complexity of the lorenzan sulfates was revealed in detailed studies by HOW, JONES, and DOYLE (1969) and HOW, JONES, and STACEY (1970) or preparations from the ray *Raja clavata* and the basking shark *Cetorhinus maximus*. The glycopeptides obtained by digestion with pronase of the jelly from both species are all sulfated and all contain the same monosaccharide components, i.e. D-galactose, D-glucosamine and D-galactosamine (probably as N-acetylated derivatives), a small amount of fucose, and the same amino acids. For each species, the glycopeptides can be represented by a spectrum ranging from molecules of high galactosamine and peptide contents and low glucosamine and sulfate contents to those of low galactosamine and peptide contents and high glucosamine and sulfate contents. The glycopeptides have the following general features: all contain N-acetylgalactosamine residues linked glycosidically to threonine residues in polypeptide sequences; the N-acetylgalactosamine residue is substituted by either a short-chain oligosaccharide (2-8 sugar units) containing D-galactose and N-acetylglucosamine, or a galactan, or both. The structures possess an interesting resemblance to that of spisulan (Fig. 9.1).

Only histochemical data are available on glycans of electroreceptors in other vertebrates. The jelly in the ampullary organ of the gymnotid *Gymnotus carapo* appears to contain neutral glycans (DERBIN and DENIZOT, 1971). However, sulfated glycans are present in the mormyromast (type II of Cordier) of *Gnathonemus petersii* (DENIZOT, 1971). It is interesting that the jelly of the shark neuromast is strongly metachromatic-staining (TESTER and KENDALL, 1968). One might expect the glycans present in this mechanoreceptor organ to show chemical affinities with glycans of electroreceptors in accord with the evolutionary relationship between mechanoreceptors of the lateral-line system and electroreceptors.

ii) Electric Organs

Electric organs occur in several genera of marine and freshwater teleosts and elasmobranchs. In its most typical form, the organ is composed of a series of flattened cells, electrocytes or electroplaques, separated by connective-tissue partitions. Each electrocyte is imbedded in gelatinous connective tissue; a large fraction of the volume of electric organs is extracellular space. Electrocytes are modified from muscle fibers, except in the sternarchid family where they are modified from nerve cells; they work on the same general principles as ordinary muscle and nerve cells. The general biology and glycosaminoglycan chemistry of electric organs have been discussed by BENNETT (1971b) and by SZIRMAI (1965b).

Electric organs of the electric eel *Electrophorus electricus* contain hyaluronate and chondroitin 6-sulfate in amounts of 0.7% and 0.1%, respectively, of the dry, defatted-tissue weight (MEYER et al., 1956). A content of uronic acid of 1.6 to 2.3% of the weight of the dry, dissected connective tissue suggests a nonuniform distribution of glycosaminoglycans in the organ. The possible presence of highly sulfated glycosaminoglycans in the electric organs of the skate *Raja clavata* is suggested by the observation of SZIRMAI (1965b) that a water extract of the fresh organ showed a molar ratio of glucosamine to galactosamine of 1:1.8, and molar ratios of hexuronic acid to hexosamine to sulfate of 1:1.5:2.4.

Several investigators (see SZIRMAI, 1965b) observed that isolated glycosaminoglycans are capable of binding pharmacologically active quaternary ammonium compounds and suggested that these glycans have important biological roles in the ion-transfer mechanisms of the electric organ. However, little additional information on this intriguing problem is available. A glycoprotein with ion-exchange properties has also been isolated from electric organs of *E. electricus* (TRAMS and LAUTER, 1964).

i) Cells that Store Glycosaminoglycans

Several types of cells synthesize and retain glycosaminoglycans within intracellular granules.

i) Leukocytes

Early suggestions of the presence of glycosaminoglycans in leukocytes were strengthened by the isolation of chondroitin 4- (6-)sulfate and probably hyaluronate from lymphocytes and granulocytes of normal subjects and of patients with myeloid leukemia (CLAUSEN and ANDERSEN, 1963). These glycans (predominantly chondroitin 4-sulfate) are present in platelets as well as in leukocytes (OLSSON and GARDELL, 1967) and are synthesized by the leukocytes and leukemic myeloid cells themselves (OLSSON et al., 1968; OLSSON, 1968). The chondroitin sulfate component is synthesized on microsomal structures in both normal and leukemic cells and stored in cytoplasmic granules (OLSSON, 1969a, b, c) as a proteoglycan molecule of low protein content and relatively low molecular weight (OLSSON, 1969d; BARBER et al., 1972).

Rabbit bone-marrow cells (over 90% granulocytes) concentrate chondroitin 6-sulfate and probably hyaluronate in the granule fraction (OLSSON, 1971). However, the nature of the glycans in rabbit polymorphonuclear leukocytes is not clear. One component appears to be hyaluronate while the sulfated component resembles chondroitin 4-(6-)sulfate chemically but is not hydrolyzed by testicular hyaluronidase, and iduronic acid is absent (FEDORKO and MORSE, 1965).

ii) Kurloff Cells

These cells are considered to be lymphocytes that accumulate in guinea-pig spleen and show characteristic metachromatic granules. DEAN and MUIR (1970) isolated from whole spleen a presumably intact proteoglycan, which consisted of chondroitin 4-sulfate linked as multiple side-chains to serine residues of a polypeptide moiety. The macromolecule, thus, closely resembles the characteristic proteoglycan of mammalian cartilage (Chap. 6). Although the spleens of sheep, pig, rat, and humans do not appear to contain cells of the Kurloff type, they were shown to yield upon extraction proteoglycans similar to those obtained from guinea-pig spleens (DEAN et al., 1971).

iii) Mast Cells

This family of cell types having cytoplasmic granules staining metachromatically with cationic dyes occurs in a great number of tissues and animal species and is associated with the biosynthesis and storage of heparin (RINGERTZ, 1965; SPIRO, 1965b). BARLOW et al. (1964) and KOTUKU et al. (1967) concluded that heparin preparations from liver, lung, small intestine, and intestinal mucosa of a variety of mammalian species are biologically and chemically similar. A "new" type of heparin isolated from lung and intestine of the finback whale *Balaenoptera physalus* by YOSIZAWA (1964) is intermediate in composition (with respect to N-acetyl and N-sulfate groups) between heparan sulfate and heparin (SCHILLER, 1966a; JACQUES et al., 1966; YOSIZAWA, 1967).

j) Tumors

A specific deviation of normal control mechanisms is demonstrated by many tumors, which produce only a single type of glycosaminoglycan (reviewed by SYLVEN, 1965). The major glycosaminoglycan present appeared to be hyaluronate in Rous chicken sarcoma, pleural and peritoneal mesotheliomas, and human pancreatic carcinoma, heparin in mastocytomas (RINGERTZ, 1965), chondroitin 4-sulfate in a rat fibrosarcoma and in one human chondrosarcoma, but chondroitin 6-sulfate in another chondrosarcoma (MEYER et al., 1956; ANDERSON et al., 1963). Dermatan sulfate is associated with fibrous structures of mixed salivary tumors whereas chondroitin 6-sulfate is found in the myxomatous regions (LOVELL et al., 1966).

A transplantable rat chondrosarcoma contained a very small amount of hyaluronate in addition to a major proteoglycan fraction composed principally of chondroitin 4-sulfate. The persistence of a small proportion of glucosamine in the proteoglycan fraction may have been due to associated glycoprotein (CHOI et al., 1971).

k) Some Other Sources

i) Body Fluids

Normal human plasma contains 0.5 to 1.0 mg total glycosaminoglycans per 100 ml. The major component appears to be chondroitin 4-sulfate, with small amounts of heparin and possibly heparan sulfate (reviewed by WINZLER, 1965). CALATRONI, DONNELL, and DiFERRANTE (1969) suggested the possible presence of keratan sulfate and also indicated that a portion of the chondroitin 4-sulfate was low-sulfated. The variable degree of sulfation of chondroitin 4-sulfate was demonstrated at the disaccharide level by MURATA et al. (1972) who detected O-S, 4-S, and diS_D (Table 7.1).

Hyaluronate of synovial fluid has been extensively investigated (SUNDBLAD, 1965; SCHUBERT and HAMERMAN, 1968).

Gastric secretion of the dog contains chondroitin 4-sulfate and sulfated sialoglycoproteins (DeGRAEF and GLASS, 1968). Glycosaminoglycans have been detected in other mucus secretions (DISCHE, 1965) and in urine (BOAS, 1965).

ii) Kidney

All of the known vertebrate glycosaminoglycans, with the exception of keratan sulfate, occur in the mammalian kidney. There is a marked variation in concentration and type of glycosaminoglycan with topographical zone, but without evident correlation with kidney functions (CASTOR and GREENE, 1968; FARBER and VAN PRAAG, 1970; KRESSE and GROSSMAN, 1970; VAN PRAAG et al., 1972; SENO et al., 1972).

iii) Liver

The mast cells (see above) of the mammalian liver synthesize and store heparin, and possibly also heparan sulfate (SPIRO, 1965b). Bovine liver contains, in addition to heparan sulfate, dermatan sulfate that is composed in part of disulfated disaccharide periods (GREGOIRE et al., 1972; see Figs. 6.1 and 7.1).

iv) Umbilical Chord (Human)

The umbilical chord consists essentially of vessels and their surrounding adventitial layer, known as Wharton's jelly. Hyaluronate is the principal glycosaminoglycan of the jelly constituting about 3% of the dry weight. Sulfated glycans are also present (SZIRMAI, 1965a) in smaller amounts, largely chondroitin 6-sulfate. A portion of the chondroitin 6-sulfate, all of the dermatan sulfate, and a sulfated glucosaminoglycan (probably heparan sulfate) present in chord are associated with the blood vessels (DANISHEVSKY and BELLA, Jr., 1966). Analysis of the disaccharide composition of the chondroitin sulfate fraction shows it to consist of 80% of 6-S, 10% of 4-S, and 10% of 0-S periods (Table 7.1; MATHEWS, unpubl.).

v) Placenta

Interest in the possible roles of glycosaminoglycans in placental permeability and immune mechanisms has led to several analytical studies. The concentration of glycoproteins and glycosaminoglycans of mouse placenta reaches a maximum at days 15 and 16, thereafter dropping rapidly one fifth of the maximum by term. Hyaluronate and chondroitin 4- (6-)sulfate are prominent, accompanied by nearly equal amounts of glycoproteins (ŠVEJCAR, 1968). Although the collagen of human placenta (portion containing the terminal chorionic villi) increases 3-fold during the 10 to 40 week gestation period, the total content of glycosaminoglycans remains constant. At 10 weeks, the glycosaminoglycans consist of about 25% of hyaluronate, 25% of dermatan sulfate, and 50% of chondroitin 6-sulfate. The last-named fraction includes a large amount of disaccharide periods lacking sulfate. At 40 weeks, the only difference in chemical composition is the replacement of the unsulfated disaccharide by 6-sulfated periods (LOVELL et al., 1967). When the entire term placenta is extracted with saline, the extract contains hyaluronate, chondroitin 4- (6-)sulfate and a small proportion of dermatan sulfate and heparan sulfate (CALATRONI and DiFERRANTE, 1969). The residual glycosaminoglycans, obtained after proteolytic digest, are mainly heparan sulfate and dermatan sulfate with a smaller proportion of chondroitin 4-sulfate. These glycans are apparently associated with blood vessels and coarse collagen fibers.

vi) Tendon

In addition to glycoprotein, bovine tendon contains hyaluronate, chondroitin 6-sulfate, and dermatan sulfate (MEYER et al., 1956). Whereas chondroitin 6-sulfate is the most abundant glycosaminoglycan in mature bovine tendon, dermatan sulfate is predominant in rabbit tendon. Regenerating rabbit achilles tendon contains, in addition, chondroitin 4-sulfate (DORNER, ANTONOPOULOS, and GARDELL, 1968). The predominant glycosaminoglycan in 4-day, regenerating tissue of rabbit tendon was hyaluronate which was replaced largely by dermatan sulfate after 4 months (DORNER, 1968). DAHMEN (1965) observed that, with embryonic development of rabbit tendon, thin (250-300 Å diameter) collagen fibrils grouped in loose bundles and randomly oriented were replaced by thicker and more tightly packed fibrils, while the concentration of glycosaminoglycans declined. Unfortunately, chemical data were not reported.

vii) Rooster Comb

Virtually all of the hyaluronate content (about 6% of the dry-tissue weight) of rooster comb and very little chondroitin 4- (6-)sulfate can be extracted by distilled water (SWANN, 1968). The remaining glycosaminoglycans, chondroitin 4-(6-sulfate) and dermatan sulfate, are contained in the "epithelial layer" and the "central core", which contains collagen bundles, blood vessels, and fat tissue (DOYLE and SZIRMAI, 1961). The small amount of chondroitin 4- (6-)sulfate that is extracted with hyaluronate from the "mucoid layer" may be associated with the fine collagen fibrils found in that region of the comb.

viii) Allantoic Fluid

A sulfated glucosaminoglycan-peptide was isolated from chick allantoic fluid by MEYER et al. (1967). The purified material bears a close resemblance to skeletal keratan sulfate.

ix) Toad Venom

Earlier histochemical indications led VIALLI et al. (1969) to isolate and partially characterize the polyanionic glycans of toad "venom" of the parotoid glands. The purified preparations were obtained in yields of 10 to 14% of the dry weight as chondroitin 4- (6-)sulfate and chondroitin 6-sulfate from adult *Bufo vulgaris (Bufo bufo)* and *Bufo viridis*, respectively. The molar ratio of sulfate to galactosamine was 1.01 and 1.4 to 1.8 for preparations from *B. vulgaris* and *B. viridis*, respectively. These results were confirmed for the *B. vulgaris* secretion, which contained nearly 20% of the dry weight as chondroitin 6-(4-)sulfate (ratio of 6-S to 4-S periods about 3) with equimolar proportions of sulfate and hexosamine residues and no indication of the presence of disulfated disaccharide periods (MATHEWS and MUZII, unpubl.). The major residual amino acid in the preparation obtained by proteolysis of the secretion was serine, apparently involved in the linkage of the glycan to polypeptide (see Chap. 6, Sect. 2a).

The composition of the parotoid secretion is quite unlike that of skin of the same species (Table 9.5). The physiological role of the glycan, which constitutes nearly 1% of the weight of the original secretion, remains a biological puzzle.

x) Lungs and Gills

Mammalian lungs contain most of the known vertebrate glycosaminoglycans and are notable for their contents of heparan sulfate and heparin (WUSTEMAN, 1972; BRIMACOMBE and WEBER, 1964). It is interesting that fish gills, which are analogous to lungs of higher vertebrates, also contain in their soft tissues, besides chondroitin sulfate, significant amounts of sulfated glucosaminoglycans of the heparan sulfate or heparin types with anticoagulant properties (WASSERMAN et al., 1972).

5. Discussion

a) Origins and Taxonomy

Complex polyanionic heteropolysaccharides are found in all animal groups, from the protists to the vertebrates, as well as in plants (see Chap. 2). It appears likely, therefore, that the biological apparatus necessary for their synthesis was established early in evolution. Subsequent developments greatly enlarged the repertoire of glycan structures available to meet the functional requirements for specific adaptations.

Identical, or very similar, glycan structures have been selected by many animal groups that are widely separated phylogenetically. The best known of these is the chondroitin sulfate-type polysaccharide chain into which sulfate groups can be introduced to varying degrees and at differing residue sites. This type of glycan chain is present in tissues of all vertebrates and also in tissues of some molluscs, annelids, arthropods, and echinoderms. Hyaluronate is widely distributed in vertebrates and is also found in some cephalochordates, arthropods, possibly in some annelids, and also in a few bacteria. Heparin is present in molluscs as well as in mammals. Occurrence of glycans of similar structure in very distantly related phyla is possibly due to processes of evolutionary convergence.

Other glycan structures appear to have a more restricted taxonomic distribution. These include keratan sulfate in vertebrates, and most of the glycans in invertebrate tissues described earlier. However, the narrowness of distribution of any one type of glycan among animal groups may be more apparent than real, since both the extent of sampling of species and the degree of structural characterization of glycans are still very limited.

b) Development and Aging

Changes in glycosaminoglycan content and composition with embryonic development, maturation, and aging are apparent for many tissues of vertebrates. However, comparable chemical data for invertebrates are not available. Changes with aging in tissues, such as skin, intervertebral disc, and arteries, seem to be inevitably accompanied by "degenerative alterations" or loss of function. Similar difficulties are encountered in interpretation of the changes in aging cartilage (Chap. 8, Sect. 5c).

Since mature vertebrate tissues possess tissue-specific glycosaminoglycan compositions, comparisons between species are meaningful only for homologous tissues. Nevertheless, a general trend of temporal alteration appears to be common to many different tissue types. In general, early developmental stages are characterized by glycosaminoglycans preponderantly composed of hyaluronate or of low-sulfated chondroitin sulfate. With increasing embryonic age these types of glycan, of low linear charge density, are partially or wholly replaced by tissue-specific sulfated glycans of greater linear charge density.

The wide range of variation within this general pattern is shown by several vertebrate tissues. In the developing avian cornea, hyaluronate, which is the only detectable glycosaminoglycan at very early stages, is virtually entirely replaced by chondroitin 4-sulfate (molar ratio of sulfate to galactosamine >1) and by keratan sulfate. Hyaluronate is the predominant glycan of newborn human aorta and is partially replaced by dermatan sulfate and heparan sulfate during maturation. During development of the tadpole, chicken, man, and hog, the initial high proportion of hyaluronate in the glycosaminoglycans of skin is reduced in favor of dermatan sulfate. Minor differences between species of the same class may be expected, but have been documented only in the mammals. It is interesting that hyaluronate is the prominent or sole component in regenerating amphibian limb bud (TOOLE and GROSS, 1971) and in regenerating tendon (DORNER, 1968), preceding the appearance of tissue-specific glycosaminoglycan patterns. Composition changes in glycosaminoglycans of nucleus pulposus of mammals are very complex as early as the second half of embryonic development. Hyaluronate is absent, but may be an important component at prior stages for which data are not available.

c) Evolution

The complexity and diversity of polyanionic glycan structures tends to obscure a fundamental functional unity of these molecules. Not only may different polyanionic glycan structures perform similar functions but the same glycan structure may be utilized in a variety of ways. For example, hyaluronate is an important glycan in a variety of tissues and organs of the vertebrate body and, as we have noted previously, is also present in tissues of many organisms from other phyla. Chondroitin sulfate is a component of cartilage and also of the secretion of toad parotoid glands. Heparin most probably plays different functional roles in liver and lung tissues and in mantle secretions but retains the same molecular properties in mammals and in molluscs.

At this point we are led to inquire whether the chemically similar or isologous, glycans present in different tissues of organisms are genetically related, i.e. indirectly homologous (see Chap. 1, Sect. 2). It is apparent that definitive answers cannot be given until the primary structures of the biosynthetic enzymes, i.e. the glycosyl transferases, sulfotransferases, etc., are known. However, the existence of widespread homologies of other types of enzymes catalyzing the same chemical reaction in different animal species (see Chap. 2) suggests the possibility that many of the functionally similar enzymes involved in polyanionic glycan synthesis will also prove to be homologous. In any case, it is likely that many, if not all, of these functionally related enzymes in different tissues of the same organism are identical. Yet, the possibility that genetic polymorphism may be found should be borne in mind. There exist ample precedents for the occurrence of functionally similar, nonidentical proteins in the same organism, including the example of the several different types of collagen.

Chapter 10

Molecular Interrelationships and Functions

"Such old antitheses as that of form and function need not, indeed, detain us, for, as WOODGER'S (WOODGER, 1929) analysis once and for all made clear, form is simply a short time slice of a single spatio-temporal entity. We have enlarged our concept of structure so as to include and recognize that in the living organism there is no spatial structure with an activity as something over against it, but that the concrete organism is a spatio-temporal structure and that this spatio-temporal structure is the activity itself." NEEDHAM (1968).

"The things of the universe are not sliced off from one another with a hatchet, neither the hot from the cold nor the cold from the hot." ANAXAGORAS (ca. 500BC)

1. General Considerations

The notion of a "spatial hierarchy" of biological organization has already been alluded to in Chap. 2. It is a paradigm (KUHN, 1970) of wide influence (LEAKE, 1969) which may be stated as follows: 1. there exists hierarchical continuity between successive integrative levels of biological organization, and 2. there are central organizing principles to be discovered (NEEDHAM, 1968).

Firstly, the search for such principles requires an appropriate methodology. It should be evident that data derived solely from the molecular level are not adequate in themselves to account for organization and function at higher biological levels. We are reminded by SIMPSON (1964) that evolutionary selection acts at the phenotype level and that problems of phylogenetic adaptation and of molecular function require a synthesis of molecular and organismal viewpoints. In a similar vein, MAYR (1964) suggests that the ideal approach in biology is a combination of an analytical with a systems approach. Thus, questions of function and evolution of connective tissue macromolecules are linked to considerations of the nature and role of higher levels of organization.

Biological adaptation is achieved in two ways - environmental plasticity and genetic specialization (DOBZHANSKY, 1951). Both processes operate at all levels of biological organization. They are expressed by the macromolecules that make up such multimolecular intracellular structures as cell membranes, chromosomes, endoplasmic reticulum, mitochondria, chloroplasts, etc., as well as those that form the large variety of extracellular structures of connective tissues. In general, the molecular building blocks appear to contain all, or most of the implicit interaction potentials required to bring about their assembly under appropriate conditions into a functional array. Amongst these macromolecules, it appears that those destined for the formation of extracellular structures are likely to exhibit environmental plasticity to a greater degree.

The capacity of a connective tissue matrix macromolecule for interaction with like and unlike macromolecules in different fashions so as to produce a variety of supramolecular structures derives from a covalent structure that is in part determined rigidly, and in part subject to variation. In a single organism or tissue, the invariant (or pauci-variant, if a small degree of polymorphism occurs) part of the structure is represented by the primary peptide sequences of the molecule. Secondary variation of covalent structure is extensive. For example, collagen polypeptide structures are varied by the presence of hydroxyl, aldehyde, and hexose groups (Table 3.9). Some of the same variations occur in elastin. Secondary structural differences are also found among proteoglycans and structural glycoproteins, with respect to degree and distribution of glycan chains attached to the polypeptide moiety, and a number of features of glycan structure, such as chain length, extent and site of sulfation, minor nonrepeating monosaccharide substituents, etc. As a consequence of the microheterogeneity, an enormous number of different macromolecules of the same type but with distinguishable secondary chemical features may be produced by the cells in various tissues of an organism, or by the same cell at different times.

The different members of each class of macromolecules, while showing gradations of properties in some respects, share other features in common. For example, different collagen molecules may produce aggregates that are indistinguishable in fine structure (Fig. 3.10). Alternatively, these invariant characteristics themselves permit multivaried expression in higher-order structures in different aqueous environments. This is illustrated by the production of different morphological forms of collagen (640 Å, 220 Å, and no periodicity) from a collagen solution by variations in pH, ionic strength, and concentration of various substances. The wide-ranging phenomena were summarized by GROSS (1956):
"1. An apparently homogeneous population of macromolecules may be precipitated from solution under the influence of physical chemical forces in the form of highly ordered structures, even when in the presence of a variety of other high molecular weight substances.
2. The morphological characteristics of the precipitated structures may vary so radically as to appear completely unrelated to each other, depending on readily controlled conditions in the solution. The conditions required for determining the type of structure may, in some cases, be no more than differences in concentration of the inducing agent.
3. The several different morphological forms may be readily dissolved and transformed one into another.
4. The specificity for structure formation may reside almost entirely in the reacting system, the macromolecular building blocks of collagen, while the inducing agents may be relatively nonspecific.
5. Identical structural types may be precipitated from solutions ranging greatly in ionic content and properties by using different manipulations.
6. Morphologically identical end products may be produced via different intermediate stages in structure formation."

This potentiality for multivaried expression is further magnified by the effects of microheterogeneities that allow for a seemingly infinite spectrum of possibilities for molecular interactions. It seems appropriate, therefore, to characterize the group of structural macromolecules as *versatile* and *polymorphogenic*. These descriptive terms need not be limited in application either to molecules of the four classes considered here or to those present in animal tissues. There is ample reason to consider their use for components of chitinous (JEUNIAUX, 1971) and cellulosic (REES, 1967; SIEGEL, 1968) structures as well.

Supramolecular structures of characteristic morphology in which either collagen or elastin are conspicuous have been more fully observed than have those featuring either polyanionic proteoglycan or structural glycoproteins. By far the greatest variety and complexity of ordering have been noted for collagenous tissues. Only a few of the large number of recognized types of collagenous structures can be discussed here in any detail. A brief survey follows.

Fibrillar forms, deriving from monomeric collagen possibly via oligomeric intermediates, are the first distinct structures observed in the electron microscope. An axial periodicity of 640 Å is commonly, but not always present. It may be entirely absent, as in many very small-diameter fibrils, in byssus threads, and in annelid cuticles, or it may be shorter, as in early stages of collagen formation in vertebrates, and in some mature tissues of invertebrates. Fibrillar diameters range from 100 Å to 4000 Å, frequently falling into a narrow distribution in some tissues. For many tissues (but not cornea), there is an increase in fibrillar diameter during embryonic development and after birth. It is assumed that fibrils increase in length as well as diameter (HARKNESS, 1961). Collagen is found also as very large-diameter fibers and threads that are sometimes visible in the light microscope. These appear to be complex structures that seem to involve other chemical components.

Collagenous organizations of even greater complexity are widely distributed throughout the Metazoa. A striking feature is the occurrence of laminated texture and cross-fibrillar orientation of fibrils at different dimensional levels, in various situations in the body, and occasionally at sites remote from the secreting cells. A number of examples have been discussed by PICKEN (1960) and by FITTON JACKSON (1968), including:

1. Coelenterates - mesoglea of sea-anemones containing crossed-fibrillar layers with fibrils of each layer oriented at about 45° to long axis of column, while in the interior of the mesogleal sheet fibers run both radially and longitudinally; shear and/or tension forces are possibly responsible for organization.
2. Echinoderms - body wall with orthogonal ply arrangements of fibrils.
3. Nematodes, nemerteans, annelids - basement membranes of epidermis and cuticles with alternating ply arrangements crossing at specific angles with respect to the longitudinal axis of the animal.
4. Teleosts and amphibia - basement membranes of embryonic epidermis with orthogonal plies of collagen fibers (Fig. 2.1); specific number of fibrils in thickness of each lamella and specific number of lamellae.
5. Vertebrates - cornea with collagen fibrils organized into sheets or lamellae in plane of cornea forming orthogonal ply; regularly spaced fibrils with narrow range of diameters, specific number of fibrils in thickness of each lamella, and specific number of lamellae; articular cartilage, dentine, and lamellar bone with development of complex fibrillar patterns and a tendency to parallel alignments of fibrils influenced by extrinsic mechanical factors as well as intrinsic factors.

At higher levels of magnitude, the fibrillar net (notably in basement membranes of nematodes and amphibia) is specifically oriented in relation to the morphological axes of the whole animal. In these instances, the body surface is contained by a system of geodetics, either as open helices or as closed elliptical annuli (PICKEN, 1960). In nematodes, the collagen fibers are parallel-ordered in sheets, with the angles between fiber directions in adjacent sheets variable with species and related to the mode of locomotion. Although the individual fibers are inextensible, the network of fibers allows for extension and shortening of the cuticle by a lazy tongs-like action (LEE, 1965). In addition to the above types of fibrillar organization, there are important histological entities with fine structures that are not disclosed by electron microscopy. The most important of these, perhaps, are the basement membrane structures reported in Table 3.4.

Elastic tissue is widespread in the mammalian body but tends to be condensed in areas subject to periodic stress, such as walls of major arteries, lungs, skin, ligamentum flavium, ligamentum nuchae of ruminants, and capsules of visceral organs. At the different sites, the elastic fibers differ both chemically and morphologically (see reviews by AYER, 1964; BANGA, 1966). Diameters vary from 2 to 3μ in ligamentum nuchae down to the limit of visibility in the light microscope (0.2μ) in other tissues. The fibers may be thinly distributed throughout connective tissue matrices and show free branching, may be tightly packed side-by-side as coarse fibers in ligamentum nuchae, or may occur in the form of densely packed lamellae, each consisting of a dense fibrillar network.

Elastic fibers are chemically complex and consist of two or more components. The main component derives from monomeric tropoelastin (Chap. 4) by crosslinking processes. During embryogenesis, microfibrils, apparently glycoprotein in nature, are closely associated with elastic fibrillar structures. Other substances, such as collagen, polyanionic glycans, lipids or lipopolysaccharides, etc., are frequently separable from elastic structures only by drastic treatments. The nature of the association, whether due to chemical bonding or to physical entrapment, has not been clarified. According to BANGA (1966), some elastic fibers are composed of smaller, helically wound elastin fibers enclosed by spiral-running collagen fibers and held together by "mucoid" substances.

Functions of connective tissues may be classified into broad categories with respect to features that are expressed more prominently in some tissues than in others (CHVAPIL, 1967). These are mechanical or supportive functions, transportive and storage functions, and metabolic, repair, and defensive functions. Other functions are noted for some highly specialized connective tissues, for example the cornea as a light-transparent tissue. However, this traditional division of connective tissue activities tends to neglect the essential "integrated" nature of the "spatio-temporal structure."

Extracellular substances and structures play their role in "spatial extention of cellular activities" (PICKEN, 1960). Thus, the cells produce the extracellular materials that provide supporting or protective substances and structures with varied mechanical properties and, remaining viable, permit continuous growth, repair, and reconstruction. From this viewpoint, certain aspects appear more prominently than others. Apart from features related to the structures and properties of the component molecules and the physiochemical processes of extracellular matrices, aspects connected with the control of metabolic and secretory activities of cells, in particular, loom large. Many questions arise in these areas of cellular activity. What part of the process of production of extracellular molecules is determined by intrinsic and what by extrinsic factors? What mediates the relation between present and future mechanical stress and the distribution and orientation of material? In what instances, if any, is the relation direct? When it is indirect, what is the effective stimulus to the system, and how does the response acquire direction and mechanical appropriateness?

The various sections of this chapter attempt to describe aspects of connective-tissue structure at successive hierarchic levels of organization. The primary level, for our purposes, is represented essentially by the macromolecular monomer in a simple, aqueous ionic environment. Secondary and immediately succeeding levels are formed by macromolecular interactions involving both like and unlike molecules in more complex environments. At the tissue level, we are concerned with processes and properties of gross structures that reflect and illuminate lower-level structures and organizations. Finally, evolutionary implications are considered.

2. Primary Level — Molecular Properties

a) Collagen

Although the tropocollagen molecule is a complex structure of peptide chains associated in a triple helical conformation, it is taken here as the monomeric extracellular unit. The physical parameters of the vertebrate molecules in solution corespond with those predicted for a protein of molecular weight near 3×10^5 in the shape of a rod, 15 Å in diameter and about 2900 Å in length (VON HIPPEL, 1967). This group of closely related proteins has been extensively studied in the last two decades. Unfortunately, other collagens and their associated macromolecules in supporting tissues of invertebrates are still too little known to warrant further discussion (see Chap. 3).

There appears to be wide agreement that water molecules bind loosely to polar groups in the collagen backbone as well as to side chains and form continuous chains along the fiber axis with the participation of water molecules not directly linked to the protein. Water contributes to the stability of the collagen structure through intramolecular hydrophobic bonds, hydrogen-bonds, and possibly also by linking polar groups. It can also stabilize and destabilize the collagen structure through interaction with specific groups. BARDELMEYER (1973) suggests that the inner hydration layer consists of 50 grams of water per gram of dry collagen.

Neutral salts may greatly affect the structure and stability of the collagen molecule. Some, such as calcium chloride, lithium bromide, and guanidinium chloride, destabilize the collagen structure, others may have little effect, and a few, such as potassium fluoride, stabilize the collagen structure. The effects are not related to the sign or magnitude of the ionic charge, or to specific ion binding to charged protein side-chains. They appear to be generally similar to the effects of salts upon the conformation of a large variety of other types of macromolecules, and are perhaps related to alterations of water structure (see reviews by VON HIPPEL, 1967; and VON HIPPEL and SCHLEICH, 1969).

In addition, neutral salts interact in a direct and specific fashion with collagen. The marked effect of microions on the isoelectric point of soluble collagen indicates that ion binding to the protein occurs (VEIS et al., 1958; BENSUSAN and HOYT, 1958). DAVISON and DRAKE (1966) found isolectric point values (pH) of 6.0, 9.6, and 11.0 in the presence of phosphate, acetate, and calcium ions, respectively. In 0.1 M Tris buffer, a collagen preparation had an isoelectric point of 8.3 (BENSUSAN and HOYT, 1958). There are striking differences among ions in respect to binding by fibrillar collagen, phosphate, pyrophosphate and sulfate ions having especially high degrees of binding compared to chloride. Although bivalent cations, such as calcium, strontium, and magnesium, are bound very little compared to sodium, these ions have a profound effect in dissolving collagen fibrils (WEINSTOCK et al., 1967). Small molecules and ions also exert profound effects, apparently due to binding, upon rates of fibrillogenesis from collagen solutions. Among these are inorganic ions, amino acids, ascorbic acid, adenosine triphosphate, and neutral molecules such as urea and formamide (GROSS and KIRK, 1958; BENSUSAN and HOYT, 1958).

b) Elastin

Monomeric elastin, tropoelastin, has been obtained to date only from copper-deficient swine and is free of the aldehydic intermediate derived from lysine side-chains (Chap. 4). As a consequence it is one metabolic stage removed from the aldehyde-containing molecule that undergoes the homologous interactions leading to the cross-linked structures characteristic of elastin. Tropoelastin has an interesting property, which it probably shares with the aldehydic precursor of elastin. This in an unusual solubility in aqueous solutions containing high concentrations of aliphatic alcohols, a property apparently due to the high proportion of hydrophobic side chains in the molecule (SANDBERG et al., 1971). At this moment, the properties of the immediate precursor of cross-linked elastin remain unknown, except for suggestions from studies of tropoelastin and of the polymer.

The binding of cations, in particular calcium, to elastin is of special interest in view of the occurrence of extensive calcification of the elastic tissue of arteries in arteriosclerosis. TOSATTI et al. (1971) observed that the maximal amount of Ca^{++} bound to elastin powder from ligamentum nuchae was about 1 μmole per mg at pH 7 and attributed the binding to electrostatic interaction with ionic groups of the protein. Phosphate was bound only in trace amounts. However, binding of calcium ions to neutral, uncharged binding sites in elastin is theoretically possible and could be an initial step in calcification (URRY, 1971). Such binding was indicated by the ability of calcium ion to effect a large conformational change in solubilized α elastin from bovine ligamentum nuchae (URRY et al., 1971) and from porcine aorta (STARCHER et al., 1973) under conditions where all sites were expected to be either protonated or neutral. Confirmation was obtained by demonstration of binding of calcium ions to a neutral polypeptide containing only peptide groups (URRY et al., 1973) and to a neutral peptide identical with a repeating sequence of elastin (URRY et al., 1974). The binding depends upon the presence of β-turn polypeptide conformations (Chap. 4, Sect. 2). Coacervates of α elastin with blocked carboxyl groups bind 0.15 μmole of calcium per mg in 24 h when phosphate ions are also present (STARCHER and URRY, 1973).

c) Polyanionic Proteoglycans

The distinctive properties of this class of macromolecules derive from the following characteristic features: 1. a flexible chain structure, 2. a large molecular domain, and 3. a large number of closely adjoining, fixed anionic groups. Extensive discussions have been given by BETTELHEIM (1970), SCHUBERT and HAMERMAN (1968), OGSTON (1970), and by LAURENT et al. (1969); OOSAWA (1971) has furnished a general theoretical analysis of the properties of polyelectrolytes.

i) Conformation and Molecular Size

The sulfated glycosaminoglycans of vertebrates, with the possible exception of heparin, occur in the native state bound as side-chains to a large polypeptide backbone (Chap. 6). Molecular weights of the proteoglycans, for molecules containing 4 to 120 side-chains, are in the range 10^5 to 4×10^6. Their greatest dimension aproaches a 'length' of 4000 Å. The glycan chains have molecular weights in the range of 5×10^3 to 10^5 daltons with contour lengths from 100 Å to 2000 Å. The one unsulfated polyanionic glycan, hyaluronate, is present in tissues with very little, or no attached peptide and has molecular weights of 4×10^3 to 10^7 with contour lengths from 100 Å to 2.5×10^5 Å. The major differences in physicochemical properties between the proteoglycan and the glycan derived from it by proteolysis arise from the much greater molecular size of the kinetic unit and the resultant greater local density of glycan chains in solutions of the former.

The proteoglycan macromolecule of vertebrate cartilage (Fig. 6.3) serves as a useful model for other sulfated proteoglycan molecules. Unlike the tropocollagen molecule, as well as most other protein molecules, it does not have a fixed conformation in solution. This indeterminacy arises from a certain degree of freedom of rotation of groups about peptide and glycosidic bond linkages. Optical studies (EYRING and YANG, 1968a) indicate no secondary structure in the polypeptide chain; light-scattering data (see Chap. 6, Sect. 2b) suggest a somewhat extended conformation requiring about 2 Å average projection on the chain axis per peptide residue, dependent upon ionic strength (PASTERNACK et al., 1973). However, proteoglycan interaction with hyaluronate suggests a requirement for higher-order structure in a segment of the peptide moiety (Chap. 6, Sect. 1b).

In polymer-chain characteristics, the linear polyanionic glycosaminoglycans resemble other flexible-chain, linear, anionic polyelectrolytes of both the naturally occuring and synthetic varieties (BETTELHEIM, 1970). The average conformation of a polyanionic glycan molecule in aqueous solution is greatly dependent upon the charge type and concentration of added electrolytes. This effect is illustrated in Fig. 10.1. In dilute aqueous solution, the polymer molecules are highly ionized. The resulting fixed negative charges cause repulsion between adjacent segments of the same chain and lead to a preference for highly extended conformations.

When neighboring charged groups of the polyanion are closely spaced along the chain, a high electrostatic potential is created. This causes a strong attraction for small cations and a repulsion of small anions. On the average, cations are distributed in solution closer to the polymer chain than are anions and hence have a proportionally greater influence on conformation. Thus, the significant parameter is cation concentration and not ionic strength, a parameter that includes contributions from both types of ions. With increasing electrolyte concentration, the polyanion binds more microcations, reducing both its electrostatic energy and V_e. Such phenomena have been observed not only for chondroitin 4-sulfate (MATHEWS, 1953) but also for dermatan sulfate and chondroitin 6-sulfate (MATHEWS, 1959), hyaluronate (BALAZS, 1958), keratan sulfate (MATHEWS, unpubl.) and heparin (LIBERTI and STIVALA, 1967).

At sufficiently high electrolyte concentrations, generally obove 0.1 M as indicated in Fig. 10.1, the viscosity parameters show only a small dependence upon salt concentration. With electrostatic interactions between neighboring fixed anionic charges effectively shielded by

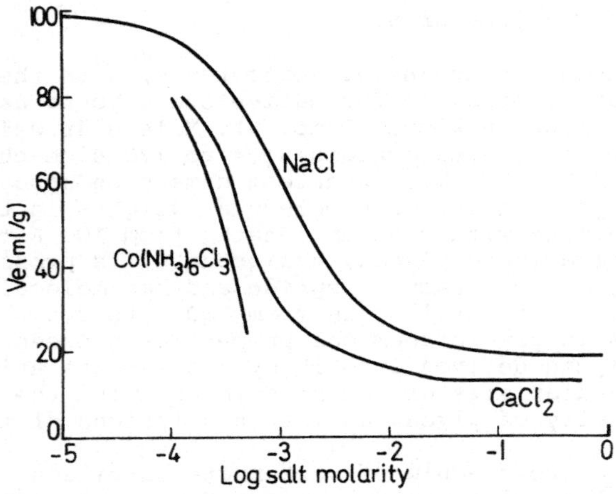

Fig. 10.1. Effect of type and concentration of added salt on polyanion conformation. Concentration of chondroitin 4-sulfate is 0.06%. The volume V_e of the equivalent hydrodynamic sphere, an approximate measure of the molecular domian, is defined here as: $V_e = 40\ [\eta]$ (with permission of the publishers; from MATHEWS, 1967c)

microions, the polyanion now exhibits the chain characteristics of an essentially uncharged polymer molecule. When intrinsic viscosity and other molecular parameters are related to various theoretical models for polymer molecules under these conditions, it is found that the polyanionic glycosaminoglycan chains possess varying degrees of stiffness (expressed in the values of the constants of the Mark-Houwink relations of Table 10.1). For example, chondroitin 6-sulfate chains possess more extended conformations than do chondroitin 4-sulfate chains.

The overall chain conformation is a dynamic reflection of molecular fine structure. Reduced chain flexibility is probably due to steric interference with free rotation by bulky groups, such as carboxyl, sulfate and acetyl, and also to a possible stabilization of secondary structure. Although considerable information is now available on the conformation of individual sugars (ELIEL et al., 1965), very little is known about specific conformational aspects of polysaccharides, especially when in solution. Data relating to the latter compounds have been derived from studies by infrared, nuclear magnetic resonance, optical rotatory dispersion, and circular dichroism techniques, with suggestions from studies of solid states, in particular, crystalline fibers. Recognition of the fact that the tissue states of polysaccharides are not comparable to either dilute solutions or to solid states has led to the investigation of conformations of polysaccharides at high concentrations and in the gel state. Our present knowledge of the conformation of neutral and of polyanionic glycans has been reviewed by REES (1967, 1972), STONE (1969), and BETTELHEIM (1970).

One might expect to find most of the hexose units of glycosaminoglycans in the C1, pyranose ring, chair conformation of minimum potential energy, a conformation that places the bulky carboxyl, N-acetyl and sulfate groups of D-glucuronic acid and D-glucosamine residues in an equatorial orientation. This geometry is less likely to involve such groups in steric interactions than the alternate axial orientations

Table 10.1. Constants[a] of Mark-Houwink relationship: $[\eta] = KM^{\alpha}$

Glycosaminoglycan	Buffer	$K \times 10^4$	α	References
Hyaluronate	0.2 M NaCl	2.28	0.816	CLELAND and WANG (1970)
Hyaluronate	0.5 M NaCl	3.18	0.777	CLELAND and WANG (1970)
Chondroitin 4-sulfate	0.2 M NaCl in 0.15 M PO_4, pH 7	3.1	0.74	MATHEWS (1956, 1959)
Chondroitin 6-sulfate	"	5.8	0.74	MATHEWS (1959, unpubl.)
Dermatan sulfate	"	3.1	0.74	MATHEWS (1956, 1959)
Chondroitin 4-sulfate[b]	0.2 M NaCl	0.050	1.14	WASTESON (1971)
Heparin	0.5 M NaCl	0.355	0.90	LIBERTI and STIVALA (1967)
Keratan sulfate (cornea)[c]	0.2 M NaCl	1.5	0.83	LAURENT and ANSETH (1961)

[a] Approximate molecular weight ranges of validity are 1×10^5 to 10^6 for hyaluronate and 5×10^3 to 1×10^5 for others.

[b] For paucidisperse preparations from ox cartilage consisting of both 4-S and 6-S repeating periods. See Table 7.1.

[c] Values estimated from data cited by authors.

which occur with the carboxyl group of L-iduronic acid residues in dermatan sulfate, heparan sulfate, and heparin. Nuclear magnetic resonance data suggest that the 1C chair conformation of L-iduronic acid residues, which places the carboxyl group in an equatorial orientation, may be energetically favored in solution (PERLIN et al., 1970; HIRANO, 1972). However, the Cl conformation is preferred in the crystalline state of dermatan sulfate (ARNOTT et al., 1973b; ATKINS and ISAACS (in press), but not of heparin (NIEDUSZYNSKI and ATKINS, 1973).

Intensive investigation of plant polysaccharides of the polyanionic type has yielded considerable data on conformation. Of particular interest here is the demonstration of secondary and tertiary order for linear glycans containing carboxyl or sulfate groups (REES, 1972). REES and collaborators have revealed that some carrageenins (sulfated galactans of seaweeds) possess a helical chain conformation and are able to gel in partial double helical structures stabilized by hydrogen bonds. Elucidation of conformational details was aided greatly by the ability to obtain good fiber diffraction patterns.

In contrast, little progress was made until recently with the glycosaminoglycans, which proved difficult to crystallize, although there were strong indications for secondary structure from a variety of experimental directions (STONE, 1969; BETTELHEIM, 1970). The first significant advance occurred when ATKINS et al. (1972) recorded excellent fiber X-ray diffraction patterns for sodium hyaluronate, which indicated an axially projected disaccharide repeating period of 9.5 Å and a rather stiff helical conformation of three disaccharide units per turn. The diffraction pattern of the free acid in fiber form was different and suggested a possible tightening of the glycan chain into a two-fold helix.

The structural organization of hyaluronate at pH 2.5 (free acid form) was clarified in further X-ray diffraction studies by DEA et al. (1973). Excellent patterns indicated the presence of a double helix consisting of two identical, antiparallel, left-handed strands with each strand having four disaccharide repeating periods per pitch length. To explain the viscoelastic properties of hyaluronic acid solutions, the authors proposed the presence of weak "cross-linking" between double helices. Sodium hyaluronate also assumed a double helical conformation (DEA et al., 1973; ARNOTT et al., 1973a) which seemed likely to persist to a small degree in solutions at physiological pH (CHAKRABARTI and BALAZS, 1973). In addition, ATKINS and SHEEHAN (1973) identified another conformational state in fibers as a four-fold single-strand helix, which they considered an intermediate between the double helix and the three-fold single-strand helix previously characterized. These observations led to the suggestion that hyaluronate in dilute solutions at neutral or physiological pH exists as a mixture of molecules in interconvertible, quasi-static, conformational states.

The first sulfated glycosaminoglycans to yield good crystalline fiber X-ray diffraction patterns (ATKINS et al., 1972; ARNOTT et al., 1973a) were chondroitin 6-sulfate (CH 6-S) and its variants from shark (CS-D) and squid (CS-E) cartilages (see Fig. 6.1). Two different forms of helical molecules of different symmetries and packing arrangements are indicated. The simplest interpretation of the data is that the two forms contain 3-fold and 8-fold helices with axial periodicities per disaccharide repeating unit of 9.5 Å and 9.8 Å, respectively, corresponding to almost fully extended chain conformations. Ionic bridges, involving counterions (NA^+), appear to be the only interactions between molecules. Although the 3-fold and 8-fold helices have similar axial periodicities, the aximuthal distribution of charged side-groups is very different in the two conformations.

Chondroitin 4-sulfate mimics the reported molecular conformations of chondroitin 6-sulfate (ISAACS and ATKINS, 1973). Dermatan sulfate occurs as 2-, 3-, or 8-fold helices in oriented crystalline films (ARNOTT et al., 1973b; ATKINS and ISAACS, in press). The evidence favors the Cl conformation of L-iduronic acid residues but does not rule out the possibility of the 1C conformation. Presently available X-ray diffraction data on heparan sulfate (ATKINS and LAURENT, 1973) and on heparin (NIEDUSZYNSKI and ATKINS, 1973) do not offer a clear choice as to conformation.

Conformations in the solution state, evidently, need not be the same as those in crystalline fibers. Other indications of a helical conformation for some glycosaminoglycans were given by earlier observations. Preparations of sodium hyaluronate of different molecular sizes undergo a reversible drop in intrinsic viscosity when the pH is increased from 7 to 12 (MATHEWS, unpubl.; SWANN, 1970). A similar drop is not observed for preparations of chondroitin 4-sulfate, chondroitin 6-sulfate, dermatan sulfate, or desulfated chondroitin sulfate. It may be noted that hyaluronate differs from desulfated chondroitin sulfate only in the configuration of C-4 of the hexosamine residue.

Insight into the nature of the transformation of hyaluronate conformation at high pH was obtained from light-scattering studies. The molecular weight (1×10^6) of a preparation from umbilical-chord hyaluronate remained unchanged at high pH, although the intrinsic viscosity decreased and the radius of gyration changed from 1120 Å to 940 Å (MATHEWS, unpubl.). SWANN (1970) had also noted that the molecular weight of rooster comb hyaluronate did not change when the pH was increased. This behavior suggests that sodium hyaluronate at pH 7, may exist in a relatively stiff, worm-like conformation, which is converted, at higher pH, into a random coil through breakage of hydrogen bonding and hydroxyl group ionization. Similar transformations were noted by RAO and FOSTER (1963) for amylose, for which helical conformations at neutral pH were proposed. However, carboxymethylamylose (RAO and FOSTER, 1965) appeared to have a random coil conformation at neutral pH.

In view of the participation of glycosaminoglycans in interactions with collagen and other molecules, the existence of molecules differing in primary, secondary, and tertiary structure acquires added biological significance. For example, the physiological functions of sulfated glycosaminoglycans may be regulated by biosynthetic variation of the degree of sulfation (STONE, 1969). Other structural variations of equal importance come to mind (Chap. 6). To these may be added the various helical conformations, each with a different azimuthal distribution of charged groups.

ii) Cation Binding

The polyanionic glycans share with other members of the class of flexible-chain, anionic polyelectrolytes the characteristic property of selective association, or binding, with small inorganic cations and with cationic groups of macromolecules. This section is concerned only with interactions with micro-ions, principally H^+, Na^+, and Ca^{++}. The theoretical analysis of polyelectrolytes presented by OOSAWA (1971) is particularly useful as a basis for interpretation of the experimental findings (see also MANNING, 1969; WELLS, 1973a, b).

The volume of solution surrounding a polyanion may be divided into four potential regions: a sharp slope and deep potential hole at charged groups; a sharp and deep potential valley along the length of the chain; a spherical trough in the apparent volume occupied by the macro-ion; and the region outside the macro-ion. Counterions are distributed in these four regions and can be defined as "bound" to the macro-ion within the first three regions. However, they are mobile in the cylindrical potential valley and in the spherical potential trough. When trapped by the sharp-potential holes, they are localized at charged groups, forming ion pairs.

The counterions in the region outside the macro-ions are free and are in equilibrium with the bound, but mobile counterions. The essential features of this equilibrium are retained by an approximation which considers this division of counterions as constituting two phases. The contribution of counterions bound as ion pairs may generally be neglected for monovalent and divalent cations. However, hydrogen ions are an exception: they can be site-bound to carboxylate ions and are in equilibrium with bound but mobile counterions.

The distribution equilibrium between bound and free counterions is given by the equation.

$$\ln (1-\beta)/\beta = \ln\phi/(1-\phi) — e_o\delta\psi/kt \qquad (10.1)$$

Where β = apparent degree of dissociation; ϕ = apparent volume concentration of macro-ions; e_o = the electronic charge; $\delta\psi$ = the electric potential difference between bound counterions and free counterions; k = the Boltzman constant; and T = the absolute temperature. For a long, rod-like macro-ion with an average distance between neighboring charges of d, the potential difference is expressed as

$$\phi\psi = -(\beta e_o/\varepsilon_o d) \ln (1/\phi) \qquad (10.2)$$

where ε_o is the dielectric constant of the solvent. The equation may be written as:

$$\ln(1-\beta)/\beta = \ln\phi/(1-\phi) + \beta Q \ln (1/\phi) \qquad (10.3)$$

where $Q = e_o^2/\varepsilon_o k\, T\, d$.

The nondimensional quantity Q is the most important parameter determining the intensity of binding of counterions. It increases with the square of the counterion valence and with 1/d, or the average linear-charge density of the macro-ion. The larger the value of Q, the greater the proportion of counterions bound to the macro-ion or, in other words, the greater the reduction in thermodynamic activity of the counterions derivable by dissociation from the macro-ion.

The small-cation-binding properties of the polyanionic glycans of biological origin may be generally related to the characteristic Q values. Thus, for monovalent cation salts of the glycans in aqueous solution at 25°, $Q = 7.2 \times 10^{-8}/d$. Estimated values of Q for hyaluronate, chondroitin sulfates or dermatan sulfate, and heparin are 0.7, 1.4, and 2.5, respectively. The predicted order of binding affinity for inorganic cations has been experimentally verified. However, actual binding affinities may be significantly influenced by the chemical structure and conformation of the glycans (REES, 1972; WILLIAMSON, 1972), the size and hydration structure of inorganic cations, and the chemical structure of organic cations. The binding properties of vertebrate and plant polyanionic glycans with respect to both inorganic and organic cations have been reviewed by BETTELHEIM (1970). The following is a critical summary of the data available to date for glycosaminoglycans: Consideration of ionization of carboxylic acid groups in polyanionic glycans falls into a special category because the phenomenon involves site-binding. The D-glucuronic acid group in hyaluronate as well as in chondroitin 4-sulfate and in chondroitin 6-sulfate has an intrinsic ionization constant pk_o of about 3.1; the L-iduronic acid group in dermatan sulfate is a weaker acid by 0.5 pk units (MATHEWS, 1961). The biological significance of this difference in acid strength is not clear; at pH over 7, both types of carboxyl groups are completely ionized.

Although negigible for H^+-binding, the structural differences among the isomeric chondroitin sulfates become significant for polyvalent-cation binding. The binding affinity for the trivalent cation $Co(NH_6)^{3+}$ (MATHEWS, 1960, 1964) is hyaluronate = corneal keratan sulfate < chondroitin 6-sulfate < chondroitin 4-sulfate < dermatan sulfate < heparin. A similar order was found by BUDDECKE and DRZENICK (1962) for Ca^{++}. The general order of binding increases with increase in linear charge density. The differences between the isomeric chondroitin sulfates and dermatan sulfate may be related to the orientations of carboxyl and sulfate groups, which are either axial or equatorial in the repeating disaccharide unit. With the assumption of the C1 conformation for chondroitin 6-sulfate, both groups are equatorial; for chondroitin 4-sulfate, the carboxyl group is equatorial and the sulfate group is axial; for dermatan sulfate, both groups are axial. Sulfate groups on uronic acid residues of the glycans enhance cation affinities.

BOYD and NEUMAN (1951) demonstrated that the binding of cations by hyaline cartilage was due to its chondroitin sulfate content. The order of increasing affinity for cartilage was $K^+ < Na^+ < Mg^{++} < Ca^{++} < Sr^{++} < Ba^{++}$ (DUNSTONE, 1959, 1960). A similar order was found for chondroitin sulfate isolated from cartilage and a nearly similar order for dermatan sulfate (DUNSTONE, 1962). A different order of affinities was indicated for heparin and heparan sulfate. MAROUDAS and EVANS (1972) found the order of affinity of human cartilage to be $Na^+ < K^+ < Ca^{++}$. Although these authors concluded otherwise, their data showed reduction of activity coefficients, or binding of counterions. The quantitative effects of the polyanionic glycans in the tissue were comparable to those shown experimentally by chondroitin sulfate in NaCl solutions (PRESTON et al., 1972; MATHEWS, unpubl.), corresponding to about 50% dissociation of sodium chondroitin sulfate. The affinity of chondroitin 4-sulfate for choline and lithium cations exceeds that for Na^+ (SIMARD and FRIEDMAN, 1970).

It should be noted that direct experimental measurement of cation binding to yield thermodynamic constants is not feasible as yet without simplifying assumptions. Calculations (MATHEWS, 1959) suggest that the chondroitin sulfates bind Na^+ in a degree equivalent to about 50% of the anionic sites of the macro-ion; sodium hyaluronate binds counterions to about 30% of the sites. A similar estimate for chondroitin sulfate was deduced by FARBER and SCHUBERT (1957), who also found that a greater percentage of Ca^{++} than Na^+ was bound. Although chondroitin sulfate shows only a small preference for binding Ca^{++} over Na^+ (URIST et al., 1968), it is possible that this component functions, in cartilage, as a store of Ca^{++} and has a specific role in tissue calcification. It may be noted that, unlike some plant glycans, chondroitin 6-sulfate does not aggregate upon binding calcium

(EYRING and YANG, 1968b) or show cooperative effects (GRANT et al., 1973).

Alginates, which are polyuronic acids containing different proportions of D-mannuronic acid and L-guluronic acid, are noteworthy for their property of forming gels upon exchange of alkali-metal ions for those of calcium or other divalent metal ions. The high affinity and, in particular, the divalent cation selectivity, are due principally to the L-guluronic acid content of the alginate and apparently also to the insoluble, gel state of the glycan (SMIDSRØD and HAUG, 1968). It is interesting to note the analogy of this affinity to the greater affinity for Ca^{++} of dermatan sulfate, as compared to chondroitin 4-sulfate, whose uronic acid components (L-iduronic and D-glucuronic acid) are the C-2 epimers of L-guluronic acid and D-mannuronic acid, respectively. The dependence of selectivity upon the gel state suggests that tertiary as well as secondary structure is related to divalent cation binding (REES, 1972; MACKIE et al., 1971; GRANT et al., 1973).

Attempts to find a biological activity for chondroitin sulfate have led to a search for components of cartilage having an especially high affinity for Ca^{++}. A report by WOODWARD and DAVIDSON (1968) that a proteoglycan preparation from cartilage showed 100% binding of Ca^{++}, whereas the isolated chondroitin sulfate chains had little affinity for this cation, was contradicted by observations by MATHEWS (1970a) and by MacGREGOR and BOWNESS (1971) showing no difference in Ca affinity between proteoglycan and isolated chondroitin sulfate. However, particular fractions from cartilage that appear to be aggregated or associated with collagen show a relatively high affinity for Ca^{++} (BOWNESS, 1962; MacGREGOR and BOWNESS, 1971; SMITH and LINDENBAUM, 1971). Their relationship to mineralization is unclear.

iii) Water Binding and Osmotic Pressure

Connective tissues usually contain water in about the same proportion as do other types of soft tissue in animals. Osmotic equilibrium of tissue water with water in 'fluid tissues', such as capillary blood and lymph, may be generally assumed. Frequently, the contribution of polyanionic glycans to tissue osmotic pressure is sufficiently great not only to influence water balance but also to introduce excess swelling pressure that must be balanced by an internal structural resistance (OGSTON, 1970). An inner osmotic pressure of about 50 milliosmolar was estimated for bovine hyaline cartilage (OGSTON and WELLS, 1972; compare WELLS, 1973b; MAROUDAS, 1973). Because of cation binding, this value is substantially lower than might be expected on the basis of ideal Gibbs-Donnan behavior. A role in maintaining mechanical flexibility and elasticity of tissues is suggested.

The polyanionic glycans are associated with "retention of water" in connective tissues. However, the terms commonly applied to this function, "water binding" and "hydration", are ambiguous and require careful evaluation (OGSTON, 1966a). We may consider the glycans to retain water by two mechanisms: 1. by direct interaction between polysaccharides and water due to polar and hydrogen bonds and 2. by osmotic mechanisms. Discussions of these topics have been given by LAURENT et al. (1969), BETTELHEIM (1970), and KATCHALSKY (1964).

A variety of physicochemical techniques indicate that the amount of water that is chemically bound to polysaccharide is less then 10% of the weight of the glycan; it is immediately evident that this is a minor proportion of the total water present in the tissue. The bulk may be allocated as follows: in part (usually small) to cells, in part (also small) bonded to intercellular components, and in part (major amount) to retention by osmotic mechanisms. The portion of tissue water held by osmotic mechanism is determined by the balance achieved between osmotic expansive forces due to dissolved components and the restraining forces due to the structural components of the tissue, principally collagen and elastic fibers. For the present, only the macromolecular components responsible for osmotic expansion will be discussed.

Polyelectrolytes obey Van't Hoff's law for osmotic pressure only at infinite dilution (TANFORD, 1961). Their behavior at finite concentrations may be described by the virial form of the ideal equation:

$$\pi = \frac{RTc}{M}(1 + Bc + Cc^2 + - - -) \tag{10.4}$$

where c is concentration and B and C are virial coeffients. Ordinarily, only the first two terms are retained for consideration. The unique properties of the highly charged glycans are revealed by an analysis of the coefficient B. The Donnan osmotic pressure, π_D, is the difference in osmotic pressure between a polyelectrolyte-salt solution and an external salt solution with which it is in equilibrium. It is given (KATCHALSKY, 1964) by the following expression for a fully neutralized polyanion that is a regular repeating polymer:

$$\pi_D = \frac{RTc_p}{1000}\left(\frac{1}{P} + \frac{c_p\,\phi_p^2}{4m_s}\right) \tag{10.5}$$

where c_p is the concentration of the polyelectrolyte in ionizable equivalents per liter, P the number of ionizable equivalents per molecule, ϕ_p the osmotic coefficient, and m_s the molarity of univalent-univalent salt. A plot of π_D/c_p (1000/RT) versus c_p, for a series of polyanion solutions in equilibrium with the same salt solution, yields a straight line the intercept of which gives the molecular weight of the polymer (1/P) and the initial slope of which, $\phi_p^2/4\ m_s$, provides a means of determining the factor ϕ_p. Thus, a value for ϕ_p of about 0.4 may be calculated from osmotic-pressure data on sodium chondroitin sulfate (MATHEWS and DORFMAN, 1953). This value is considerably reduced from the ideal value ($\phi = 1$) as a consequence of counterion binding.

The relative contribution of the counterions to the osmotic pressure, compared to 1 for the polyanion chain, is measured by the coefficient B of Eq. 10.4, which is evaluated as $P^2\phi_p^2/4\ Mm_s$. It increases with increase in linear charge density (P/M) and in P at fixed salt-concentration, compensated by an expected decrease in ϕ_p with increase in P/M. At the high concentrations of polymer that occur in such tissues as cartilage, the small-ion contribution becomes a preponderant determinant of osmotic pressure. Thus, the net effect of a polyanionic glycan is to combine the high equivalent contribution to osmotic pressure of a low-molecular-weight component with the low diffusivity of a high-molecular-weight solute. A high degree of Donnan type of salt exclusion is also characteristic of highly charged polyanions. Indeed, solutions of polyanions are liquid analogs of solid cation exchangers and possess comparable dynamic as well as static properties (PRESTON et al., 1972; PRESTON and SNOWDEN, 1972).

The simplified treatment just described requires expansion under several conditions. When the polyanion is of very large molecular size, excluded-volume effects between like molecules (TANFORD, 1961) result in increased osmotic pressures, reflected in an increase of the measured coefficient B. Other large-molecular-size components, present simultaneously with the polyanion will exert greater than expected osmotic pressures due to excluded-volume effects between unlike molecules. Such effects have been demonstrated for connective-tissue components (LAURENT et al., 1969; OGSTON, 1966a). In some cases, more than two terms of the virial equation are required to fit osmotic pressure data even at moderate concentrations of solute.

Sorption of solvent by gels should be mentioned also, because this behavior resembles osmotic swelling. However, the phenomenon is not associated - particularly in the case of electrically neutral gels - with appreciable reduction by the polymer-chain backbone of the activity of water relative to the outside phase and hence does not produce a significant osmotic effect (OGSTON, 1966b). Thus, it is unlikely to be a physiologically important factor in determining the water content of tissues.

iv) Molecular Exclusion

The phenomenon of molecular exclusion has been discussed for polymers in general by TANFORD (1961) and for polyanionic glycans by OGSTON (1970), LAURENT (1970), and LAURENT et al. (1969). It is essentially a steric phenomenon, which may produce large physical chemical effects in solutions, and is ideally dependent only upon the sizes and concentrations of the interacting molecules. Since the polyanionic-glycan molecules of connective tissue are flexible linear chains that possess a random coil-like conformation in solution, they occupy a much larger volume or domain, than would a molecule of comparable size in the shape of a solid sphere of the same specific molecular volume. As a consequence, the glycan chain will resist the penetration of its volume by other molecules (like and unlike) to a degree dependent upon their sizes and shapes.

For example, a molecule of sodium hyaluronate of molecular weight 10^7 occupies a volume approximated by a sphere of 1 μm diameter corresponding to a volume of about 5 liters per gram. Consequently, at a concentration of only 0.2 mg per ml, the molecules of the glycan, without overlap, occupy the entire volume of the solution. At higher concentrations, such as commonly occur *in vivo*, interpenetration of glycan molecules must occur, leading to a continuous network, or felt, of entangled molecular chains. A protein molecule like albumin, which has a more compact shape and a much smaller domain than the glycan but is much larger than salt ions, is only partially excluded from the domain of the sodium hyaluronate molecule. It may also be noted that most sulfated glycans exist in the tissues as moieties of very large proteoglycan molecules. The latter and their aggregates also produce large excluded-volume effects.

Excluded-volume phenomena may be categorized broadly as equilibrium, dynamic or transport, and gross or tissue effects. They are expressed in various ways: osmotic pressures are nonideal and increased; large and diffusible molecules (but not very small molecules) are unequally partitioned between different tissue compartments; excluded molecules show decreased solubilities and may aggregate or precipitate when glycan concentrations are raised. Exclusion results in a lowering of diffusion coefficients. Bulk flow rates and permeability coeficients of solvents, and distribution and diffusion coefficients of solutes are reduced by the presence of a closely tangled network of glycan chains (MAROUDAS, 1970; McCABE, 1972); under flow conditions, excluded molecules may be retarded or filtered off. Systems of excluded molecules and entangled molecular chains exhibit non-Newtonian behavior and gel-like properties (BALAZS and GIBBS, 1970). When combined with gross network of collagen and elastin fibrils, they confer properties of elasticity, viscoelasticity, resistance to compression, etc. upon connective tissues.

d) Structural Glycoproteins

The molecular properties of the structural glycoproteins described in Chapter 5 have been investigated in a few instances only. High contents of acidic residues, such as sialic, phosphoric, sulfuric, aspartic and glutamic acid, confer upon these compounds the properties of anionic polyelectrolytes of moderate charge at physiological pH. Cation-binding characteristics have been noted for the sialoprotein from bovine bone (HERRING, 1969, 1972; CHIPPENFIELD, 1970) and for a glycoprotein from cartilage (VITTUR et al., 1972a, b).

3. Macromolecular Interactions

Secondary and higher levels of spatial organization of matrices acquire their characteristic temporal stability through interactions involving macromolecules. Primary (homotypic) macromolecular interactions may be defined as those involving the formation of small, oligomeric-type aggregates or units from like macromolecules. Secondary (heterotypic) interactions are those leading to further aggregation and short-range ordered structures and may involve more than one type of macromolecule or unit aggregate. Tertiary and higher interactions, commonly involving heterologous macromolecules, produce long-range ordered structures which may be visualized at the electron-microscopic or higher dimensional level.

a) Primary (Homotypic) Interactions

i) Collagen

The elementary unit taken here is the final product of limited proteolysis of procollagen (Chap. 3, Sect. 11). Primary interaction may produce an oligomeric unit of four to six collagen molecules in a staggered arrangement (Chap. 3, Sect. 3). Evidence for such microfibrillar structures is limited to mammalian skin collagen as yet. The suggestion that collagen molecules in simple salt solution may under certain conditions give rise to a regular supermolecular structure akin to that produced by liquid crystalline substances is particularly intriguing (PRIVALOV et al., 1971).

ii) Elastin

The presumptive elementary unit for elastin is the tropoelastin molecule (Chap. 4), which has been isolated in an unoxidized state only from tissues of copper-deficient animals. The formation of filamentous arrays from tropoelastin (Chap. 4, Sect. 4) suggests a specific alignment of molecules during aggregation of the type probably necessary for interchain cross-link formation.

iii) Polyanionic Proteoglycans

Homotypic aggregation does not appear to occur with the proteoglycan from bovine nasal cartilage (Chap. 6, Sect. 1). However, the situation with the proteoglycan from nucleus pulposus, which is rich in keratan sulfate (HASHIMOTO and LUDOWIEG, 1968), requires clarification.

iv) Structural Glycoproteins

Data on self-aggregation of structural glycoproteins are meager (VITTUR et al., 1972a, b).

b) Secondary (Heterotypic) Interactions

In general, heterotypic interactions will be grouped in classes according to the predominant macromolecular component involved.

i) Collagen

The process by which collagen solutions form "native-type" fibrils under a variety of physical and chemical environmental conditions has been extensively studied (see comprehensive review by CASSEL, 1971). Although the major questions relating to biological control of fibril initiation, growth, and orientation are still unanswered, much has been learned of the basic properties of the complex systems from *in vitro* experimentation. Apparently, other types of macromolecules are not essential for the self-aggregation of collagen molecules, but they do profoundly modify the process *in vitro*. It is probable that such common macromolecular constituents of matrices as proteoglycans and structural glycoproteins, in addition to possessing other functions, also have a biological role in fibrillogenesis. Interactions of these constituents with collagen will be discussed shortly. We are concerned more immediately with systems presumably containing collagen as the sole macromolecular component.

Major difficulties in interpretation and comparison of data obtained by many investigators arise from the heterogeneities inherent in most, if not all, solutions of collagen. These solutions, prepared and purified to different degrees, differ in several respects. They invariably contain different proportions of monomeric, dimeric, and higher-order forms of collagen, are cross-linked to different degrees, and vary with respect to bound anions and cations, traces of associated lipids (RABINOWITZ and SHAPIRO, 1972), peptides and other small molecules. In addition, the solutions are rarely stable, even at low temperatures. As a consequence, the properties of fibril-forming systems deduced from observations on *in vitro* systems may not be wholly reliable indices of the behavior of living systems.

Formation of native-type fibrils occurs spontaneously upon raising the temperature of collagen solutions, usually to 37°, as an endothermic reaction. It has been suggested that the process is largely entropy-driven and that it involves disruption of the water structure surrounding collagen molecules. Isothermal studies have demonstrated that fibrillogenesis occurs in two steps: a lag or induction period, followed by a period of rapidly increasing precipitation. The kinetics have been described in terms of a nucleation process and a growth process. The number of nuclei increases during a lag phase, which is followed by a growth phase during which fibrils increase in length and diameter by accretion of collagen molecules.

The interactions of collagen with polyanionic proteoglycans are of particular importance, and considerable effort has been expended to elucidate their nature (see reviews by CASSEL, 1971; MATHEWS, 1970b; JACKSON and BENTLEY, 1968). Difficulties encountered by some workers (MASHBURN and HOFFMAN, 1967; BANGA, 1969; KUBOTA et al., 1969) in completely separating proteoglycan from collagen in extracts from vertebrate cartilage have led to the suggestion of a covalent linkage between the two classes of macromolecules. However, STEVEN et al. (1969) obtained complete separations from extracts of human intervertebral disc with special chromatographic procedures. GLIMCHER et al. (1969) found that more than 96% of the proteoglycan of lathyritic chick cartilage could be readily obtained free of collagen by mild procedures; much lower yields were obtainable from normal chick cartilage. It would appear, therefore, that most of the proteoglycan of cartilage is physically enmeshed in a gel network of collagen fibrils (see below), in addition to being reversibly immobilized by electrostatic interactions (MATHEWS, 1970b). However, the possibility exists that a small percentage of molecules are covalently united (HOFFMAN and MASHBURN, 1970; KOBAYASHI and PEDRINI, 1973).

The numerous studies directed toward clarifying the molecular mechanisms of interaction of collagen and proteoglycans have produced much confusion and only modest progress. Many problems stem from unrecognized chemical and physical chemical variations in the collagen solutions (see above) and in the polyanion glycans and proteoglycans, and also from differences in experimental methodology. However, there is major agreement on many aspects of interaction phenomena. The following are the most reliable conclusions:

Polyanionic glycans and proteoglycans interact at physiological pH with collagen in solution and in fibrous form by electrostatic mechanisms dependent upon ionic composition and strength, linear charge density and length of glycan chains, and also upon molecular structure and size of proteoglycans (MATHEWS, 1970b; ÖBRINK and WASTESON, 1971; ÖBRINK, 1971, 1973a, b).

Interaction requires the undenatured state of collagen (MATHEWS, 1965b).

Excluded-volume (molecular-entanglement) effects are important and operate simultaneously and mutually with electrostatic factors (MATHEWS and DECKER, 1968b; ÖBRINK, 1971, 1973a, b).

Interaction influences the process of native-type collagen fibril formation in solutions, variously retarding or accelerating the lag phase and the growth phase, and influencing the diameters of resultant fibrils (MATHEWS and DECKER, 1968b; ÖBRINK, 1973b; VON KÜHN, GRASSMAN, and HOFMANN, 1959; CASSEL, 1971; MUTHIAH and KÜHN, 1973).

Proteoglycans interact more effectively than their constituent glycan chains (MATHEWS and DECKER, 1968b). Dermatan sulfate - proteoglycans are unusually reactive under some conditions (LOWTHER, TOOLE, and HERRINGTON, 1970; ÖBRINK, 1973a, b).

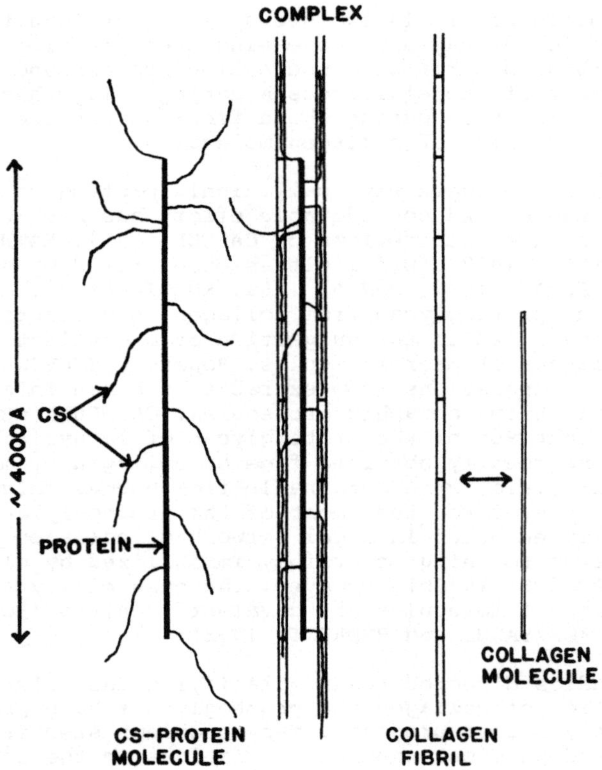

Fig. 10.2. A schematic representation of interaction between collagen and the basic unit of the principal proteoglycan of vertebrate cartilage. For purposes of clarity, the figure shows only a small fraction of the total number of chondroitin sulfate (CS) side-chains, which occur as doublet pairs (Fig. 6.3). Reproduced with permission of the Biochemical Journal; from MATHEWS (1965b)

The model shown in Fig. 10.2 emphasizes the major role of electrostatic forces in the interaction of proteoglycans with collagen. It illustrates the initial stages of one type of spatial orientation of interactants that may occur in connective tissues and suggests that the chondroitin sulfate chains are aligned parallel to the microfibril axis. In dilute solution, maximum interaction results from parallel alignment of the protein core of the proteoglycan also. Further growth of the interaction complex may then occur by accretion of additional fibrils, collagen molecules, or proteoglycan molecules. Finally, proteoglycan molecules themselves may form self-aggregates by lateral association with the participation of noncollagenous proteins (MATHEWS and LOZAITYTE, 1958). A role for such a noncollagenous, aggregation-promoting protein was suggested by JACKSON and BENTLEY (1968). Subsequently, LOWTHER and NATERAJAN (1972) showed that chondroitin sulfate proteoglycan that is aggregated by addition of "glycoprotein", unlike the monomeric proteoglycan, had no delaying effect on the rate of fibril formation from collagen solutions. This effect is possibly related to the reduction upon aggregation of the excluded volume of the proteoglycan and the number concentration of kinetic units. In general, the character of the interaction will depend upon the chemical structure and molecular parameters of proteoglycans, collagens, and other participating molecules that are species- and tissue-variable and also change during development, growth, and aging. More complex steric relationships between macromolecules than the one shown in Fig. 10.2 are possible under other states of connective-tissue matrices (see Sect. 3c below). Strikingly analogous behavior is shown by fibril-forming systems containing monomeric cellulose precursor and sodium carboxymethylcellulose (BEN-HAYYIM and OHAD, 1965).

Collagen is involved in complex interaction with platelets (see Chap. 3, Sect. 11). In this phenomenen, the glucosylgalactosylhydroxylysine residue of α1-CB5 appears to have a specific role (KATZMAN et al., 1973).

ii) Elastin

Data on aggregation of monomeric elastin are not available. Whether the glycoprotein microfibrils associated with embryonic development of elastic fibrils (Chap. 4) have a specific role in early stages of the aggregation process is unknown.

iii) Polyanionic Glycans and Proteoglycans

Polyanionic polyelectrolytes interact significantly not only with collagen but also with other biological macromolecules. The interactions may result either in large aggregates of proteoglycan or in both soluble and precipitated products in other cases.

The noncovalent aggregation of cartilage proteoglycan has already been described (Chap. 6, Sect. 1b) for preparations from bovine nasal septum. This phenomenon may be widespread among the homologous vertebrate macromolecules since it is exhibited also by proteoglycan of shark cartilage (MATHEWS, 1962); more extensive investigation is needed. Disulfide bonds are involved in the aggregation of proteoglycan from whale nucleus pulposus (HASHIMOTO and LUDOWIEG, 1968).

Interactions of polyanionic glycans with a variety of noncollagenous proteins have been reported. However, in many cases a clear distinction has not always been possible between electrostatic and steric factors. This has been particularly true for interactions involving high-molecular-weight hyaluronate, which may be associated with serum globulins in the vitreous body (VARGA and BALAZS, 1954). BETTELHEIM et al. (1966) found no evidence for electrostatic interaction between albumin and polyanionic glycans at physiological pH and ionic strength and hence concluded that steric factors only were involved (LAURENT, 1970; PRESTON et al., 1973). This conclusion was disputed by NIEDERMEIER and GRAMLING (1969).

On the other hand, electrostatic factors appear to be clearly involved in the binding of sulfated glycosaminoglycans to human plasma lipoproteins (IVERIUS, 1972) and other blood components (BERNFELD, 1966). Dermatan sulfate is an especially effective interactant; Ca^{++} appears to be essential for the formation of the complexes (SRINIVASAN et al., 1970; WALTON, 1973). Interactions of glycosaminoglycans occur also with amyloid fibrils (PENNOCK, 1968), diquaternary amines (EHRENPREIS and FISHMAN, 1960; HASSON, 1962), and with lysozyme in cartilage (ECONOMU, 1971). Glycosaminoglycans may cause adhesion of platelets. However, there is disagreement as to the relative effectiveness of the different glycosaminoglycans and proteoglycans in producing this phenomenon (MUIR and MUSTARD, 1968; LYMAN et al., 1971). The shift in circular-dichroism spectrum of poly-L-lysine produced by chondroitin 6-sulfate suggested to GELMAN et al. (1972) that the polypeptide was forced to adopt an α-helical conformation. Similar interactions were observed for other glycosaminoglycans with basic polypeptides and with collagen (GELMAN and BLACKWELL, in press; GELMAN, GLASER, and BLACKWELL, in press).

iv) Structural Glycoproteins

Glycoproteins appear to be bound covalently, probably by disulfide bonds, to basement membrane (Chap. 3, Sect. 10). The linkage to insoluble structures is apparently noncovalent in other supporting tissues, i.e in bovine cortical bone (HERRING and KENT, 1963), shark (MOCZAR and MOCZAR, 1970) and bovine cornea (ROBERT and DISCHE, 1963; DISCHE and ROTHSCHILD, 1971), bovine skin (WOLFF et al., 1971), and bovine tendon (ANDERSON and JACKSON, 1972). In bovine cornea and tendon the noncovalent linkage appears to be to collagen. Interaction with collagen, as indicated by experiments with fibril-forming systems, is strong with dentine sialoprotein, which contains about 6% of phosphate ester, and weak with bone sialoprotein, which contains 20% of sialic acid but little ester phosphate (MATHEWS and DECKER, 1968b). Data relating to glycoproteins from other tissues are given in Chapter 5.

c) Tertiary (Complex Heterotypic) Interactions

Interactions that produce large structures are included implicitly in the term tertiary interactions. They result in two or more condensed phases at some specified dimensional level and are commonly associated with the presence of two or more kinds of macromolecules. Our primary interest is in those cases in which collagen or elastin are the sole or predominant components of one of the phases.

Elementary fibrillar assemblies comprising one phase may have preferred and regular orientations over comparatively large distances and at several levels. This is exemplified in Fig. 2.1 by amphibian basement membrane, where order is present at the level of a single fibril of indefinite length, at the level of a plane of indeterminate size, and also in a direction perpendicular to the long axes of fibrils. Preferred orientations may be evident at macrodimensional levels, as in tendon. Or collagen fibrils may be nonordered, as in regions of hyaline cartilage, and form a randomized network (see Sect. 1 above for other examples).

The fibrillar phase (or phases, if complex) is usually embedded in a ground substance, which may contain additional phases such as cells, granules, vesicles, mineral, etc. The ground substance, which appears amorphous in the electron microscope, is generally considered to be "structureless." However, WEISS (1956) has postulated a lamellar organization for the ground substance demonstrated in Fig. 2.1. Although the cells, and possibly the other phases of a tissue, participate in producing and maintaining the ordered fibrillar structures, we shall be mainly concerned with the ground substance with which the fibrils are in most immediate contact and with two classes of macromolecules only. The vast bulk of our information bearing upon tertiary interactions of collagen and elastin with ground-substance components relates to polyanionic proteoglycans and structural glycoproteins. Analogous fibrillar systems of plants were discussed in Chap. 2.

i) Collagenous Structures

Collagen is closely associated with polyanionic glycans and proteoglycans, and possibly also with anionic glycoproteins in supporting tissues of a wide range of animal species. The apparent absence, or very low concentration, of the associated components in the collagenous mesoglea of two species of coelenterates (KATZMAN and JEANLOZ, 1970d; BOCQUET et al., 1972) has raised the question of an absolute biological requirement for these classes of ground-substance macromolecules in the formation and stabilization of collagenous structures. However, it may be observed that there is no lack of such ancillary macromolecules in the mesoglea of the other phylum of diploblastic mesozoa, the Porifera. One may speculate that structural requirements peculiar to some, or even all coelenterates are met either by very small amounts of anionic carbohydrate-containing macromolecules or by other classes of macromolecules that have not as yet been detected.

The association of polyanionic glycans with collagen is indicated in experiments aimed at purification of fibrils. Thus, differential extraction reveals that dermatan sulfate proteoglycan is tightly associated with collagen fibers in bovine dermis while hyaluronate is easily removed from the ground substance (TOOLE and LOWTHER, 1966; see also Chap. 9, Sect. 4a). Also, uronic- acid and amino sugar-containing glycans closely accompany spongin fibers when they are purified from marine sponges (GROSS et al., 1956).

Fine localization is possible by the use of special staining methods with the electron microscope. Polyanionic glycans of unknown structure can be demonstrated on the surfaces of collagen fibers and of spongin in the Demospongiae, *Haliclona* (GARRONE, 1969), *Chondrilla* and *Hippospongia* (THINEY and GARRONE, 1970). It is significant that *Chondrilla* contains a mixture of glycans consisting of several neutral sugars, hexosamine, uronic acid and ester sulfate (MATHEWS, unpubl.). Polyanionic glycans are also found closely associated with the outer leaflet of the plasma membrane of all cell types in the just described sponges (GARRONE et al., 1971). It was suggested (GARRONE and PAVANS DE CECCATHY, 1970) that the polyanionic-glycan coat has a role in membranar fibrillogenesis by lophocytes, which are specialized mesogleal cells responsible for collagen synthesis (GARRONE, 1971; PAVANS DE CECCATHY and GARRONE, 1971).

Polyanionic glycans not only coat the surface of the mammalian collagen fibril but also fill the interior space of the fibril, between the small-diameter microfibrils or filaments (KUHNKE, 1958). Recent studies by BOUTEILLE and PEASE (1971) and PEASE and BOUTEILLE (1971) of fresh, unfixed collagen from aorta adventitia of the rat revealed a complex substructure of collagen filaments, having a mean diameter of 30 to 35 Å, that are arranged in a helical pattern within collagen fibrils, probably always spiraling in a single direction. The filaments are separated from one another by a space-filling matrix believed to consist, at least in part, of a macromolecular carbohydrate component.

The problem of defining the steric relationships between collagen fibrils and various macromolecular interactants has been studied by a variety of methods including molecular localization by electron-microscopic procedures. The use of plane-polarized illumination in the light microscope shows that stretched-fibrous precipitates of collagen, containing chondroitin sulfate and stained with toluidine blue, are pleochroic, i.e the oriented fibrous mass has a pale blue color when the plane of vibration of the electric vector is parallel to the fiber axes and a deep purple-red (metachromatic) color when the fiber axes and plane of vibration are at right angles (NEMETH-CSÓKA, 1960, 1961). DE CAMPOS VIDAL (1963) demonstrated a similar pleochroic behavior for sections of bovine tendon. Although the electronic-optical properties of the dye when bound electrostatically to polyanionic glycans are yet understood fully (STONE and BRADLEY, 1967; STONE, 1967), a reasonable interpretation of his data was presented by DE CAMPOS VIDAL (1963), viz. that the toluidine molecules were oriented with their major axes perpendicular to the major axes of the tendon collagen fibers. He also concluded that the glycan chains were bound to the surfaces of the collagen fibers and, because of their extended conformation, were most probably aligned with the collagen fiber axes or coiled about them in a spiral fashion (MELLO and VIDAL, 1973).

Experimental support for this suggested parallel organization was obtained by MATHEWS (1965b, 1970b) and MATHEWS and DECKER (1968b). An analysis of the energetics and stereochemistry of the interacting macromolecules (see above) led to a proposed model (Fig. 10.2) for the initial stages of one type of steric relationship, involving proteoglycans applicable to collagen molecules and to very thin macrofibrils. JACKSON and BENTLEY (1968) elaborated this model further.

Quite different steric relationships were suggested by SERAFINI-FRACASSINI, SMITH and collaborators in a series of electron-microscopic studies of various connective tissues. The proteoglycan in bovine nasal cartilage was visualized by staining the tissue with bismuth nitrate in 0.1 M nitric acid in either an aqueous medium or in one consisting mainly of acetone (SERAFINI-FRACASSINI and SMITH, 1966). The proteoglycan appeared as a linear series of dark, bismuth containing, particles, regularly spaced about 50 Å apart (the model of the proteoglycan shown in Fig. 6.3 suggests that each dark particle represents two glycan chains), the backbone-polypeptide chain was not seen, since the procedure does not stain protein. On many of the largest collagen fibrils in cartilage sections the chains of dark particles appeared to lie perpendicularly across the fibril, overlaying a and b bands. Thus, the polypeptide backbone orientation was also transverse. However, no information on the glycan-chain orientation could be obtained. Similar observations were made on articular cartilage (SMITH et al., 1967). Since the characteristic steric relationship was noted only in the main matrix but not in the pericellular zone, it was suggested that a minimum fibril diameter of 500 to 600 Å was required to accommodate the proteoglycan molecules in the perceived manner. Indeed, in the epiphyseal plate cartilage of the rabbit, where collagen fibrils are 80 to 200 Å in diameter, a tangential orientation was observed (SMITH, 1970). Large, unstructured aggregates containing proteoglycan may be related to the matrix granules observed by ANDERSON and SAJDERA (1971) and associated with the calcification process. Relationships in nucleus pulposus are much more complex (SMITH and SERAFINI-FRACASSINI, 1968). In the vitreous, hyaluronate is weakly bound to small-diameter vitrosin fibrils (SMITH and SERAFINI-FRACASSINI, 1967).

Macromolecular interrelationships are particularly interesting in the corneal stroma where collagen fibrils are highly organized in parallel arrays that are organized in lamellae, with very nearly orthogonal relationships between directions of arrays in adjacent lamellae (MAURICE, 1969). For this tissue, SMITH and FRAME (1969) suggested that the protein cores of the proteoglycan are not wrapped around individual collagen fibrils as in cartilage but follow an angular course through the tissue, running at right angles to a number of collagen fibrils. However, LANGHAM et al. (1969) and FARREL and HART (1969) developed good theoretical models based upon the schema of Fig. 10.2 to account for the stability of the quasi-ordered array of parallel collagen fibrils in a corneal lamellae. By use of a new electron-dense stain, HODSON and MEENAN (1969) showed that the glycan chains did not act as bridges between fibrils but were bound to their surfaces. An explanation of the feature of alternating orthogonality of a series of corneal lamellae has not yet been offered. DISCHE (1970a) has suggested that a structural sulfated sialoglycoprotein forms part of the interfibrillar gel and is a factor in the developmental organization of the corneal stroma. Ruthenium-red staining of synovial tissue revealed several different configurations of sulfated proteoglycans and collagen (HIGHTON et al., 1968; MYERS et al., 1969). Both parallel and transverse relationships between the extended polypeptide backbone of the proteoglycan and the collagen fibril were noted.

In many intercellular matrices - for example, vitreous, cartilage, and Wharton's jelly - no long-range ordering of fibrillar elements can be detected. In these cases, the mechanistic model suggested (FESSLER, 1960) is that of a relatively coarse and open network of fibers (collagen and elastin) in which a much finer network of glycan and proteoglycan chains is trapped. Water may be assumed to be in its normal state (CLIFFORD and CHILD, 1971). Such a model is a basis for explaining the mechanical and osmotic (swelling) properties of connective tissues (see discussions by OGSTON, 1970; BALAZS, 1969; SCHUBERT and HAMERMAN, 1968; BARRETT, 1968). Statistical thermodynamic calculations of experimental models considered as random gel networks yield values of mechanical and osmotic parameters in good agreement with direct measurements on tissues (PRESTON and MEYER, 1971; MEYER, COMPER, and PRESTON, 1971).

ii) Elastic Structures

The elastic fiber is a multicomponent system organized in a complex fashion (BANGA, 1966). Several levels of organization have been revealed; relationships of elastin to microfibrillar components were described earlier (Chap. 4, Sect. 4). Bovine ligamentum nuchae fibers of about 6 μm diameter are a composite assembly of interwoven, branched and thread-like entities of increasing diameter: 30 to 100 Å filaments and 1000 to 2000 Å fibrils (GOTTE et al., 1972). Nondegradative extractions and autoclaving at neutral pH yield preparations that can be purified further only by very prolonged extractions or by the use of such hydrolytic reagents as strong alkali and proteolytic enzymes (GOTTE et al., 1963; JOHN and THOMAS, 1972). The latter treatments yield glycoproteins containing hexosamines, glucose, galactose and mannose and a protein rich in aspartic and glutamic acids. This highly polar protein, incompletely removed by hot alkali may be an integral component of aortic elastin (KEELEY and LA BELLA, 1972a). A high percentage of sulfur, not accounted for by the content of sulfur-containing amino acids, in 2 M NaCl extracts of autoclaved elastin suggests the possible presence of sulfated glycans (GOTTE, SERAFINI-FRACASSINI, and MORET, 1963).

iii) Polyanionic Glycans and Proteoglycans Structures

The properties of polyanionic glycans in gel and in concentrated solution states have been discussed by BETTELHEIM (1970). Ordered structures may be produced by the diffusion of polyvalent cations into the glycan solutions. This method of producing "ionotropic gels" was extensively studied by THIELE and coworkers (1962, 1964, 1967) who obtained, in general, five different structures: solution, striation, lenses, capillaries, and membranes. A general resemblance between these gels and such biological structures as eye lenses, cornea, aorta, skin, etc. was noted. Structures grossly resembling the biological ones were also obtained by allowing various cations, such as calcium, to diffuse into extracts obtained from various tissue layers. However, as BETTELHEIM (1970) commented, it is unlikely that the rebuilt ionotropic gels corresponded to the original gel structures in the different parts of the original tissue. ROSENBERG et al. (1970) observed ordered aggregates of the proteoglycan of bovine nasal cartilage in the electron microscope.

iv) Structural Glycoprotein Structures

Like the polyanionic glycans and proteoglycans, glycoproteins in a supporting tissue are constituents ancillary to the fibrous proteins. Connective-tissue structures composed primarily of glycoproteins have not been found.

4. The Tissues

The broad goal of this section is to describe the structural organization and properties of connective tissue in the dimensional range between the electron-microscopic and gross-anatomic levels. Mechanical functions are of primary importance (HARKNESS, 1968; CHVAPIL, 1967; VIIDIK, 1973). In many connective tissues, the cellular component is a very minor proportion of the total mass and probably contributes little to structural properties. For these tissues, our attention centers upon the intercellular matrix macromolecules. However, the cells cannot be overlooked for they are not only the biosynthetic sources of the matrix macromolecules but may also have essential roles in the organization, degradation and remodeling of matrix structures, and even (especially in highly cellular tissues) mechanical functions. In spite of many recent advances in knowledge, the precise nature and extent of the cellular contributions are still relatively obscure. For this reason, the exposition to follow is restricted to those tissues (or discrete regions of tissues) for which the cellular fraction is small, the matrix structure is well described, and the intercellular macromolecules are best known. Of necessity, numerous important and fascinating structural tissues of invertebrates and vertebrates (PICKEN, 1960) must be neglected.

a) Skin

The skin is at the boundary, or interface, of the organism and its physicochemical environment. As such, it is involved in a vast number of physical and chemical transactions of both exo- and intra-organismic character. Two intimately associated problems arise: firstly, what are the biologically important mechanical functions of skin, and secondly, what are the contributions of the different components of skin to its mechanical properties: HARKNESS (1971) has given a detailed analysis of these questions for mammals with particular emphasis upon the macromolecular components of the dermis layer, which is of major importance in most member species.

Observation of animals in their natural habitat is essential to the acquisition of knowledge of the biological functions of skin. Information so gained may then be related to laboratory studies of mechanical properties and to the chemical nature of skin components. The main mechanical function of mammalian dermis is protection against a large variety of potentially damaging environmental contacts. These include energetic impacts of many types, including those from stones, plant thorns, predator teeth, claws, and horns, penetrating structures of parasites, etc. Collagen comprises the bulk (70-80% of the dry weight) of the dermis. The fibers run in many directions and are responsible for the resistance to shear and tensile strength of the layer. Systematic organization of fibers may exist but is difficult to detect. Elastin is a minor component and is commonly under tension in the normal state of the skin. Other components of the intercellular matrix of the dermis include polyanionic glycans and proteoglycans (Chap. 9), structural glycoproteins (Chap. 5), plasma proteins, water, and electrolytes.

Collagen is the main material providing ultimate tensile strength for skin. Its contribution to physical properties of skin has been described by ELDEN (1968) and GIBSON and KENEDI (1970). Hyaluronate may have some lubricant role between collagen fibers. Dermatan sulfate proteoglycan is found in close association with collagen and may possibly contribute to mechanical function. Interpretation of experimental results in the light of theoretical models leads to the conclusion that the major factors contributing to the tensile strength of dermis are the nature and degree of the intermolecular cross-linking of collagen (HARKNESS, 1971). The important cross-links are possibly those of the Schiff-base type. Excessive cross-linking, such as can occur with increased age, may contribute to weakening of collagenous structures.

Scaleless skin of mutant chicks lacks feathers. This defect is associated with the absence of a fibrous collagenous network, but is not due to a failure to synthesize sufficient collagen (GOETINCK and SEKELLICK, 1972). The relationship of glycosaminoglycans to this failure of collagen to be organized into a lattice is currently under investigation (GOETINCK, priv. comm.).

b) Cornea

The cornea is divided classically into five layers lying parallel to its surfaces, from without inwards: epithelium, Bowman's membrane, stroma, Descemet's membrane, and endothelium. Details of structure, physiology, and biochemistry are given by MAURICE (1969) and MAURICE and RILEY (1970). The stroma, which makes up the bulk of the cornea, is divided into sheets or lamellae of collagenous material lying parallel to the surface. In the electron microscope, the lamellae are seen to contain long collagen fibrils of uniform diameter running parallel to each other and approximately at right angles to one another in alternate layers. The majority of the cells in the stroma are modified fibroblasts known as keratocytes. The intercellular matrix represents 97 to 98% of the volume of the tissue and consists of 78% of water, 15% of collagen, 5% of other proteins, 1% of glycosaminoglycans, and 1% of salts.

The stroma functions as a supporting structure adapted for the transmission of a high percentage of incident light of visible wavelengths (MAURICE, 1969). Its transparency is dependent upon the presence of the quasi-regular, long-range, fibrillar order, which can be disturbed by swelling of the stroma upon uptake of water. Considerable interest attaches, therefore, to the developmental processes which bring about this supramolecular order as well as to the mechanisms involved in its maintenance.

The development of the fine structure of the chick cornea has been described by HAY and REVEL (1969) and also, in a less extensive fashion, by FITTON JACKSON (1968). The primary corneal stroma is formed from secretions by cells of the ectodermal epithelia (HAY and DODSON, 1973), which remain active in producing glycosaminoglycans (MEIER and HAY, 1973) and orthogonally arranged collagen fibrils until stage 28 (5.5 days of incubation). Thereafter, mesenchymal cells destined to become corneal fibroblasts invade the acellular stroma and take over the secretory task (CONRAD, 1970a, b) until at 19 days the structure of the cornea of an adult is achieved. It is conceivable that the epithelial cells actively influence the orientation of collagen fibrils during the stage of formation of the primary stroma and that the corneal stroma is completed by accretion of the secretory products of fibroblasts upon the initial template (TRELSTAD and COULOMBRE, 1971). However, fibroblasts in mass culture are capable of generating self-maintaining, parallel arrays and also orthogonal multilayers, stabilized by secreted collagen (ELSDALE and BARD, 1972). Alternatively, no immediate intervention by cells is required and the long-range order results from the spontaneous assembly of macromolecules (TRELSTAD, 1973). In either case, it is likely that the macromolecules involved possess to a considerable degree the capacity for highly specific interactions (see above; also COULOMBRE and COULOMBRE, 1972).

The maintenance of the structure of the mammalian cornea requires the constant action of a fluid transport system dependent upon the metabolic activity of endothelial cells. When the activity of these cells is inhibited, the excised cornea is capable of rapidly absorbing water and swelling to many times its normal volume. For the last 20 years or so, i.e since recognition of the important role of polyanionic glycans in the passive swelling process, the interpretation of swelling has been based upon the comparison of tissue properties with theoretical properties of polyelectrolyte gel models of natural and synthetic polymers (MAURICE, 1969; KATCHALSKY, 1964). This approach to the problem often produced oversimplified interpretations of experimental results and led to disagreements. Early workers, and more recently HODSON (1971), suggested that cation binding by polyanionic glycans is negligible and that swelling occurs because the Donnan potential that is associated with the charged glycans is unopposed by structural cross-linking. However, others found appreciable monovalent cation binding (GREEN et al., 1971) and concluded that the Donnan contribution to swelling pressure is sensitive to the molecular organization of the stromal ground substances (FRIEDMAN and GREEN, 1971).

The contributions of specific chemical structures and conformations of the matrix macromolecules appear to be important. For example, the occurence of a highly sulfated keratan sulfate is associated with the rise in transparency of stroma during chick corneal development; in corneal scars, normal collagen fibrils are replaced by coarse fibrils of varied diameters when accompanied by dermatan sulfate; stromal transparency and normal hydration is correlated with the presence of the normal proportions of keratan sulfate and chondroitin 4-sulfate (ANSETH and FRANSSON, 1970).

Significant variations occur among species. Elasmobranch corneal stroma differs from the homologous tissue of mammals and other vertebrates in failing to swell, a phenomenon attributed to the presence of sutural fibers. At normal hydration, dogfish cornea exhibits no swelling pressure and therefore has no need for an active dehydrating mechanism (TOLPIN et al., 1969). Unlike mammalian stroma, the elasmobranch tissue contains highly sulfated chondroitin sulfate as well as keratan sulfate and a preponderance of chondroitin sulfate over keratan sulfate (MATHEWS, 1969; PRAUS and GOLDMAN, 1970a, b).

c) Vitreous

The adult vitreous body fills the internal cavity between the lens and the retina and serves to maintain ocular shape and transmit light. A thorough review of the anatomy, physiology and biochemistry of this tissue was recently given by BERMAN and VOADEN (1970). The reader is referred to this work for full details.

The vitreous consists approximately of 99% of water and 0.1% of macromolecules, the remainder being low-molecular-weight solutes. The important structural macromolecules are vitrosin (about 0.01%) and hyaluronate (0.001-0.08%) of high molecular weight. Both macromolecules are inhomogeneously distributed throughout the vitreous, and the average contents of these components are highly variable with species, development, and aging, even among individuals of the same species.

Vitrosin is a basement membrane-like collagen (Chap. 3, Sect. 10) and may be specific for the tissue. It is observed as very thin fibrils of less than 300 Å width without clearly distinguishable axial periodicity; subfilaments of 15 to 25 Å width have also been noted. The protein forms a thinly dispersed, three-dimensional fibrillar network within which the hyaluronate molecules are distributed. Although there is some evidence for weak electrostatic interactions (WOODIN and BORUCHOFF, 1955) and ordered relationships (SMITH and SERAFINI-FRACASSINI, 1967) between vitrosin fibrils and hyaluronate, the properties of the vitreous are probably best interpreted in terms of the random gel network model (Sect. 3c). According to this view, the structure of the vitreous is stabilized as a consequence of molecular entanglement, frictional factors, and excluded volume effects within the system of two interpenetrating gel networks, one formed by vitrosin fibrils and the other by hyaluronate molecules.

d) Cartilage

Cartilage performs its principal mechanical function as a resilient tissue, resisting forces of shear and compression and, usually to a lesser degree, tension. It appears in many anatomical locations and is designated as hyaline, elastic, or fibrous depending upon the relative proportions of the ground substance, collagen, elastic fibers, and cells present. Apart from minor components, the mass of the extracellular matrix in most vertebrate species consists of very fine collagen fibrils and proteoglycosaminoglycans; the ratio of cell mass to total organic content is very low, in the range of 0.01 to 0.10 for some forms of adult hyaline cartilage. Comprehensive discussions of the various types of cartilage in vertebrates were given by MAXIMOW (1927), SCHAFFER (1930), and LUBOSCH (1938) and of mammalian articular cartilage by SOUTEYRAND-BOULENGER (1971) and FREEMAN (1973); more specialized presentations with emphasis on the biochemistry and matrix structure of mammalian cartilage were given by SCHUBERT and HAMERMAN (1968), BARRETT (1968), and SOKOLOFF (1969).

The best model for interpreting the properties of cartilage appears to be a three-dimensional, random network of collagen fibrils with an interspersed gel network of proteoglycan molecules. No directional ordering of fibrils is detected in resting cartilage, except in such special regions as the surface of articular cartilage. Many of the distinctive properties of adult cartilage derive from the very high content (4-8% of the wet-tissue weight) of proteoglycan, unique to this tissue; collagen contents also are high (10-20%). The effect of a decreased amount or loss of proteoglycan is dramatically seen in the loss of turgor of tracheal cartilage in inherited chondrodysplasia of the mouse (SEEGMILLER et al., 1972) and of ear cartilage of rabbits injected with papain (THOMAS, 1964). Alterations of conformation and interactions of proteoglycans that are induced by the replacement of sodium ions by divalent and trivalent cations results in increased deformability and loss of elasticity (SOKOLOFF, 1963). It is not surprising, therefore, that complete degradation of the chemical structure of proteoglycans destroys interlocked tensile stresses dependent upon the osmotic-expansive contributions of these molecules (FRY and ROBERTSON, 1967).

Chondroitin sulfate and keratan sulfate are inhomogeneously distributed within a cartilage tissue and may be present in low concentrations at surface regions (SZIRMAI et al., 1967; STOCKWELL and SCOTT, 1967; BALAZS et al., 1966; EISENSTEIN et al., 1971; MAROUDAS et al., 1969; MAROUDAS, 1972). It is likely, therefore, that mechanical properties also vary throughout a tissue, since the creep modulus (stiffness) of articular cartilage in different individuals correlates positively with average content of glycosaminoglycan (KEMPSON et al., 1970). Composition appears associated with function, since the content of sialic acid (structural glycoprotein?) is correlated with weight-bearing load (TRANCREDI and PECORELLI, 1966), as is glycosaminoglycan content (MATTHEWS, 1952; CONVERY and AKESON, 1969; OLAH and KOSTENSZKY, 1972).

The particular mechanical properties of hyaline cartilage represented by a resilient resistance to forces of shear and compression depend in large measure upon the resistance to fluid flow of the complex network. This resistance is a function primarily of the interfibrillar ground substance and to a much lesser degree of the fibrillar network (OGSTON, 1970); permeability to fluid flow is correlated strongly in an inverse manner with glycosaminoglycan concentration (MAROUDAS, 1968; MAROUDAS et al., 1969). In this tissue glycosaminoglycan concentration has a predominant role in determining mechanical parameters. Joint lubrication may be facilitated by a polarity that is induced by loading and is possibly related to piezoelectric and streaming-potential phenomena (BASSETT and PAWLEK, 1972). The streaming-potential effects are dependent upon the presence of glycosaminoglycans (MAROUDAS, 1968).

The high concentration of glycosaminoglycans, and hence of fixed negative charge, has important consequences for the electrolyte composition of cartilage fluids. In this respect, the tissue resembles an ion-exchange resin and, with respect to most of its properties, can be described quantitatively by means of ion-exchange theory (MAROUDAS, 1968; HANDLER-BERNICH et al., 1972). The diffusion coefficients of small ions, such as Na^+, K^+, Cl^- and $SO_4^{=}$, are reduced in bovine nasal cartilage (MAROUDAS, 1968) and in calf rib cartilage (KAWALEC et al., 1972) to about 50% of their values in aqueous solution; however, the diffusion coefficient of water is little affected (MAROUDAS, 1973).

The full significance of these observations for the electrolyte physiology of connective tissues remains to be explored (MANERY, 1966).

Mineralization of cartilage and other supporting tissues commonly occurs in intimate relationship to matrix components, for example, vesicles in cartilage, highly ordered collagen fibers in bone, elastic fibers of aorta, or specialized proteins in dentine and in various invertebrate tissues. Although these processes have been investigated for more than half a century, there exist as yet no satisfactory biochemical and biophysical explanations of the phenomena at the molecular level. Thus, a detailed discussion here can serve our purposes little. Nevertheless, these intriguing problems are of great biological importance; the reader is referred to review articles by ALCOCK (1972), BACHRA (1970), SCHIFFMAN et al. (1970), EASTOE (1971), JOPE (1971), WILBUR and SIMKISS (1968), GLIMCHER and KRANE (1968), TRAVIS et al. (1969), LAVINE and ISENBERG (1964) and MOSS (1964). Data on structural glycoproteins of bone (Chap. 5), polyanionic glycans of vertebrate bone and calcified cartilage (Chap. 7), and polyanionic glycans of invertebrate mineralized tissues (Chap. 9) have been discussed earlier. HERRING (1972) has discussed possible roles of these components in mineralization.

Recent observations implicate specific structural features of matrix macromolecules in the mechanism of calcification. Analyses of collagen isolated from different regions of the long bones of 3-week-old chicks showed that the growth plate and cartilagenous regions contained $[\alpha 1(II)]_3$-type collagen, while ossifiying regions contained the $[\alpha 1(I)]_2\alpha 2$-type collagen (TOOLE et al., 1972; LINSENMEYER et al., 1973). The bone-type collagen in osteoid contained much more hydroxylisine and carbohydrate than did bone collagen from mature chickens. TOOLE et al. (1972) suggested that the high degree of glycosylation in osteoid collagen prevented calcification by steric interference with apatite crystal nucleation in these regions (see GLIMCHER and KRANE, 1968). Cartilage-type collagen is likewise unable to seed apatite crystals by virtue of its high content of glycosylated hydroxylysine residues and possibly also because of various primary structural features. In addition, noncalcified regions of epiphyseal cartilage (in the rat) appear to contain a specific proteoglycan inhibitor of calcification, which is diminished or absent from calcified regions (PITA et al., 1970). The isomeric shift of the position of ester sulfate in chondroitin sulfate proteoglycan that occurs with calcification of cartilage was noted earlier (Chap. 7, Sect. 5). Cartilage contains a tissue-specific proteoglycan in addition to a tissue-specific collagen (PALMOSKI and GOETINCK, 1972; LEVITT and DORFMAN, (1974). It is probable that these two types of macromolecules are specifically adapted to produce hyaline cartilage matrices.

5. Extrinsic Factors Influencing Function

Connective-tissue cells possess, as do all other types of cells in multicellular organisms, mechanisms for growth and maintenance in an appropriate environment as well as for control of internal processes. A universal paradigm in biology embodies the homeostatic aspect of biological reactions to the external environment. Since the prime function of these reactions is the construction and maintenance of adaptive support structures, it is not surprising to find that connective-tissue cells are especially responsive to external mechanical factors. Recognition of the importance of this type of reactivity should not imply depreciation of other types of environmental influences - either of important extrinsic chemical factors such as hormones (DONGHERTY and BERLINER, 1968), vitamins (GOULD, 1968c), nutritive metabolites, inorganic ions (CHVAPIL, 1967), etc., or of various matrix-degrading enzymes such as collagenases (SEIFTER and HARPER, 1971; NORDWIG, 1971; WOESSNER, 1968), hyaluronidases (MEYER, 1972); elastase (HARTLEY and SHATTON, 1971), lysomal enzymes (WOESSNER, 1968), etc.

a) Mechanical Factors - Physical Feedback

Time-dependent mechanical forces acting upon connective tissue produce tensive, compressive, and shear stresses which result in alterations of structure and shape at various dimensional levels. Both cells and matrix undergo changes. WEISS (1959, 1960, 1961, 1962, 1968) has done much to clarify the complex interrelationships between the cell and its physical environment under such circumstances. For example, when the fibrin network in which fibroblasts are growing is placed under tension and stretched, the cells become greatly elongated and align themselves along the lines of stress. Collagenous fibrils appear along these lines (WEISS, 1959), suggesting that the tensile forces have stimulated fibroblasts to an increased rate of biosynthesis of collagen and to a polarized secretion of this tension-bearing matrix component. The response of the cells has been to counteract the effect of the imposed stress.

In this connection, it is interesting to note that the growth of mammalian tendon is greatly influenced by mechanical stress (ELLIOT, 1965). Removal of mechanical stress, as in joint immobilization, may alter patterns of biosynthesis of glycosaminoglycans (AKESON et al., 1973). Pressure and nonpressure areas of cartilage differ greatly in glycosaminoglycan content (STUHLSATZ, 1973). It is noteworthy that cartilage of aquatic mammals has an unusually high content of hyaluronate (see Chap. 7, Sect. 4d). It appears that mechanical factors have a significant role in determining the regional composition of glycosaminoglycan of cartilage matrices, both at the gross-anatomic level between different cartilage tissues in the same organism and at the microscopic level within a cartilagenous structure (Chap. 8, Sect. 4e).

Mechanical forces have a determining influence also on the type of connective tissue produced by cells in culture. A cellular outgrowth from a bone explant will form bone, cartilage, or fibrous tissue according to whether the conditions of growth are compaction or stretch in the presence of a high or low concentration of O_2 (BASSETT, 1964). The role of piezoelectricity produced by the action of mechanical forces upon collagen and bone crystals in these phenomena has been reviewed by BASSETT (1971). The participation of collagen and ground-substance components in the complex series of events of chick skeletal morphogenesis and endochondral ossification has been discussed by FITTON JACKSON (1968).

Arteries fulfill mechanical functions maintained by elastic fibers, collagen, and smooth muscle cells, which are aligned circumferentially in concentric layers (see Chap. 9, Sect. 4d). These blood vessels, like many other connective tissues, are adapted to the continually changing mechanical forces to which they are subjected. GLAGOV (1972) has discussed the relationships of mechanical stresses imposed by blood pressure and flow to the size, architecture and composition of normal mammalian arterial walls. The role of mechanical forces in determining forms and structures of arteries and of connective-tissue systems generally was analyzed by RODBARD (1970, who presented a broad hypothesis with supporting data. In essence, mechanical stress and the associated feedback mechanisms induce tissue changes which result in the elimination of the effects of these stresses on the tissues. Some of the inducing tensorial forces, which may also appear in combinations, are listed below with the corresponding arterial tissue responses.

Absence of stress produces ground substance.
Tension produces collagen.
Rate of change of tension produces elastin.
Acceleration of tension produces muscle.
Excess contracting tension produces muscle hypertrophy.
Tension during relaxation produces muscle elongation.
Compression produces osteoid.
Compression and tension produces calcification and bone formation.
Rate of change of compression produces cartilage.
Rate of change of tension (compression) produces elastic cartilage.
Tension and a high rate of change of compression produces fibrocartilage.

b) Matrix Macromolecules - Chemical Feedback

It is evident that in the creation of an intercellular matrix of a particular chemical composition and molecular structure the connective-tissue cells must regulate the temporal sequence of both biosynthesis and degradation of matrix components. Questions of major importance are: What are the relative roles in biosynthesis of (a) fixed internal programming through information storage and (b) regulation through feedback mechanisms involving the chemical microenvironment? Also, how are chemical- and mechanical-feedback mechanisms coupled? Data relevant to these questions, as well as to related questions pertinent to cell differentiation, cell - cell interactions, tissue development, etc., are still quite new and permit only very tentative conclusions to be drawn (SLAVKIN, 1972). For example, difficulties of interpretation arise in cases where the relative contributions of physical, chemical and other factors cannot be readily evaluated. This situation is illustrated by the observation that chondrocytes form collagen type II (cartilage type) when in organ culture, but form collagen type I (skin type) when in monolayer culture (LAYMAN et al., 1972).

i) Biosynthesis of Matrix

GREER et al. (1968) observed that the rate of synthesis as well as the rate of degradation of cartilage proteoglycan in different regions of the epiphyseal growth plate of the rabbit were inversely related to the concentration of proteoglycan in each zone. They concluded that a feedback mechanism may be operative in such a manner that the cartilage cell responds to decreasing amounts of proteoglycan by increasing its synthetic rate. Similar conclusions were drawn from the effect on biosynthesis of partial depletion by papain (BOSMANN, 1968) and by hyaluronidase (FITTON JACKSON, 1970; HARDINGHAM et al., 1972) of the proteoglycan content of embryonic chick tibiae in culture. Under these conditions, only a small increase in the rate of synthesis of collagen was observed. Removal of collagen by collagenase increased rates of synthesis of both collagen and proteoglycan (FITTON JACKSON, 1970). Since both macromolecules are necessary for matrix formation, it seems likely that the syntheses of collagen and proteoglycans are coupled (PROCKOP et al., 1964; ROKOSOVÁ-ČMUCHALOVÁ and BENTLEY, 1968). However, the biosynthetic mechanisms are known to be under separate controls (BHATNAGER and PROCKOP, 1966) in a cell that is capable of synthesizing both types of macromolecules (LINDNER, 1963; CONRAD,

1970b) concurrently (SMITH, 1972). Also, collagen synthesis appears normal in cartilage of mutant embryo chicks that are defective for synthesis of proteoglycan in cartilage but not in other tissues (MATHEWS, 1967b).

In seeming contradiction to the conclusions described above are a number of observations that suggest that proteoglycans may stimulate their own biosynthesis. However, it should be noted that, in the cases to be cited the phenomenon is seen under special conditions. For example, the formation of cartilage matrix was stimulated when chondrocytes were cultured under compaction and low oxygen tension (BASSETT and HERRMAN, 1961). Cartilage, particularly of the hyaline type, is apparently adapted to a low-oxygen environment (PAWALEK, 1969). With thyroxine present in the culture medium, low oxygen tension markedly stimulated the rate of synthesis of chondroitin sulfate by embryonic chick chondrocytes when grown in mass monolayer culture but not when grown as clones (PAWALEK, 1969). When ascorbic acid was an added component of the culture medium, low oxygen tension greatly increased the rate of production of both chondroitin sulfate proteoglycan and collagen by embryonic chick chondrocytes in suspension culture (NEVO et al., 1972). Since, in this instance, both capsule formation and cell aggregation are favored, it appears likely that retention of the proteoglycan in the immediate vicinity of the cell had stimulated biosynthesis.

Evidence for a positive feedback effect was obtained by NEVO and DORFMAN (1972), who observed that removal of the proteoglycan resulted in a decline in its rate of synthesis. Furthermore, the addition of exogenous proteoglycan produced a stimulation of the rate of synthesis of proteoglycan but not of collagen, particularly under conditions of low oxygen tension. A similar stimulation was produced by various sulfated glycosaminoglycans and by dextran sulfate, but not by hyaluronate, suggesting a requirement for a polyanionic glycan of high linear charge density. The authors suggested that the synthesis of proteoglycan may be regulated by surface receptors sensitive to their chemical environment. In this connection it is interesting to note that collagen synthesis by embryonic chick tibiae is stimulated by addition of polyinosinic-polycytidylic acid to the culture medium. Accordingly, WADELL (1971) suggested that, in developing mesenchymal systems, the regulation of protein synthesis may be related to the presence of macromolecular polyelectrolytes. Unfortunately, observations on proteoglycans were not included in her study. Biosynthesis of the cationic protein, lysozyme, by chondrocytes may also be regulated by a feedback mechanism (KUETTNER et al., 1972).

ii) Interactions between Cells

Of the very great number of different phenomena that are included in this topic (WEISS, 1958), only examples that indicate the possible involvement of components of connective tissue matrices will be discussed here. There are numerous indications in the early literature (BALAZS and JACOBSON, 1966) that proteoglycans, collagen, and other macromolecules, alone or in various combinations, influence cell aggregation, differentiation, and tissue development. The insoluble matrix of acid-decalcified bone stimulated mesenchymal cells to differentiate into cartilage or into bone (URIST et al, 1973; REDDI and HUGGINS, 1972). However, the "morphogenetic" substances have not been characterized yet. LIPPMAN (1968) suggested that proteoglycans, acting via adsorption on cell surfaces, have a role in the regulation of cell division.

The glycosaminoglycans may function in the aggregation of cells and in other developmental processes. For example, factors that promote cell aggregation by several types of chick embryo and mammalian cell cultures are synthesized by the same cells, and all these factors appear to be hyaluronate (PESSAC and DEFENDI, 1972). Presumably, the high-polymeric glycan can bind to receptor sites on some cell surfaces and act as a ligand. However, it interferes with cell - cell interactions in lymphocyte cultures (DARZYNKIEWICZ and BALAZS, 1971) and in trypsinized cultures of precartilage cells (TOOLE, 1973). Hyaluronate appears to have a specific developmental role in the developing chick embryonic cornea. It is associated with the stages during which the primary stroma, which contains fine collagenous fibrils, is highly swollen and is being invaded by mesenchymal cells. At about day 10, the hyaluronate is removed by hyaluronidase and replaced by the sulfated glycans characteristic of adult corneal stroma (TOOLE and TRELSTAD, 1971). An analogous series of events appears to occur in the notochord of the developing chick embryo (O'CONNELL and LOW, 1970), in the regenerating newt limb (TOOLE and GROSS, 1971), and in healing wounds (BENTLEY, 1969). In all these situations, hyaluronate may be involved in the regulation of collagen fibril formation and as a substratum for migration of cells. The collagenous and (unknown) polysaccharide components of the mesoglea in Hydra also appear to function as a medium for cell movement (SHOSTAK et al., 1965). Polyanionic glycans are present in cell coats of invertebrate and vertebrate eggs and have possible roles in fertilization and development (Chap. 9, Sect. 1). Mediation of cell aggregation by proteoglycans appears to be a primitive characteristic of animal cells since the phenomenon is also exhibited by sponges (see Chap. 9, Sect. 2b).

The interactions between neighboring tissues in the developing embryo that lead to differentiation and organogenesis have been studied over many years. Although the nature of the mechanisms involved is obscure, it is likely that these mechanisms depend upon substances intimately associated with the surfaces of cells and the interfaces between tissues. From investigations of inductive interactions between epithelium and mesenchyme in particular, it appears that both collagen and proteoglycans may be involved as mediators in morphogenetic events (GROBSTEIN, 1969; FITTON JACKSON, 1968; TRELSTAD, 1973). Recent studies of epithelial - mesenchymal interactions in mouse embryo submandibular glands by BERNFIELD and BANERJEE (1972) and BERNFIELD et al. (1972) indicate that salivary morphogenesis is dependent upon the presence at the epithelial surface of proteoglycans of the chondroitin 4-sulfate or chondroitin 6-sulfate type, and possibly also hyaluronate. To account for the distribution of proteoglycan as well as of collagen fibers near the epithelial surface, the authors suggested that the proteoglycans at the sites of new cleft formation initiate the fibrogenesis of collagen which ultimately acts to stabilize the epithelial morphology. KOSHER et al. (1973) observed that a preparation of proteoglycan from embryonic chick cartilage stimulated *in vitro* chondrogenesis of chick somites.

The surface of mammalian cells is known to possess a wide variety of complex carbohydrates in addition to other components (KRAEMER, 1971a). Recent observations indicated that the polyanionic glycosaminoglycans or proteoglycans are common to many cells, independent of their specialization for other functions (COOK and STODDARD, 1973). Thus, heparan sulfate had been shown by KRAEMER (1971b, c) to be both an intracellular and a surface component of Chinese hamster cells, as well as of a diploid mouse-embryo lung strain and six other established mammalian cell lines. Chondroitin 4-sulfate, chondroitin 6-sulfate, and dermatan sulfate have been identified by SUZUKI et al. (1970) as synthetic products in cultures of cell lines of both fibroblastic and nonfibroblastic origin. The glycans appeared to be major constituents of the cell surface of these cells. Other sulfated glycans were present but were not identified. They might have included heparan sulfate. It is possible that collagen interacts with the anionic groups of these cell-surface glycans.

Although the cells of the various lines studied by KRAEMER (1971b, c) differ widely in species of origin, morphology, ability to grow in culture, and retention of ability to synthesize specific types of other glycosaminoglycans, they all retain the ability to synthesize heparan sulfate to a similar degree. For this reason, KRAEMER (1971c) suggested that the capability to synthesize heparan sulfate is vital for some general life process. In this connection, it should be noted that, although collagen is also synthesized by a large variety of cells of both fibroblastic and nonfibroblastic origin, the rates of synthesis vary over a range of 10^4 for the different types of cultures cells and that lymphoma cells and lymphocytes fail to synthesize detectable amounts of the protein (GREEN and GOLDBERG, 1968).

6. Conclusions

An attempt has been made in this chapter to order our knowledge of connective tissue macromolecules in terms of organizational levels that possess hierarchic relationships. The results, which are limited to the lowest levels of hierarchic structure and are summarized in Table 10.2, raise many questions for which there are no ready answers. It is evident that the *boundaries* between levels do not necessarily coincide with physical boundaries between recognizable structural entities. Thus, some secondary-level structures that are destined for extracellular location may originate within the cell, which itself maintains other secondary (and higher-order) structures as permanent intracellular features. The cell membrane becomes less of a wall and more of a communication structure, especially when one sees the role of the extracellular matrix as a "spatial extension of cellular activities" (PICKEN, 1960).

Major difficulties and complexities arise when we attempt to define the boundary conditions between hierarchic levels and to generate information about one hierarchic level from a description of units at a subordinate level. These flow primarily from the condition that living systems necessarily possess functional as well as structural hierarchies (PATTEE, 1969), and that knowledge of functions of units at one level can be obtained only from observation of properties and behavior of systems at superordinate levels. For example, the functions of molecular units in the developing corneal stroma can be understood only in terms of phenomena at the level of the complex supramolecular network of fibrils, interfibrillar ground substance, and cells at the next higher level. Thus, one needs to explore not only *horizontal* interactions between units within a single level but also *vertical* interactions between levels (GERARD, 1969).

GUTMAN (1969) argues that an understanding of the genesis of hierarchical structure requires a clarification of the relationship of structure to function and of organic whole to its parts within the context of the history of living organisms, including not only ontogeny but also phylogeny. He views hierarchical order as arising from the dominance of function over structure "on the assumption of a

Table 10.2. Structure of matrix and of major matrix components

Site	Level	Collagen	Elastin	Structural Glycoprotein Proteoglycan
INTRACELLULAR	PRIMARY			
	Sublevel: gene action	1 to 3 polypeptides	polypeptide	polypeptide
	Sublevel: enzyme action	hydroxylation glycosylation aldehyde formation	hydroxylation aldehyde formation	glycosylation epimerization sulfation
		procollagen (triple helix) limited proteolysis		
EXTRACELLULAR	SECONDARY	limited proteolysis intramolecular bonding oligomeric assembly fibril assembly intermolecular bonding	helical assembly (?) intramolecular bonding intermolecular bonding fibril assembly	ordered (?) aggregates (assoc. glycoprotein; proteoglycan)
	TERTIARY AND HIGHER hetero-associations, network, long-range fiber pattern	fiber (assoc. glycoprotein, proteoglycan, elastin, mineral) fiber pattern	fiber (assoc. proteoglycan, glycoprotein, collagen) fiber array	molecular order (assoc. collagen, elastin)

primacy of the whole over the parts in living organisms and that development of parts from a central whole is the result of progressive individualization and specialization, i.e. a delegation of function to specialized structures." Thus, the molecular (and cellular) components of connective-tissue matrices possess physiological functions by virtue of their interactions and interrelationships in higher level structures. Since it is not the basic macromolecules alone which have evolved but also their functions, the study of molecular evolution will acquire biological relevance only within the context of functional interrelationships between unit structures.

Adaptation of organisms and of their parts arises as a consequence of the interplay of two processes, which may be termed (a) genetically determined prefittedness and (b) direct physiological adjustment (WEISS, 1949). Genetically determined primary structures of macromolecules have changed during evolution. However, rigid determination of structure has been moderated by the availability of tissue-specific variants of collagen and possibly of other proteins, and of protein moieties of macromolecular components of connective-tissue matrices. It is possible that the more important evolutionary changes, at least in vertebrates, have occurred in mechanisms controlling biosynthesis and functional interrelationships rather than in structural genes. The evolutionarily conservative character of the partnership between collagen and proteoglycans is attested to by observations on 450-million-year-old graptolites (TOWE and URBANEK, 1972; see Chap. 3, Sect. 13c).

The great significance of the process of direct physiological adjustment is perhaps more clearly illustrated in the connective tissues than in other tissues. It is revealed, for example, in the regulation of supramolecular structures *via* variation of a large number of secondary chemical features of secreted macromolecules and, generally, in the responsiveness and functional adaptation of cells and tissues to chemical and physical environmental factors.

Chapter 11

Conclusion

"Appearances are a glimpse of the unseen." ANAXAGORAS, ca. 500 BC

Supporting tissues of multicellular organisms are greatly diverse in outward form but serve similar physical and chemical functions. They possess extracellular matrices having structures that reflect common principles of organization at various dimensional levels. Mechanical needs are met typically by systems of microfibrils composed of protein and polysaccharide subunits and embedded in a ground substance containing proteins, polyanionic proteoglycans, and polysaccharides. Although the matrix macromolecules of both plants and animals are broadly related in functional properties and in mechanisms of biosynthesis from simple precursor molecules, they vary considerably in chemical structure.

Chemical diversity is greatly reduced in the animal kingdom to a basic plan involving principally collagen. More specifically for the metazoa, and particularly for vertebrates, collagen and the polyanionic proteoglycans represent an accessible series of clearly homologous "informational" macromolecules that are well suited as subjects for investigations of the molecular evolution of connective tissue. The basic chemical structural data and the methodologies for acquiring such data are rapidly developing for these classes of molecules. Although progress in elucidating the structures of homologous glycoproteins and elastins is now extremely slow, we may reasonably expect that, with time, complete structural characterization of these molecules will be achieved.

However, molecular structural information on matrix components will be insufficient in itself to produce true understanding of evolving biological organizations. It is evident that many of the structural features of these macromolecules serve functions that are associated with intermolecular interactions and with the formation of multimolecular structures. Thus, we need to learn more of the organization of connective tissues at dimensional levels higher than the molecular one as well as of structure-functional relationships between hierarchical levels. In the words of MAYR (1964): "Evolution of higher organisms must be studied both as an elaboration of systems and through the analysis of the elementary units."

Determination of the primary covalent structures of the structural proteins and protein moieties of complex proteins can yield data related directly to variation in the structural genes. Assuredly, this course has great intrinsic significance. However, it may yield fewer rewards than investigation of genetic control mechanisms and of biosynthetic enzymes, particularly in vertebrates, and especially for collagens and proteoglycans. The universal distribution of the homologous compounds of both classes in tissue-specific patterns in all vertebrate classes suggests that primitive vertebrates possessed the complete genetic apparatus for controlled synthesis of these connective tissue macromolecules.

For example, structural polymorphism at the structural gene level is evident for collagen, is likely for structural glycoprotein, but has yet to be clearly established for proteoglycans and elastin. However, even in the case of collagen, which is the best known component of the group, one cannot readily choose between two alternate explanations: the types of collagens now known to be present in higher vertebrates either arose by gene duplication during vertebrate evolution or were initially present in the earliest vertebrates. In view of the origins of the protein in invertebrates and the probable early appearance of polymorphism in animal evolution, the latter explanation appears to be the more plausible hypothesis. This position is reinforced when one takes into account the probable equal antiquity of the functional association of collagen with structural glycoproteins and proteoglycans in the formation of polymorphic, higher-order structures.

Although evolution of the structural genes that determine the primary polypeptide structures of extracellular macromoles has been extensive, amino acid sequences and charged-group patterns have been strongly retained. This conservative aspect of protein evolution is best documented in collagen molecules, which have rigid requirements for formation of the triple-helix conformation and for ordered self-assembly into fibrils, fibers, and fibril networks in both invertebrates and vertebrates. Structural conservatism is also characteristic of the proteoglycans in vertebrates. Since the biological functions of both classes of molecules are expressed primarily through mutual interactions within connective tissue matrices, one might expect the phylogenetic histories of these molecules to possess a certain degree of interdependence. However, such types of genetic relationship have yet to be demonstrated for collagen and proteoglycans, although they are likely to exist for all matrix components, including glycoproteins and elastin.

Thus, the available evidence suggests that evolution of connective tissue in the metazoa, and more particularly in the vertebrates, might be considered principally in terms of changes in regulatory systems. Operating at the molecular level, such systems would influence rates of synthesis of polypeptide chains and of their secondary chemical modifications. They would also control the rates of synthesis of glycan chains having different monosaccharide residues, the lengths of glycan chains, and the introduction of sulfate groups.

For example, one may consider how phylogenetic change in regulation accounts for the variations in polyanionic glycan biochemistry of cartilage among vertebrates. Within the class Cyclostomata, which is most representative of the earliest members of the vertebrates, the hagfish appear to have retained a primitive state reflecting an invertebrate ancestry. Specialization in the other group of cyclostomes, the lamprey, led to segregation of the parent types of chondroitin 6-sulfate and chondroitin 4-sulfate within cartilage and notochord, respectively. Activities transferring sulfate residues to positions other than the 4 or 6 positions were very greatly reduced. However, the capacity to introduce sulfate groups into uronic acid residues was maintained in the chondrichthyes, although greatly decreased in those members permanently adapted to fresh water.

The uronic acid-sulfating activity is effectively repressed in all other extant vertebrates with the exception of the coelacanth and fish with acellular bone. In these instances, as in the case of the chondrichthyes, one cannot clearly distinguish between retention of an ancestral capacity by direct descent and the reemergence of a primitive enzyme activity that had been repressed in immediate ancestral forms. However, this equivocation seems to be avoidable for mammals (and probably for their tetrapod ancestors also) where this sulfation activity is effectively repressed in cartilage of young adults.

Additional problems are introduced by the aging process. The appearance of products that indicate low levels of enzyme activities for uronic acid sulfation and for introduction of a second sulfate group into galactosamine residues of chondroitin sulfate in old human cartilage suggests that biosynthetic control mechanisms may change in senescence.

Evolutionary modification of regulatory systems accounts also for the species differences in biochemical patterns of embryonic development of cartilage. The changes that have occurred during phylogeny appear to be relatively moderate, in accord with the 'primitive' character of cartilage and the generally conservative nature of developmental sequences in vertebrates. For example, the principal alterations that have occurred in developmental patterns of proteoglycans appear to involve the glycan moieties; the characteristic architecture of the entire macromolecule has been strongly retained. Furthermore, only modest changes in developmental patterns of glycan chemistry have been made during vertebrate evolution, concerning primarily the temporal sequences of relative activities of glycosyltransferases and sulfotransferases. The broad trend in developmental patterns with evolution has been toward increased sulfation of glycan chains, with a shift from sulfation of C-6 positions toward sulfation of C-4 positions in galactosamine residues. Amino acid sequences have undoubtedly changed also. However, it is likely that such alterations in proteoglycan, and in cartilage-specific types of collagen as well, will prove to be minor ones.

Mutation of control mechanisms may be responsible for many inherited disorders of connective tissue that show disturbances of biosynthesis of proteoglycans or glycoproteins or which show deviations from normal in the secondary and noncoded structural aspects of collagen and elastin molecules. Special kinds of deviation from normal control seem to occur in tumors that show distinctive patterns of biosynthesis of proteoglycans and collagen.

Regulation of biosynthesis of extracellular molecules implies regulation also of the organization of cells and matrix components. Indeed, the high degree of histogenic plasticity that is characteristic of connective tissues derives largely from the capacity for fine modulation of the structures of matrix macromolecules. One cannot fail to note that the range of variation of structure consistent with retention of principal biological function is very much greater for these molecules than for enzymes, for example. It is principally through secondary chemical modification of primary direct-coded structures that a very great number of different molecular forms is achieved by each organism. The resultant potential for molecular interaction in a variety of different ways confers upon each class of structural macromolecule an impressive versatility in morphogenesis.

The expression of this 'polymorphogenic' capacity is subject not only to inherited determinants but also to a variety of extrinsic factors, including physical forces. Responsiveness to environmental influences appears to be an essential property of supporting tissues. It is an important element in the phylogenesis of structures, in embryonic developmental mechanisms, and in mechanisms of physiological adjustment during maturation, aging, and pathological conditions.

Addendum

Chapter 2

Chemical analogies among structural components of plants and animal were noted for the mannan-proteins of yeast cell walls (BALLOU and RASCHKE, 1974) and animal connective tissue proteoglycans. The probable conversion of D-glucuronic acid to L-iduronic acid at the polymer level of heparin (Chap. 6) is paralleled by the enzymatic conversion of polymannuronic acid to a mixed polymer containing guluronic acid residues in brown algae (MADGWICK et al., 1973). The chitin system has been reviewed by RUDALL and KENCHINGTON (1973).

Chapter 3

It is probable that a five-stranded microfibril is the basis for the ordered substructure of collagen fibrils (DOYLE et al., 1974; PIEZ, in press). The patterns of bands in the reconstituted native-type-fibril and in the long spacing forms of calf-skin collagen match each other and correspond with the distribution of charged residues in the known α1 chain sequence (CHAPMAN, 1974).

NOWACK and NORDWIG (1974) have verified previous suggestions that sea anemone collagen is composed of three identical α chains.

The fibrils of the inner layers of the egg capsule wall of the shark, *Scyliorhinus caniculus*, consist of collagen in close association with other proteins that have a high tyrosine content. They show a unique fine structure by electron microscopy (KNIGHT and HUNT, 1974). The collagens of sturgeon cartilage and notochord (see Sect. 10c; MILLER and MATHEWS, 1974) and frog, tadpole, and newt skin (MAILMAN et al., 1974) have been characterized. Chick lens and vitreous collagen appear to be mainly $(\alpha 1)_3$, possibly type III (TRELSTAD and KANG, 1974). The CNBr peptides from human α1 (III) have a unique amino acid composition. The two cysteinyl residues of the chain which provide for interchain disulfide bonding occur in a collagen-like sequence near the COOH-terminus (CHUNG et al., 1974).

Collagen biosynthesis has been extensively reviewed (BORNSTEIN, 1974; SCHOFIELD and PROCKOP, 1973). The synthesis of collagen in nonfibroblastic cells was confirmed by use of cloned cell lines (LANGNESS and UDENFRIEND, 1974).

An evolutionary interpretation of comparative structures of collagen emphasizes the necessity for considering supermolecular, tissue organization, and species ecology (BAIRATI, 1972). Cuticulin from *A. lumbricoides* contains about 75 mole % of proline and yields polypeptide fragments that form polyproline II-type helices in solution (KANAYA and FUJIMOTO, 1973). A collagen-like sequence of about 78 residues is present in a polypeptide chain of Cl_q of human complement. It contains hydroxylysine that is in part glycosylated (REID, 1974).

Evidence for a high degree of homology of α1 (I) chains is now supported by extensive amino acid sequence data. Thus, only five inter-species amino acid differences were noted between the 149 residue sequences of α1-CB3 peptides from rat and calf (BUTLER et al., 1974), whereas a sequence identity level of only 80% was noted for the first 75 residues of this sequence when the α1 (I) chain was compared with the α1 (II) chain from calf (BUTLER et al., 1974). Compared to sequences such as α1-CB3 from the helical portion of the collagen molecule, the C-terminal, nonhelical sequences from chicken, rabbit, man, rat, and calf possess homology to a lesser degree (RAUTERBERG, 1973). The α2-CB2 peptides derived from the internal helical region of collagen from skin of calf-rabbit, pig, and man show a comparative high degree of sequence variability (FIETZEK et al., 1974).

With increasing embryonic development type III collagen is largely replaced by type I collagen in human skin (EPSTEIN, Jr., 1974) and type I collagen is replaced by type II collagen in chick cartilage (SEYER and VINSON, 1974).

Chapter 4

Additional data on amino acid sequences in tryptic peptides of tropoelastin led GRAY et al. (1973) to propose a new model for elastin, which, like the liquid drop model, emphasizes hydrocarbon-water interactions, but views the monomeric units as fibrillar rather than globular. The major crosslinks, desmosine and isodesmosine, link only two chains. The principal site of hydration is the peptide chain rather than amino acid side chains. This 'oiled coil' structure is a broad coil in which the glycines occupy exterior positions exposed to solvent while the hydrophobic residues of other amino acids are buried. HOEVE and FLORY (1974) find this model unacceptable. They interpret the elastic properties of elastin as indicating a network structure of random chains like that in a typical rubber. The branched fibers of alkali-purified elastin reveal at high resolution arrays of slender (35-40 Å diameter) filaments arranged rougly parallel to the fiber axis (GOTTE et al., 1974).

Chapter 5

Glycoproteins from bovine aorta (RADHAKRISHNAMURTHY and BERENSON, 1973), chick allantoic fluid (CHOI and ULEGER, 1974), and from chicken egg shell membranes and oviduct (PICARD et al., 1973) have been characterized further.

Chapter 6

Skeletal keratansulfate contains the following non-repeating residues per chain: one mannose, three N-acetylneuraminic acid, one N-acetylgalactosamine, and one or less fucose and is bound to protein by two types of linkage, one of which is an alkali-sensitive glycosidic bond between N-acetylgalactosamine and threonine (HOPWOOD and ROBINSON, 1974a, b). The properties of heparin were reviewed by EHRLICH and STIVALA (1973). The process of formation of iduronic acid residues from glucuronic acid residues at the polymer level in heparin is linked to sulfation of hydroxyl groups (HÖÖK, LINDAHL, and IVERIUS, 1974).

The interaction of polymeric hyaluronate with a specific region of the polypeptide chain (HEINEGÅRD and HASCALL, 1974) in proteoglycan subunits is competitively inhibited by oligosaccharides of hyaluronate not smaller in size than decasaccharide (HASCALL and HEINEGÅRD, 1974). Most of the hyaluronate of pig cartilage (about 0.7% of the total uronic acid) is associated with proteoglycan aggregates (HARDINGHAM and MUIR, 1974). EISENSTEIN et al. (1973) suggest that epiphyseal cartilage contains two pools of proteoglycan one of which may be covalently bound to collagen. The nutritionally-essential element silicon may act as crosslinking agent for glycosaminoglycans in connective tissue (SCHWARZ, 1974; CARLISLE, 1974).

Chondroitin sulfate disaccharide periods that are disulfated on the galactosamine residue were reported to constitute a surprisingly high proportion (about 20%) of the total glycosaminoglycan disaccharide periods of rabbit articular cartilage, or produced by chondrocytes in culture (SRIVASTAVA et al., 1974). SENO et al. (1974) have determined the structure of a novel disaccharide from glycosaminoglycans of cartilage of the king crab, *Tachypleus tridentatus*, a species belonging to the arthropods. This disaccharide is a variant of diS_D (see Table 7.1) and has an ester sulfate residue on C-4 of the amino sugar. The suggested abbreviation is diS_K.

Chapter 8

Cultures of cartilage of six week old human embryos incorporate ester sulfate primarily into the C-6 position compared to the C-4 position of chondroitin sulfate. A significant amount of sulfate that appears in heparan sulfate at six weeks is greatly reduced by eight weeks (LASH et al., 1974).

Chapter 9

The distribution of heparin in vertebrate tissues was discussed by EHRLICH and STIVALA (1973). The carbohydrates of molluscan tissues (GOUDSMIT, 1972) and the biochemistry and structure of the molluscan shell (GREGOIRE, 1972) were reviewed. The pedal gland of a marine slug secretes a high molecular weight sulfated proteoglycan (TRENCH, 1973). The glycosaminoglycans of bovine leucocytes consist of 75-80% chondroitin 4-sulfate, 10-15% hyaluronate, and 10% heparan sulfate (MURATA et al., 1974). Sulfated glycosaminoglycans of various embryonic mouse tissues include undersulfated chondroitin sulfate and heparan sulfate (SHAPIRO and SHERMAN, 1974). The glycosaminoglycans of two solid rat tumors consist of a mixture of hyaluronate, heparan sulfate, dermatan sulfate, and chondroitin 4-sulfate. The major component was heparan sulfate (70%) in one tumor and hyaluronate (57%) in the other tumor (KURODA et al., 1974).

Chapter 10

ARNOTT et al. (in press) have now shown that all the ordered hyaluronate systems studied by X-ray diffraction consist of single helical chains. Bovine cartilage proteoglycan shows a conformational transition in quanidinium hydrochloride solutions (PASTERNACK et al., 1974). Interactions of cationic polypeptides with bovine cartilage proteoglycan and with derived fragments produce conformational changes in the polypeptide backbone, possibly dependent upon the distribution of glycan chains (GELMAN et al., 1974). Differences in interactions of acid glycans and proteoglycans with cationic polypeptides are probably paralleled by differences in their interactions with collagen in tissues (GELMAN and BLACKWELL, 1974) and may control collagen fibril diameter and length (see PIEZ, in press). The collagenous filaments of horny sponges have a complex 3-layered organization in which microfibrils of collagen of about 60 Å diameter are prominent (GARRONE et al., 1973).

Collagen has an important role in the process of homeostasis in extracellular tissues (DAVISON, 1973) and especially in corneal morphogenesis (HAY, 1973; MEIER and HAY, 1974). BASSETT (1971) has given a comprehensive discussion of the effect of force on skeletal tissues. A direct effect of cyclical stress on biosynthesis of collagen, elastin, and glycosaminoglycans by smooth muscle cells in culture was shown (LEUNG et al., in press). Heparan sulfate of high molecular weight (135,000) may regulate cell-environment interactions (KRAEMER and SMITH, 1974). Two species of proteoglycan are produced in differing proportions by chondrocytes at different stages of differentiation (KUMATA et al., 1974).

References

AARONSON, S. (197o): Molecular evidence for evolution in the algae: A possible affinity between plant cell walls and animal skeletons. Ann. N. Y. Acad. Sci. 175, 531-54o.

ABELSON, P.H. (1956): Paleobiochemistry. Sci. Am. 195, 83-9o.

AGUILAR, J.H., JACOBS, H.G., BUTLER, W.T., CUNNINGHAM, L.W. (1973): The distribution of carbohydrate groups in rat skin collagen. J. Biol. Chem. 248, 51o6-5113.

AKESON, W.H., WOO, S.L.Y., AMIEL, D., COUTTS, R.D., DANIEL, D. (1973): Connective tissue response to immobility. Biochemical changes in periarticular connective tissue of the immobilized rabbit knee. Clin. Orthop. Relat. Res. 93, 356-362.

AKIYAMA, F., SENO, N., ANNO, K. (1971): Mucopolysaccharides from skin of hagfish. Presented at the 44th General Meeting of the Japanese Biochemical Soc., Oct. 4, 1971, Sendai. Seikagaku 43, 554.

ALCOCK, N.W. (1972): Calcification of cartilage. Clin. Orthop. Relat. Res. 86, 287-311.

ALLISON, W.S., KAPLAN, N.O. (1964): The comparative enzymology of triosephosphate dehydrogenase. J. Biol. Chem. 239, 214o-2152.

ALLEN, A.K., NEUBERGER, A. (1973): The purification and properties of the lectin from potato tubers, a hydroxyproline-containing glycoprotein. Biochem. J. 135, 3o7-314.

ALSTON, R.E. (1967): Biochemical systematics. Evol. Biol. 1, 197-3o5.

ALSTON, R.E., TURNER, B.L. (1963): Biochemical Systematics. Englewood Cliffs, N.J.: Prentice Hall.

AMPRINO, R. (1938): Studi sul significato meccanico delle fibrille collagene della cartilagine jalina dell'uomo. Z. Zellforsch. 28, 734-76o.

AMPRINO, R., BAIRATI, A. (1933): Studi sulle trasformazioni delle cartilagini dell' uomo nell'accrescimento e nella senescenza. I. Cartilagini jalini. Z. Zellforsch. 2o, 134-2o5.

AMPRINO, R., BAIRATI, A. (1934a): Studi sulle trasformazioni delle cartilagini dell' uomo nell'accrescimento e nella senescenza. II. Cartilagini elastiche. Z. Zellforsch. 2o, 489-522.

AMPRINO, R., BAIRATI, A. (1934b): Studi sulle trasformazioni delle cartilagini dell' uomo nell'accrescimento e senescenza. III. Cartilagini fibrose. Z. Zellforsch. 21, 448-482.

ANDERSEN, S.O. (1964): The cross-links in resilin identified as dityrosine and trityrosine. Biochim. Biophys. Acta 93, 213-215.

ANDERSEN, S.O. (1971): Resilin. In: Comprehensive Biochemistry (eds. M. FLORKIN, E.H. STOTZ), Vol. 26 C, pp. 633-567. Amsterdam: Elsevier.

ANDERSON, B.S., ODELL, T.T. (196o): Changes in rat cartilage mucopolysaccharide with age and radiation. J. Gerontol. 15, 249-252.

ANDERSON, C.E., LUDOWIEG, J., EYRING, E.J., HORWITZ, B. (1963): Ultrastructure and chemical composition of chondrosarcoma. J. Bone Joint Surg. 45-A, 753-764.

ANDERSON, H.C., SAJDERA, S.W. (1971): The fine structure of bovine nasal cartilage. J. Cell Biol. 49, 65o-663.

ANDERSON, J.C., JACKSON, D.S. (1972): Isolation of glycoproteins from bovine achilles tendon and their interaction with collagen. Biochem. J. 127, 179-186.

ANDREW, W. (1959): Textbook of Comparative Histology, p. 32. Oxford: University Press.

ANFINSON, C.B. (1959): The Molecular Basis of Evolution. New York: Wiley.

ANNO, K., KAWAI, Y., SENO, N. (1964): Isolation of chondroitin from squid skin. Biochim. Biophys. Acta 83, 348-349.

ANNO, K., SENO, N., MATHEWS, M.B., YAMAGATA, T., SUZUKI, S. (1971): A new dermatan polysulfate, chondroitin sulfate H, from hagfish notochord. Biochim. Biophys. Acta 237, 173-177.

ANSETH, A. (1961a): Studies on corneal polysaccharides. III. Topographic and comparative biochemistry. Exp. Eye Res. 1, 1o6-115.

ANSETH, A. (1961b): Glycosaminoglycans in the developing corneal stroma. Exp. Eye Res. 1, 116-121.

ANSETH, A., FRANSSON, L.-A. (197o): The importance of the polysaccharides for the normal function of the corneal stroma. Acta Ophthal. 48, 455-464.

ANSETH, A., LAURENT, T.C. (1962): Polysaccharides in normal and pathologic corneas. Invest. Ophthal. 1, 195-2o1.

ANTONOPOULOS, C.A., ENGFELDT, B., GARDELL, S., HJERTQUIST, S.-O., SOLHEIM, K. (1965): Isolation and identification of the glycosaminoglycans from fracture callus. Biochim. Biophys. Acta 1o1, 15o-156.

ARAMBOURG, C. (1958a): Classe des aconthodieus. Classe des placodermes. In: Traité de Zoologie (ed. P.-P. GRASSÉ), Vol. 13, pp. 1984-2oo9. Paris: Masson.

ARAMBOURG, C. (1958b): Sous-classe des crossoptérygiens. Formes fossile. In: Traité de Zoologie (ed. P.-P. GRASSÉ), Vol. 13, pp. 2541-2552. Paris: Masson.

ARAMBOURG, C., BERTIN (1958a): Classe des chondrichthyens. In: Traité de Zoologie (ed. P.-P. GRASSÉ), Vol. 13, pp. 2o1o-2o67. Paris: Masson.

ARAMBOURG, C., BERTIN (1958b): Super-ordres des holostéens et des halécostomes. In: Traité de Zoologie (ed. P.-P. GRASSÉ), Vol. 13, pp 2173-22o3. Paris: Masson.

ARAMBOURG, C., GUIBE, J. (1958): Sous-classe de dipneustes. In: Traité de Zoologie (ed. P.-P. GRASSÉ), Vol. 13, pp 2522-254o. Paris: Masson.

ARMITAGE, P.M., CHAPMAN, J.A. (1971): New fibrous long spacing form of collagen. Nature New Biol. 229, 151-152.

ARMSTRONG, W.G., TARLO, L.B.H. (1966): Amino-acid components in fossil calcified tissues. Nature 21o, 481-482.

ARNOTT, S., GUSS, J.M., HUKINS, D.W.L., MATHEWS, M.B. (1973a): Mucopolysaccharides: comparison of chondroitin sulfate conformations with those of related polyanions. Science 18o, 743-745.

ARNOTT, S., GUSS, J.M., HUKINS, D.W.L., MATHEWS, M.B. (1973b): Dermatan sulfate and chondroitin 6-sulfate conformations. Biochem. Biophys. Res. Commun. 54, 1377-1383.

ASHHURST, D.E. (1968): The connective tissue of insects. Ann. Rev. Entomol. 13, 45-74.

ASHURST, D.E., COSTIN, N.M. (1971a): Insect mucosubstances. I. The mucosubstances of developing connective tissue in the locust, *Locusta migratoria*. Histochem. J. 3, 279-295.

ASHURST, D.E., COSTIN, N.M. (1971b): Insect mucosubstances. II. The mucosubstances of the central nervous system. Histochem. J. 3, 297-31o.

ASHURST, D.E., COSTIN, N.M. (1971c): Insect mucosubstances. III. Some mucosubstances of the nervous system of the wax-moth *(Galleria mellonella)* and the stick insect *(Carausius morosus)*. Histochem. J. 3, 379-387.

ASTBURY, W.T. (1938): X-ray adventures among the proteins. Trans. Faraday Soc. 34, 377-388.

ATKINS, E.D.T. (1972): Molecular shape of connective tissue mucopolysaccharides as a function of aging. Scand. J. Clin. Lab. Invest. 29, Suppl. 123,5.

ATKINS, E.D.T., GAUSSEN, R., ISAAC, D.H., NANDANWAR, V., SHEEHAN, J.K. (1972): X-ray fiber diffraction patterns from chondrotin-6-sulfate. J. Polym. Sci., pt B., 1o, 863-865.

ATKINS, E.D.T., ISAAC, D.H. (1973): X-ray diffraction studies on the connective tissue polysaccharides. Molecular conformations of dermatan sulfate. J. Mol. Biol. 8o, 773-779.

ATKINS, E.D.T., LAURENT, T.C. (1973): X-ray diffraction patterns from chondroitin 4-sulphate, dermatan sulphate and heparan sulphate. Biochem. J. 133, 6o5-6o6.

ATKINS, E.D.T., PHELPS, C.F., SHEEHAN, J.K. (1972): The conformation of the mucopolysaccharides. Biochem. J. 128, 1255-1263.

ATKINS, E.D.T., SHEEHAN, J.K. (1973): Hyaluronates: relation between molecular conformations. Science 179, 562-564.

ATTWOOD, M.M., ZOLA, H. (1967): The association between chitin and protein in some chitinous tissues. Comp. Biochem. Physiol. 2o, 993-998.
AYER, J.P. (1964): Elastic tissue. Int. Rev. Conn. Tissue Res. 2, 33-1oo.
BACCETTI, B. (1967): Collagen of the earthworms. J. Cell Biol. 34, 885-891.
BACH, G., BERMAN, E.R. (197o): Characterization of a sialoglycan isolated from cattle retina. Ophthalmic Res. 1, 257-258.
BACHRA, B.N. (197o): Calcification of connective tissue. Int. Rev. Conn. Tissue Res. 5, 165-2o8.
BAILEY, A.J. (1968): The nature of collagen. In: Comprehensive Biochemistry (eds. M. FLORKIN, E.H. STOTZ), Vol. 26 B, Extracellular and Supporting Structures, pp. 297-423. New York: Elsevier.
BAILEY, A.J. (197o): Comparative studies on the nature of the cross-links in the collagen of various fish tissues. Biochim. Biophys. Acta 221, 652-656.
BAILEY, A.J. (1971): Comparative studies on the nature of the cross-links stabilizing the collagen fibres of invertebrates, cyclostomes and elsmobranchs. FEBS Letters 18, 154-158.
BAILEY, A.J., SHIMAKOKOMAKI, M.S. (1971): Age related changes in the reducible crosslinks of collagen. FEBS Letters 16, 86-88.
BAKER, J.R., CIFONELLI, J.A., MATHEWS, M.B., RODÉN, L. (1969): Mannose-containing glycopeptides from keratosulfate. Federation Proc. 28, 6o5.
BAKER, J.R., JOHNSON, A.H. (1973): Digestion of chondroitin sulfate proteoglycan by trypsin and chymotrypsin. Federation Proc. 32, 676 abs.
BAKER, J.R., RODÉN, L., STOOLMILLER, A.C. (1972): Biosynthesis of chondroitin sulfate proteoglycan. Xylosyl transfer to Smith-degraded cartilage proteoglycan and other exogenous acceptors. J. Biol. Chem. 247, 3838-3847.
BALAZS, E.A. (1958): The physical chemistry of hyaluronic acid. Federation Proc. 17, 1o86-1o93.
BALAZS, E.A. (1965): Amino sugar-containing macromolecules in the tissues of the eye and ear. In: The Amino Sugars (eds. E.A. BALAZS, R.W. JEANLOZ), Vol. 2A, pp. 4o1-46o. New York: Academic Press.
BALAZS, E.A. (1969): Some aspects of the aging and radiation sensitivity of the intercellular matrix with special regard to hyaluronic acid in synovial fluid and vitreous. In: Aging of Connective and Skeletal Tissue (eds. A. ENGEL, T. LARSSON), pp. 1o7-134. Stockholm: Nordiska Bokhandelus Förlag.
BALAZS, E.A. (197o): Preface. In: Chemistry and Molecular Biology of the Intercellular Matrix (ed. E.A. BALAZS), Vol. 1, pp. xxiii-xxv. New York: Academic Press.
BALAZS, E.A., BLOOM, G.D., SWANN, D.A. (1966): Fine structure and glycosaminoglycan content of the surface layer of articular cartilage. Federation Proc. 25, 1813-1816.
BALAZS, E.A., GIBBS, D.A. (197o): Rheological properties and biological function of hyaluronic acid. In: Chemistry and Molecular Biology of the Intercellular Matrix (ed. E.A. BALAZS), pp. 1241-1253. New York: Academic Press.
BALAZS, E.A., JACOBSON, B. (1966): Interaction of amino sugar and amino sugar-containing macromolecules with viruses, cells, and tissues. In: The Amino Sugars (eds. E.A. BALAZS, R.W. JEANLOZ), Vol. 2B, pp. 361-395. New York: Academic Press.
BALAZS, E.A., LAURENT, T.C., LAURENT, V.B.G., De ROCHE, M.H., BUNNEY, D.M. (1959): Studies on the structure of the vitreous body. VIII. Comparative biochemistry. Arch. Biochem. Biophys. 81, 464-479.
BALAZS, E.A., ROGERS, H.J. (1965): The amino sugar-containing compounds in bones and teeth. In: The Amino Sugars (eds. E.A. BALAZS, R.W. JEANLOZ), Vol. 2A, pp. 281-3o9. New York: Academic Press.
BALDUINI, C., BROVELLI, A., CASTELLANI, A.A. (197o): Biosynthesis of glycosaminoglycans in bovine cornea. The effect of uridine diphosphate xylose. Biochem. J. 12o, 719-723.
BALDUINI, C., PALLAVICINI, G., CASTELLANI, A.A. (197o): Identification of the hexosamine linked to hydroxyamino acid in the protein-polysaccharide linkages. Ital. J. Biochem. 19, 253-261.
BALDWIN, E. (1948): An Introduction to Comparative Biochemistry. Cambridge: University Press.

BALIAN, G., CLICK, E.M., HERMODSON, M.A., BORNSTEIN, P. (1972): Structure of rat skin collagen α1-CB8. Amino acid sequence of the hydroxylamine-produced fragment HA2. Biochemistry 11, 3798-38o6.

BANGA, I. (1966): Structure and Function of Elastin and Collagen. Budapest: Akadémiai Kiadó.

BANGA, I. (1969): The structure of protein-polysaccharides and their role in the function of connective tissue fibres. Acta Physiol. Acad. Sci. Hung. 35, 355-363.

BANNISTER, D.W., BURNS, A.B. (1972): Pepsin treatment of avian skin collagen. Effects on solubility, subunit composition and aggregation properties. Biochem. J. 129, 677-681.

BARBER, A.J., KAESER-GLANZMANN, R., JAKABOVA, M., LUESCHER, E.F. (1972): Characterization of a chondroitin 4-sulfate proteoglycan carrier for heparin neutralizing activity (platelet factor 4) released from human blood platelets. Biochim. Biophys. Acta 286, 312-329.

BARDACK, D., ZANGERL, R. (1971): Lampreys in the fossil record. In: The Biology of Lampreys (eds. M.W. HARDISTY, I.C. POTTER), Vol. 1, pp. 67-84. New York: Academic Press.

BARDELMEYER, G.H. (1973): Electrical conduction in collagen. II. Some aspects of hydration. Biopolymers 12, 23o3-23o7.

BARKER, S.A. (1963): Polysaccharides of bacteria, moulds, yeasts, and protozoa. In: Comprehensive Biochemistry (eds. M. FLORKIN, E.H. STOTZ), Vol. 5, pp. 246-261. New York: Elsevier.

BARKER, S.A., CRUIKSHANK, C.N.D., WEBB, T. (1965): Mucopolysaccharides in rat skin. Part 1. Isolation and identification. Carbohyd. Res. 1, 52-61.

BARKER, S.A., KENNEDY, J.F., CRUIKSHANK, C.N.D. (1969): Identification of acidic mucopolysaccharides from human skin by radioactive precursors. Carbohyd. Res. 1o, 65-7o.

BARKER, S.A., KENNEDY, J.F., SOMERS, P.J. (1969): Acidic mucopolysaccharides of human skin. Carbohyd. Res. 1o, 57-63.

BARLOW, G.H., COEN, L.J., MOZEN, M.M. (1964): A biological, chemical, and physical comparison of heparin from different mammalian species. Biochim. Biophys. Acta 83, 272-277.

BARNES, M.J., CONSTABLE, B.J., MORTON, L.F., KODICEK, E. (1971): Hydroxylysine in the N-terminal regions of the α1- and α2-chains of various collagens. Biochem. J. 125, 433-437.

BARNES, M.J., PARTRIDGE, S.M. (1968): Isolation and characterization of a glycoprotein from human thoracic aorta. Biochem. J. 1o9, 883-896.

BARRETT, A.J. (1968): Cartilage. In: Comprehensive Biochemistry (eds. M. FLORKIN, E.H. STOTZ), Vol. 26B, pp. 425-474. Amsterdam: Elsevier.

BARRINGTON, E.J.W. (1965): The Biology of Hemichordata and Protochordata. San Francisco: Freeman.

BARTOS, F., LEDVINA, M. (1971): Age changes in cross-links in rat skin elastin. Arch. Dermatol. Forsch. 241, 134-14o.

BASSETT, C.A.L. (1971): Biophysical principles affecting bone structure. In: Biochemistry and Physiology of Bone, 2nd Ed. (ed. G.H. BOURNE), Vol. 3, pp. 1-76. New York: Academic Press.

BASSETT, C.A. (1964): Environmental and cellular factors regulating osteogenesis. In: Bone Biodynamics (ed. H.M. FROST), pp. 233-234. Boston: Little, Brown.

BASSETT, C.A.L., HERRMANN, I. (1961): Influence of oxygen concentration and mechanical factors on differentiation of connective tissues in vitro. Nature 19o, 46o-461.

BASSETT, C.A.L., PAWLEK, R.J. (1972): Electrical behavior of cartilage during loading. Science 178, 982-983.

BAUER, C. (1934): Über die Beziehungen zwischen Zelle und Interzellularsubstanz in embryonalem Bindegewebe und die Lehre von den mesenchymalen Keimlagen im erwachsenen Organismus. Z. mik. Anat. Forsch. 35, 362-424.

BAXTER, E., MUIR, H. (1972): The antigenicity of cartilage proteoglycans: the relationship of the antigenic determinants present in Smith degraded and intact proteoglycans. Biochim. Biophys. Acta 279, 276-281.

BEAR, R.S. (1952): The structure of collagen fibrils. Adv. Prot. Chem. 7, 69-16o.

BECKER, U., TIMPL, R., KÜHN, K. (1972): Carboxyterminal antigenic determinants of collagen from calf skin. Localization within discrete regions of the nonhelical sequence. Eur. J. Biochem. 28, 221-231.

BECKS, H., EVANS, H.M. (1953): Atlas of the Skeletal Development of the Rat. Berkeley: Am. Institute of Dental Medicine.

De BEER, G. (1958): Embryos and Ancestors, 3rd Ed. London: Oxford University Press.

BELLAMY, D., JONES, I.C. (1961): Studies on *Myxine glutinosa*. I. The chemical composition of the tissues. Comp. Biochem. Physiol. 3, 175-183.

BENDITT, E.P., ERIKSEN, N. (1972): Chemical similarity among amyloid substances associated with long standing inflammation. Lab. Invest. 26, 615-625.

BEN-HAYYIM, G., OHAD, I. (1965): Synthesis of cellulose by *Acetobacter xylinum*. VIII. On the formation and orientation of bacterial cellulose fibrils in the presence of acidic polysaccharides. J. Cell Biol. 25(2), 191-2o7.

BENMAMAN, J.D., LUDOWIEG, J.J., ANDERSON, C.E. (1969): Glucosamine and galactosamine distribution in human articular cartilage: relationship to age and degenerative joint disease. Clin. Biochem. 2, 461-464.

BENNETT, M.V.L. (1971a): Electroreception. In: Fish Physiology (eds. W.S. HOAR, D.J. RANDALL), Vol. 5, pp. 493-574. New York: Academic Press.

BENNETT, M.V.L. (1971b): Electric organs. In: Fish Physiology (eds. W.S. HOAR, D.J. RANDALL), Vol. 5, pp. 347-491. New York: Academic Press.

BENNINGHOFF, A. (1925): Form und Bau der Gelenkknorpel in ihren Beziehungen zur Funktion. Z. ges. Anat. 76, 43-63.

BENSUSAN, H.B., HOYT, B.L. (1958): The effect of various parameters on the rate of formation of fibers from collagen solutions. J. Am. Chem. Soc. 8o, 719-724.

BENTLEY, J.P. (1966): Adaptations of amphibia to arid environments. Science 152, 619-623.

BENTLEY, J.P. (1969): Mycopolysaccharide synthesis in healing wounds. In: Repair and Regeneration (eds. J.E. DUNPHY, W. VAN WINKLE), pp. 151-16o. New York: McGraw-Hill.

BERENSON, G.S., RADHAKRISHNAMURTHY, B., DALFERES, Jr., E.R., SRINIVASAN, S.R. (1971): Carbohydrate macromolecules and atherosclerosis. Human Path. 2, 57-79.

BERENSON, G.S., RADHAKRISHNAMURTHY, B., FISKIN, H.F., DESSAUER, H., ARQUEMBOURG, P. (1966): Individuality of glycoproteins in human arota. J. Atherosclerosis Res. 6, 214-223.

BERG, R.A., PROCKOP, D.J. (1973): The thermal transition of a non-hydroxylated form of collagen. Evidence for a role for hydroxyproline in stabilizing the triple-helix of collagen. Biochem. Biophys. Res. Commun. 52, 115-12o.

BERMAN, E.R. (1969): Mucopolysaccharides (glycosaminoglycans) of the retina: identification, distribution and possible biological role. Bibl. Ophthalmol. No. 79, 5-31.

BERMAN, E.R. (197o): Proteoglycans of bovine corneal stroma. In: Chemistry and Biology of the Intercellular Matrix (ed. E.A. BALAZS), Vol. 2, pp. 879-886. New York: Academic Press.

BERMAN, E.R., BACH, G. (1968): The acid mucopolysaccharides of cattle retina. Biochem. J. 1o8, 75-88.

BERMAN, E.R., VOADEN, M. (197o): The vitreous body. In: Biochemistry of the Eye (ed. C.N. GRAYMORE), pp. 373-471. New York: Academic Press.

BERNFELD, P. (1966): Interaction of polyanions with blood components. In: The Amino Sugars (eds. E.A. BALAZS, R.W. JEANLOZ), Vol. 2B, pp. 252-266. New York: Academic Press.

BERNFIELD, M.R., BANERJEE, S.D. (1972): Acid mycopolysaccharide (glycosaminoglycan) at the epithelial-mesenchymal interface of mouse embryo salivary glands. J. Cell Biol. 52, 664-673.

BERNFIELD, M.R., BANERJEE, S.D., COHEN, R.H. (1972): Dependence of salivary epithelial (morphology and branching morphogenesis upon acid mucopolysaccharide protein (proteoglycan) at the epithelial surface. J. Cell Biol. 52, 674-689.

BERRIL, N.J. (1955): The Origin of the Vertebrates. London: Oxford Press.

BERTIN, L. (1958): Tissues squelettiques. In: Traité de Zoologie (ed. P.-P. GRASSÉ), Vol. 13, pp. 532-55o. Paris: Masson.

BETTELHEIM, F.A. (197o): Physical chemistry of acidic polysaccharides. In: Biological Polyelectrolytes (ed. A. VEIS), pp. 131-2o9. New York: Dekker.

BETTELHEIM, F.A., LAURENT, T.C., PERTOFT, J. (1966): Interaction between serum albumin and acidic polysaccharides. Carbohyd. Res. 2, 391-4o2.

BHATNAGER, R.S., PROCKOP, D.J. (1966): Dissociation of the synthesis of sulfated mucopolysaccharides and the synthesis of collagen in embryonic cartilage. Biochim. Biophys. Acta 13o, 383-392.

BIGELOW, H.B., SCHROEDER, W.C. (1948): Sharks. In: Fishes of the Western North Atlantic, Vol. 1 (Part 1), pp. 59-576. New Haven: Mem. Sears Found. Mar. Res., Yale Univ.

BIGELOW, H.B., SCHROEDER, W.C. (1953): Sawfishes, guitarfishes, skates and rays. In: Fishes of the Western North Atlantic, Vol. 1 (Part 2), pp. 1-514. New Haven: Mem. Sears Found. Mar. Res., Yale Univ.

BIGELOW, H.B., SCHROEDER, W.C. (1961): *Carcharhinus nicaraguensis*, a synonym of the bull shark, *C. leucas*. Copeia: 359.

BLANQUET, R., LENHOFF, H.M. (1966): A disulfide-linked collagenous protein of nematocyst capsules. Science 154, 153-154.

BLUMENKRANTZ, N., PROCKOP, D.J. (197o): Variations in the glycosylation of the collagen synthesized by chick embryo cartilage. Effects of development and several hormones. Biochim. Biophys. Acta 2o8, 461-466.

BOAS, N.F. (1965): Amino sugar-containing compounds in urine. In: The Amino Sugars (eds. E.A. BALAZS, R.W. JEANLOZ), Vol. 2A, pp. 95-114. New York: Academic Press.

BOCQUET, J., PAJOL, J.P., ROLLAND, M., BOUILLON, J., COPPOIS, G. (1972): Absence de mucopolysaccharides acides dans la mésoglée de *Rhizostoma pulmo* L. (Scyphoméduse). C.R. Acid. Sci. (Paris), Sér. D, 274, 1o1-1o4.

BOILLY, B. (197o): Présence de collagène dans le proventricole des syllidiens (annélides polychétes). Z. Zellforsch. 1o3, 265-281.

BOLLET, A.J., NANCE, J.L. (1966): Biochemical findings in normal and osteoarthritic articular cartilage. II. Chondroitin sulfate concentration and chain length, water and ash content. J. Clin. Invest. 45, 117o-1177.

BONNER, J., VARNER, J.E. (1965): Plant Biochemistry. New York: Academic Press.

BOREL, J.P., DESANTI, M., CLOGNE, M., BOCHET, C. (197o): Constitution de 6 glycoproteins associées au collagène dans le derme de lapin. C.R. Acad. Sci. (Paris) 27o, 2714-2716.

BORNSTEIN, P. (1967): Comparative sequence studies of rat and tendon collagen. I. Evidence for incomplete hydroxylation of individual prolyl residues in normal proteins. Biochemistry 6, 3o82-3o93.

BORNSTEIN, P., EHRLICH, H.P., WYKE, A.W. (1972): Procollagen: conversion of the precursor to collagen by a neutral protease. Science 175, 544-546.

BORNSTEIN, P., KANG, A.H. (197o): Comparative biochemical studies of collagen. In: Chemistry and Molecular Biology of the Intercellular Matrix. (ed. E.A. BALAZS), Vol. 1, pp. 99-1o7. New York: Academic Press.

BORNSTEIN, P., von der MARK, K., WYKE, A.W., EHRLICH, H.P., MONSON, J.M. (1972): Characterization of the pro-α1 chain of procollagen. J. Biol. Chem. 247, 28o8-2813.

BORNSTEIN, P., NESSE, R. (197o): The comparative biochemistry of collagen: the structure of rabbit skin collagen and its relevance to immunochemical studies of collagen. Arch. Biochem. Biophys. 138, 443-45o.

BORNSTEIN, P., PIEZ, K.A. (1964): A biochemical study of human skin collagen and the relation between intra- and intermolecular cross-linking. J. Clin. Invest. 43, 1813-1823.

BOSMANN, H.B. (1968): Cellular control of macromolecular synthesis: rates of synthesis of extracellular macromolecules during and after depletion by papain. Proc. Roy. Soc. B. 169, 399-425.

BOSMANN, H.B. (1971): Platelet adhesiveness and aggregation: The collagen:glycosyl, polypeptide:N-acetylgalactosaminyl and glycoprotein:galactosyltransferases of human platelets. Biochem. Biophys. Res. Commun. 43, 1118-1124.

BOSMANN, H.B., EYLAR, E.H. (1968): Attachment of carbohydrate to collagen. Isolation, purification, and properties of the glycosyl transferase. Biochem. Biophys. Res. Commun. 3o, 89-94.

BOSMANN, B., JACKSON, J.J. (1968): Glycoprotein structure: the carbohydrate of bovine corneal collagen. Biochim. Biophys. Acta 17o, 6-14.

BOSTRÖM, H. (1969): General metabolic changes in connective tissue with age. In: Aging of Connective and Skeletal Tissues (eds. A. ENGEL, T. LARSSON), pp. 2o1-212. Stockholm: Nordiska Bokhandelns Förlag.

BOUTEILLE, M., PEASE, D.C. (1971): The tridimensional structure of native collagenous fibrils, their proteinaceous filaments. J. Ultrastr. Res. 35, 314-338.

BOWES, J.H., ELLIOT, R.G., MOSS, J.A. (1958): The extraction of soluble protein from skin by alkaline solutions. In: Recent Advances in Gelatin and Glue Research (ed. J.H. BOWES), pp. 71-75. New York: Pergamon Press.

BOWNESS, J.M. (1962): Calcium binding by chondroitin sulfate associated with collagen. Biochim. Biophys. Acta 58, 134-136.

BOYD, E.S., NEUMAN, W.F. (1951): The surface chemistry of bone. V. The ion-binding properties of cartilage. J. Biol. Chem. 193, 243-251.

BOYDEN, A. (1969): Homology and analogy. Science 164, 455-456.

BRANDT, K.D., MUIR, H. (1969): Characterization of protein-polysaccharides of articular cartilage from mature and immature pigs. Biochem. J. 114, 871-876.

BRANDT, K.D., MUIR, H. (1971): Heterogeneity of protein-polysaccharides of porcine articular cartilage. Biochem. J. 123, 747-755.

BRANDT, K., TSIGANOS, C.P., MUIR, H. (197o): Immunology of pig cartilage proteoglycans. In: Chemistry and Molecular Biology of the Intercellular Matrix (ed. E.A. BALAZS), Vol. 3, pp. 1579-159o. New York: Academic Press.

BRAY, B.A., LEEBERMAN, R., MEYER, K. (1967): Structure of human skeletal keratosulfate. The linkage region. J. Biol. Chem. 242, 3373-338o.

BREEN, M., JOHNSON, R.L., SITTIG, R.A., WEINSTEIN, H.G., VEIS, A. (1972): The acidic glycosaminoglycans in human fetal development and adult life: cornea, sclera, and skin. Conn. Tissue Res. 1, 291-3o3.

BRIMACOMBE, J.S., WEBER, J.M. (1964): Mucopolysaccharides. Chemical structure, distribution and isolation. New York: Elsevier.

BROWN, F., STEIN, W.D. (196o): Balance of water, electrolytes, and nonelectrolytes. In: Comparative Biochemistry (eds. M. FLORKIN, H.S. MASON), Vol. 2, pp. 4o3-47o. New York: Academic Press.

BROWN, Jr., G.W., BROWN, S.G. (1967): Urea and its formation in coelacanth liver. Science 155, 57o-573.

BRUNS, R.R., GROSS, J. (1973): Band pattern of the segment-long-spacing form of collagen. Its use in the analysis of primary structure. Biochemistry 12, 8o8-815.

BRUNS, R.R., TRELSTAD, R.L., GROSS, J. (1973): Cartilage collagen: a staggered substructure in reconstituted fibrils. Science 181, 269-271.

BRYSK, M.M., CHRISPEELS, M.J. (1972): Synthesis and secretion of hydroxyproline-containing macromolecules in carrots. 6. Isolation and partial characterization of a hydroxyproline-rich cell wall glycoprotein and its cytoplasmic precursor. Biochim. Biophys. Acta 257, 421-432.

BRYSON, V., VOGEL, H.J. (eds.) (1965): Evolving Genes and Proteins. New York: Academic Press.

BUDDECKE, E., DRZENIEK, R. (1962): Stabilitätskonstanten der Calciumkomplexe von sauren Mucopolysacchariden. Hoppe-Seylers Z. Physiol. Chem. 327, 49-64.

BUDDECKE, E., SZIEGOLEIT, M. (1964): Isolierung, chemische Zusammensetzung und altersabhängige Verteilung von Mucopolysacchariden menschlicher Zwischenwirbelscheiben. Hoppe-Seylers Z. Physiol. Chem. 337, 66-78.

BURSON, S.L., FAHRENBACH, M.J., FROMMHAGEN, L.H., RICCARDI, B.A., BROWN, R.A., BROCKMAN, J.A., LEWRY, H.V., STOKSTAD, E.L.R. (1956): Isolation and purification of mactins, heparin-like anticoagulants from mollusca. J. Am. Chem. Soc. 78, 5874-5878.

BUTLER, W.T. (1968): Partial hydroxylation of certain lysines in collagen. Science 161, 796-798.

BUTLER, W.T. (197o): Chemical studies on the cyanogen bromide peptides of rat skin collagen. The covalent structure of α1-CB5, the major hexose-containing cyanogen bromide peptide of α1. Biochem. 9, 44-5o.

BUTLER, W.T. (1972): The structure of α1-CB3, a cyanogen bromide fragment from the central portion of the α1 chain of rat collagen. The tryptic peptides from skin and dentin collagens. Biochem. Biophys. Res. Commun. 48, 154o-1548.

BUTLER, W.T. (1973): High level of hydroxylysine in dentin collagen. Ala. J. Med. Sci. 1o, 1o3-1o6.

BUTLER, W.T., CUNNINGHAM, L.W. (1966): Evidence for the linkage of a disaccharide to hydroxylysine in tropocollagen. J. Biol. Chem. 241, 3882-3888.

CAIN, G.D. (197o): Collagen from the giant Acanthocephalan *Macrocanthorhynchus hirudinaceus*. Arch. Biochem. Biophys. 141, 264-27o.

CAIN, G.D., FAIRBAIRN, D. (1971): Protocollagen proline hydroxylase and collagen synthesis in developing eggs of *Ascaris lumbricoides*. Comp. Biochem. Physiol. 4o, 165-179.

CALATRONI, A., DI FERRANTE, N. (1969): The glycosaminoglycans of human term placenta. Carbohyd. Res. 1o, 535-548.

CALATRONI, A., DONNELLY, P.V., DI FERRANTE, N. (1969): The glycosaminoglycans of human plasma. J. Clin. Invest. 48, 332-343.

CAMPO, R.D. (197o): Protein-polysaccharides of cartilage and bone in health and disease. Clin. Orthop. Rel. Res. 68, 182-2o9.

DE CAMPOS VIDAL, B. (1963): Pleochroism in tendon and its bearing to acid mucopolysaccharides. Protoplasma 56, 529-536.

CANDLISH, J.K., HOLT, F.J. (1971): Proteoglycans of fowl cortical and medullary bone. Comp. Biochem. Physiol. B 4o, 283-293.

CARMICHAEL, D.J., DODD, C.M. (1973): An investigation of the phosphoprotein of the bovine dentin matrix. Biochim. Biophys. Acta 317, 187-192.

CASSEL, J.M. (1971): Aggregation phenomena of collagen. In: Biophysical Properties of Skin (ed. H.R. ELDEN), pp. 63-1oo. New York: Interscience.

CASTELLANI, A.A., BONFERONI, B., RONCHI, S., FERRI, G., MALCOVATI, M. (1962): Light chondromucoprotein from metaphysis cartilage of new-born pigs. Ital. J. Biochem. 11, 187-195.

CASTEX, M.N. (1967): Fresh water venomous rays. In: Animal Toxins, pp. 167-176. New York: Pergamon Press.

CASTOR, C.W., GREENE, J.A. (1968): Regional distribution of acid mucopolysaccharides in the kidney. J. Clin. Ivest. 47, 2125-2132.

CALDWELL, C.B., HENKART, P., HUMPHRIES, T. (1973): Physical properties of sponge aggregation factor. A unique proteoglycan complex. Biochemistry 12, 3o51-3o55.

CHAKRABARTI, B., BALAZS, E.A. (1973): Optical properties of hyaluronic acid. J. Mol. Biol. 78, 135-141.

CHAPMAN, D.M. (1966a): Evolution of the scyphistoma. In: The Cnidaria and Their Evolution. (ed. W.J. REES), pp. 51-117. New York: Academic Press.

CHAPMAN, D.M. (1966b): The structure and functions of the mesoglea. In: The Cnidaria and Their Evolution (ed. W.J. REES), pp. 147-168. New York: Academic Press.

CHAPMAN, G.E., CAMPBELL, I.D., McLAUCHLAN, K.A. (197o): NMR investigation of deuterated collagen. Nature 225, 639-641.

CHIPPENFIELD, A.R. (197o): Calcium ion binding by bone mucosubstances. Biochem. J. 118, 36P.

CHOI, H.U., MEYER, K., SWARM, R. (1971): Mucopolysaccharide and protein-polysaccharide of a transplantable chondrosarcoma. Proc. Natl. Acad. Sci. US 68, 877-879.

CHRISPEELS, M.J. (197o): Biosynthesis of cell wall protein: sequential hydroxylation of proline, glycosylation of hydroxyproline and secretion of the glycoprotein. Biochem. Biophys. Res. Commun. 39, 732-737.

CHUNG, E., MILLER, E.J. (1974): Collagen polymorphism: characterization of molecules with the chain composition $[\alpha 1\ (III)]_3$ in human tissues. Science 183, 12oo-12o1.

CHURCH, R.L., TANZER, M.L., PFEIFFER, S.E. (1973): Collagen and procollagen production by a clonal line of Schwann cells. Proc. Natl. Acad. Sci. US 7o, 1943-1946.

CHVAPIL, M. (1967): Physiology of Connective Tissue. London: Butterworths.

CHVAPIL, M., BOUCEK, M., EHRLICHOVA, E. (197o): Differences in the protocollagen hydroxylase activities from Ascaris muscle and hypodermis. Arch. Biochem. Biophys. 14o, 11-18.

CHVAPIL, M., HURYCH, J. (1968): Control of collagen biosynthesis. Int. Rev. Conn. Tissue Res. 4, 67-196.

CIFONELLI, J.A. (1968): Reaction of heparitin sulfate with nitrous acid. Carbohyd. Res. 8, 233-242.

CIFONELLI, J.A. (197o): Chemical Interrelationship of heparan sulfate and heparin. In: Chemistry and Molecular Biology of the Intercellular Matrix (ed. E.A. BALAZS), Vol. 2, pp. 961-967. New York: Academic Press.

CIFONELLI, J.A., KING, J. (1972): The distribution of 2-acetamido-2-deoxy-D-glucose residues in mammalian heparin. Carbohyd. Res. 21, 173-186.
CIFONELLI, J.A., KING, J. (197o): The isolation and properties of heparitin sulfate from human umbilical chord tissues. Biochim. Biophys. Acta 215, 273-279.
CIFONELLI, J.A., MATHEWS, M.B. (1972a): A comparison of the chemical structures of mactins with mammalian heparins. Conn. Tissue Res. 1, 121-13o.
CIFONELLI, J.A., MATHEWS, M.B. (1972b): Structural studies of spisulan: A mucopolysaccharide from clams. Conn. Tissue Res. 1, 231-241.
CIFONELLI, J.A., MATHEWS, M.B. (1966): The reaction of mactins with nitrous acid. Federation Proc. 25, 742.
CIFONELLI, J.A., RODÉN, L. (1968): The isolation of heparin and heparitin sulfate from hog skin. Biochim. Biophys. Acta 165, 553-554.
CIFONELLI, J.A., SAUNDERS, A., GROSS, J.I. (1967): Keratan sulfate fractions from bovine and human tissues. Carbohyd. Res. 3, 478-485.
CLARK, C.C., BORNSTEIN, P. (1972): Cyanogen bromide cleavage of guinea pig skin collagen. Isolation and characterization of peptides from the α1 and α2 chains. Biochemistry 12, 1468-1474.
CLARK, C.C., VEIS, A. (1972): High molecular weight α chains in acid-soluble collagen and their role in fibrillogenesis. Biochemistry 11, 494-5o2.
CLARK, E., VON SCHMIDT, K. (1965): Sharks of the central gulf coast of Florida. Bull. Marine Sci. 15, 13-83.
CLAUSEN, J., ANDERSEN, V. (1963): Acid mucopolysaccharides in human leucocytes. Clin. Chim. Acta 8, 5o5-512.
CLELAND, R.L. (197o): Ionic polysaccharides. IV. Free-rotation dimensions for disaccharide polymers. Comparison with experiment with hyaluronic acid. Biopolymers 9, 811-824.
CLELAND, R.L., WANG, J.L. (197o): Ionic polysaccharides. III. Dilute solution properties of hyaluronic acid fractions. Biopolymers 9, 799-81o.
CLICK, E.M., BORNSTEIN, P. (197o): Isolation and characterization of the cyanogen bromide peptides from the α1 and α2 chains of human skin collagen. Biochemistry 9, 4699-47o6.
CLIFFORD, J., CHILD, T.F. (1971): Water in protein and polysaccharide gels. Proc. 1st Eur. Biophys. Congr. (ed. E. BRODA), Vol. 3, pp. 461-469. Vienna: Wiener Med. Akad.
COHEN, A.S. (197o): Chemical and immunological characterization of two components of aymloid. In: Chemistry and Molecular Biology of the Intercellular Matrix (ed. E.A. BALAZS), Vol. 3, pp. 1517-1535. New York: Academic Press.
COHEN, P.P. (1966): Biochemical aspects of metamorphosis: transition from ammonotelism to ureotelism. Harvey Lect. Ser. 6o, 119-154.
COHEN, P.P. (197o): Biochemical differentiation during amphibian metamorphosis. Science 168, 533-543.
COLVIN, J.R. (1972): The structure and biosynthesis of cellulose. In: CRC Critical Reviews in Macromolecular Science, pp. 47-81. Cleveland: Chemical Rubber Co.
COMFORT, A. (1964): The Process of Aging. New York: Am. Library of World Literature.
COMFORT, A. (197o): Basic research in gerontology. Gerontologia 16, 48-64.
CONCHIE, J., PERCIVAL, E.G.V. (195o): Fucoidin. Part II. The hydrolysis of a methylated fucoidin prepared from *Fucus vesiculosus*. J. Chem. Soc. 827-832.
CONRAD, G.W. (197oa): Collagen and mucopolysaccharide biosynthesis in the developing chick cornea. Devel. Biol. 21, 292-317.
CONRAD, G.W. (197ob): Collagen and mucopolysaccharide biosynthesis in mass cultures and clones of chick corneal fibroblasts *in vitro*. Devel. Biol. 21, 611-635.
CONVERY, F.R., AKESON, W.H. (1969): Ester sulfate and collagen content of articular cartilage from human femoral heads. Surg. Forum 15, 435-437.
COOK, G.M.W., STODDARD, R.W. (1973): Surface Carbohydrates of the Eukaryotic Cell. New York: Academic Press.
COOKE, A.S., BALCH, D.A. (197o): The distribution and carbohydrate composition of the organic matrix in hen egg shell. Br. Poult. Sci. 11, 353-365.
COOMBS, J., VOLCANI, B.E. (1968): Studies on the biochemistry and fine structure of silica-shell formation in diatoms. Chemical changes in the wall of *Navicula pelliculosa* during its formation. Planta 82, 28o-292.

COOPER, A. (1971): Thermal stability of tropocollagens - are hydrogen bonds really important? J. Molec. Biol. 55, 123-127.
COULOMBRE, A.J., COULOMBRE, J.L. (1972): Corneal development. IV. Interruption of collagen excretion into the primary stroma with L-azetadine-2-carboxylic acid. Dev. Biol. 28, 183-19o.
COX, R.W., GRANT, R.A., HORNE, R.W. (1967): The structure and assembly of collagen fibrils. I. Native-collagen fibrils and their formation from tropocollagen. J. Royal Micro. Soc. 87, 123-142.
COX, B.A., STARCHER, B.C., URRY, D.W. (1973): Coacervation of α-elastin in fiber formation. Biochim. Biophys. Acta 317, 2o9-213.
COX, B.A., STARCHER, B.C., URRY, D.W. (1974): Coacervation of tropoelastin results in fiber formation. J.Biol. Chem. 249, 997-998.
CUSTOD, T.F., YOUNG, I.J. (1968): Cat brain mucopolysaccharides and their *in vivo* hyaluronidase digestion. J. Neurochem. 15, 8o9-813.
DAHMEN, G. (1965): Beobachtungen bei der Reifung des Bindegewebes. Z. Rheumaforsch. 24, 332-341.
DALFERES, Jr., E.R., RUIZ, H., KUMAR, V., RADHAKRISHNAMURTHY, B., BERENSON, G.S. (1971): Acid mucopolysaccharides of fatty streaks in young male aortas. Atheroscler. 13, 121-131.
DANIEL, J.F. (1934): The Elasmobranch Fishes, 3rd. Ed. Berkeley: University of California Press.
DANISHEVSKY, I., ABRAHAM, G. (1971): Isolation and purification of chondroitin 6-sulfate protein from umbilical chord. Prep. Biochem. 1, 133-14o.
DANISHEVSKY, I., BELLA, Jr., A. (1966): The sulfated mucopolysaccharides from human umbilical chord. J. Biol. Chem. 241, 143-146.
DARZYNKIEWICZ, Z., BALAZS, E.A. (1971): Effect of connective tissue intercellular matrix on lymphocyte stimulation. Exp. Cell Res. 66, 113-123.
DAVIDSON, E.A., SMALL, W. (1963): Metabolism *in vivo* of connective tissue mucopolysaccharides. I. Chondroitin sulfate C and keratosulfate of nucleus pulposus. Biochim. Biophys. Acta 69, 445-452.
DAVIDSON, E.A., SMALL, W., PERCHEMLIDES, P., BAXLEY, W. (1961): Age-dependent metabolism of connective tissue polysaccharides. Biochim. Biophys. Acta 46, 189-19o.
DAVISON, P.F., DRAKE, M.P. (1966): The physical characterization of monomeric tropocollagen. Biochemistry 5, 313-321.
DAVISON, P.F., LEVINE, L., DRAKE, M.P., RUBIN, A., BUMP, S. (1967): The serological specificity of tropocollagen telopeptides. J. Exp. Med. 126, 331-346.
DEA, I.C.M., MOORHOUSE, R., REES, D.A., ARNOTT, S., GUSS, J.M., BALAZS, E.A. (1973): Hyaluronic acid: A novel, double helical molecule. Science 179, 56o-562.
DEAN, M.F., MUIR, H. (197o): The characterization of a protein-polysaccharide isolated from Kurloff cells of the guinea pig. Biochem. J. 118, 783-79o.
DEAN, M.F., MUIR, H., MARSHALL, A.H.E., REVELL, P.A., VERNON-ROBERTS, B. (1971): Proteoglycans from sheep, pig, rat and human spleens having chemical resemblances to that in Kurloff cells. FEBS Letters 16, 183-185.
DEASY, C.L. (1972): Relationship between the amino acid compositions of α1- and α2-chains of collagens. Biochim. Biophys. Acta 278, 142-146.
DEFRETIN, R. (1971): The tubes of polychaete annelids. In: Comprehensive Biochemistry (eds. M. FLORKIN, E.H. STOTZ), Vol. 26C, pp. 713-747. Amsterdam: Elsevier.
DEGENS, E.T., SPENCER, D.W. (1967): Paleobiochemistry of molluscan shell proteins. Comp. Biochem. Physiol. 2o, 553-579.
DeGRAEF, J., GLASS, G.B.J. (1968): Chondroitin sulfate A and sulfated glycoproteins in dog gastric secretion from the fundus. Gastroenter. 55, 584-593.
DEHM, P., LIMINEZ, S.A., OLSON, B.R., PROCKOP, D.J. (1972): A transport form of collagen from embryonic tendon: Electron microscopic demonstration of an NH_2-terminal extension and evidence suggesting the presence of cystine in the molecule. Proc. Natl. Acad. Sci. US 69, 6o-64.
DEHM, P., PROCKOP, D.J. (1973): Biosynthesis of cartilage collagen. Eur. J. Biochem. 35, 159-166.
DELAUNAY, A. (1966): Connective tissue - in the past. In: Biochemie et Physiologie du Tissu Conjonctif (ed. Ph. COMTE). Lyon: Societé Orimeco et Imprimerie du Sud. Est à Lyon.

DE METO, R.H., LIN, Y., NARASIMHULU, S. (1967): The biosynthesis of mactin. Comp. Biochem. Physiol. 2o, 581-591.
DENISON, R.H. (1956): A review of the habitat of the earliest vertebrates. Fieldiana Geol. 2, 359-457.
DENISON, R.H. (1963): The early history of the vertebrate calcified skeleton. Clin. Orthopaed. 31, 141-152.
DENISON, R.H. (1967): Ordovician vertebrates from western United States. Fieldiana Geol. 16, 131-192.
DENIZOT, J.-P. (1971): Etude histochimique des mucopolysaccharides du mormyromaste (type II de Cordier) chez *Gnathonemus petersii* mormyridés. Histochem. 28, 3o5-315.
DERBIN, C., DENIZOT, J.-P. (1971): Ultrastructure de l'organe ampullaire de *Gymnotus carapo* (Gymnotidae); origine et nature des mucopolysaccharides. Z. Zellforsch. 113, 531-543.
DESMUKH, K., NIMNI, M.E. (1972): Identification of stable intermolecular crosslinks present in reconstituted native collagen fibers. Biochem. Biophys. Res. Commun. 46, 175-182.
DESMUKH, K., NIMNI, M.E. (1973): Isolation and characterization of cyanogen bromide peptides from the collagen of bovine articular cartilage. Biochem. J. 133, 615-622.
DeVINCENTIIS, M., MATERAZZI, G. (1964): Indagine citochimica sui mucopolisaccaridi delle cellule testacee di ovociti di *Ciona intestinalis L.* nel corso dell'ovogenesi. Rivista Biol. 58, 3o1-325.
DI BENEDETTA, C., BRUNNGRABER, E.G., WHITNEY, G., BROWN, B.D., ARO, V. (1969): Compositional patterns of sialofuco hexosaminoglycans derived from rat brain glycoproteins. Arch. Biochem. Biophys. 131, 4o4-413.
DI FERRANTE, N. (197o): The immunological properties of hyaluronidase-treated bovine proteoglycans. In: Chemistry and Molecular Biology of the Intercellular Matrix (ed. E.A. BALAZS), Vol. 3, pp. 1551-1561. New York: Academic Press.
DI FERRANTE, N., DONNELLY, P.V., GREGORY, J.D., SAJDERA, S.W. (197o): Antigenic determinants of two components of proteoglycan complex from bovine cartilage. FEBS Letters 9, 149-151.
DISCHE, Z. (1965): Amino sugar-containing compounds in mucuses and in mucous membranes. In: The Amino Sugars (eds. E.A. BALAZS, R.W. JEANLOZ), Vol. 2A, pp. 115-14o. New York: Academic Press.
DISCHE, Z. (197oa): Collagen of embryonic type in the vertebrate eye and its relation to carbohydrates and subunit structure of tropocollagen. Dev. Biol. Suppl. 4, 164-194.
DISCHE, Z. (197ob): Biochemistry of connective tissue of the vertebrate eye. Int. Rev. Conn. Tissue Res. 5, 2o9-279.
DISCHE, Z., ROTHSCHILD, C. (1971): A complex between short chain collagen and a sulfated sialoglycosaminoglycan in the bovine corneal stroma. Federation Proc. 3o, 1223 abs.
DIXON, G.H. (1966): Mechanisms of protein evolution. Essays Biochem. 2, 147-2o4.
DOBERENZ, A.R., WYCKOFF, R.W.G. (1967): Fine structure in fossil collagen. Proc. Nat. Acad. Sci. USA 57, 539-541.
DOBZHANSKY, T. (1951): Genetics and the Origin of Species, 3rd Ed. New York: Columbia University Press.
DOEZEMA, P. (1969): Carbohydrates and carbohydrate metabolism of echinoderms. In: Chemical Zoology (eds. M. FLORKIN, B.T. SCHEER), Vol. 3, pp. 1o1-122. New York: Academic Press.
DORFMAN, A., HO, P.-L. (197o): Synthesis of acid mucopolysaccharides by glial tumor cells in tissue culture. Proc. Nat. Acad. Sci. USA 66, 495-499.
DORNER, R.W. (1968): Changes in glycosaminoglycan composition associated with maturation of regenerating rabbit tendon. Arch. Biochem. Biophys. 128, 34-39.
DORNER, R.W., ANTONOPOULOS, C.A., GARDELL, S. (1968): Application of cetylpyridinium-complex elution methods to analysis of glycosaminoglycans of tendon. Biochim. Biophys. Acta 158, 336-343.
DONGHERTY, T.F., BERLINER, D.L. (1968): The effects of hormones on connective tissue cells. In: Treatise on Collagen (ed. B.S. GOULD), Vol. 2A, pp. 369-394. New York: Academic Press.

DOYLE, J. (1967): The lorenzan sulphates. Biochem. J. 1o3, 325-33o.
DOYLE, J. (1968): The lorenzan sulphates: A comparative study. Comp. Biochem. Physiol. 24, 479-485.
DOYLE, J. (1969): Aging changes in cartilage from *Squalus Acanthias* L. Comp. Biochem. Physiol. 25, 2o1-2o6.
DOYLE, J., SZIRMAI, J.A. (1961): Chondroitin sulfate and collagen in the connective tissue of the rooster comb. Biochim. Biophys. Acta 5o, 582-584.
DRACH, P. (1948): Embranchment des cephalocordés. In: Traité de Zoologie (ed. P.-P. GRASSÉ), Vol. II, pp. 931-1o37. Paris: Masson.
DUNSTONE, J.R. (1959): Some cation-binding properties of cartilage. Biochem. J. 72, 465-473.
DUNSTONE, J.R. (196o): Ion-exchange reactions between cartilage and various cations. Biochem. J. 77, 164-17o.
DUNSTONE, J.R. (1962): Ion-exchange reactions between acid mucopolysaccharides and various cations. Biochem. J. 85, 336-351.
DUNSTONE, J.R. (1967): Acid glycosaminoglycans of pig aorta. Experientia 23, 344-345.
EAKIN, R.M., WESTFALL, J.A. (1962): Fine structure of the notochord of amphioxus. J. Cell Biol. 12, 646-651.
EASTOE, J.E. (1967): Composition of collagen and allied proteins. In: Treatise on Collagen, Vol. 1, Chemistry of Collagen (ed. G.N. RAMACHANDRAN), pp. 1-72. New York: Academic Press.
EASTOE, J.E. (1971): Dental enamel. In: Comprehensive Biochemistry (eds. M. FLORKIN, E.H. STOTZ), Vol. 26C, pp. 785-834. Amsterdam: Elsevier.
ECONOMU, J.S. (1971): Observations on a possible association between acid mucopolysaccharides and lysozyme. J. Surg. Onc. 3, 89-95.
EHRENPREIS, S., FISHMAN, M.M. (196o): The interaction of quaternary ammonium compounds with chondroitin sulfate. Biochim. Biophys. Acta 44, 577-585.
EHRLICH, H.P., BORNSTEIN, P. (1972): Further characterization of a pro-α2 chain. Biochem. Biophys. Res. Commun. 46, 175o-1756.
EISENSTEIN, R., SARGENTE, N., KUETTNER, K.E. (1971): Organization of extracellular matrix in epiphyseal growth plate. Am. J. Pathol. 65, 515-528.
ELAM, J.S., GOLDBERG, J.M., RADIN, N.S., AGRANOFF, B.W. (197o): Rapid axonal transport of sulfated mucopolysaccharide proteins. Science 17o, 458-46o.
ELDEN, H.R. (1968): Physical properties of collagen. Int. Rev. Conn. Tissue 4, 283-348.
ELIEL, E.L., ALLINGER, N.L., ANGYAL, S.J., MORRISON, G.A. (1965): Conformational Analysis. New York: Wiley (Interscience).
ELKAN, E. (1968): Mucopolysaccharides in the anuran defense against desiccation. J. Zool. (London) 155, 19-53.
ELLIOT, D.H. (1965): Structure and function of mammalian tendon. Biol. Rev. 4o, 392-421.
ELSDALE, T., BARD, J. (1972): Cellular interactions in mass cultures of human diploid fibroblasts. Nature 236, 152-155.
ENGFELDT, B., HJERPE, A. (1972): Glycosaminoglycans of dentine and predentine. Calcif. Tissue Res. 1o, 152-159.
ENOMOTO, N., NAKAGAWA, H., MATSUDA, K., TOMIYASU, Y. (1966): Sialic acids in the external mucous substance and the cartilage of fishes. Bull. Japan Soc. Sci. Fisheries 32, 442-445.
EPSTEIN, E.H., SCOTT, R.D., MILLER, E.J., PIEZ, K.A. (1971): Isolation and characterization of the peptides derived from soluble human and baboon skin collagen after cyanogen bromide cleavage. J. Biol. Chem. 246, 1718-1724.
ESTES, Z.E., FAUST, R.M. (1964): Studies on the mucopolysaccharides of the greater wax moth, *Galleria mellonella* (Linnaeus). Comp. Biochem. Physiol. 13, 443-452.
EYRE, D.R., GLIMCHER, M.J. (1973a): Evidence for glycosylated crosslinks in body-wall collagen of the sea cucumber, *Thyone briareus*. Proc. Soc. Exp. Biol. Med. 144, 4oo-4o3.
EYRE, D.R., GLIMCHER, M.J. (1973b): Analysis of a crosslinked peptide from calf bone collagen: evidence that hydroxylysyl glycoside participates in the cross-link. Biochem. Biophys. Res. Commun. 52, 663-671.
EYRING, E.J., YANG, J.T. (1968a): Conformation of protein-polysaccharide complex from bovine nasal septum. J. Biol. Chem. 243, 13o6-1311.

EYRING, E.J., YANG, J.T. (1968b): Viscosity and optical activity of chondroitin sulfate C. Biopolymers 6, 694-7o1.
FABIANEK, J., HERP, A., PIGMAN, W. (1965): Comparative study of acid mucopolysaccharides in dermal connective tissues of some mammals. Comp. Biochem. Physiol. 14, 21-28.
FARBER, S., Van PRAAG, D. (197o): Composition of glycosaminoglycans (mucopolysaccharides) in rabbit renal papillae. Biochim. Biophys. Acta 2o8, 219-226.
FARBER, S.J., SCHUBERT, M. (1957): The binding of cations by chondroitin sulfate. J. Clin. Invest. 36, 1715-1722.
FARRELL, R.A., HART, R.W. (1969): On the theory of the spatial organization of macromolecules in connective tissue. Bull. Math. Biophys. 31, 727-76o.
FEDORKO, M.E., MORSE, S.I. (1965): Isolation, characterization, and distribution of acid mucopolysaccharides in rabbit leucocytes. J. Exp. Med. 121, 39-48.
FESSLER, J.H. (196o): A structural function of mucopolysaccharides in connective tissues. Biochem. J. 76, 124-132.
FESSLER, J.H., FESSLER, L.I. (1966): Electron microscopic visualization of the polysaccharide hyaluronic acid. Proc. Nat. Acad. Sci. USA 56, 141-147.
FIETZEK, P.P., REXRODT, F.W., HOPPER, K.E., KÜHN, K. (1973): The covalent structure of collagen. 2. The amino-acid sequence of α1-CB7 from calf-skin collagen. Eur. J. Biochem. 38, 396-4oo.
FIETZEK, P.P., REXRODT, F.W., WENDT, P., STARK, M., KÜHN, K. (1972a): The covalent structure of collagen. Amino-acid sequence of peptide α1-CB6-C2. Eur. J. Biochem. 3o, 163-168.
FIETZEK, P.P., WENDT, P., KELL, I., KÜHN, K. (1972b): Covalent structure of collagen. Amino acid sequence of α1-CB3 from calf skin collagen. FEBS Letters 26, 74-76.
FISCHER, E., KLUJBER, L. (197o): Protein und sulfatiertes Mucopolysaccharid in den Botryoidgranula des Pferdeegels (*Haemopsis sanguisuga*, L.). Acta Histochem. 38, 241-249.
FITCH, W.M. (197o): Distinguishing hemologous from analogous proteins. Systematic Zool. 19, 99-113.
FITCH, W.M., MARGOLIASH, E. (1967): Construction of phylogenetic trees. Science 155, 279-284.
FITTON JACKSON, S. (1964): Connective tissue cells. In: The Cell (ed. J. BRACHET, A.E. MIRIKY), Vol. 6, pp. 387-52o. New York: Academic Press.
FITTON JACKSON, S. (1968): The morphogenesis of collagen. In: Treatise on Collagen (ed. B.S. GOULD), Vol. 2B, pp. 1-66. New York: Academic Press.
FITTON JACKSON, S. (197o): Environmental control of macromolecular synthesis in cartilage and bone: Morphogenetic response to hyaluronidase. Proc. Roy. Soc. B. 175, 4o5-453.
FLEISCHMAJER, R., PERLISH, J.S., BASHEY, R.I. (1972): Human dermal glycosaminoglycans and aging. Biochim.Biophys. Acta 279, 265-275.
FLETCHER, T.C., GRANT, P.T. (1968): Glycoproteins in the external mucous secretions of the plaice, *Pleuronectes platessa*, and other fishes. Biochem. J. 1o6, 12P.
FLOOD, P.R., GUTHRIE, D.M., BANKS, J.R. (1969): Paramyosin muscle in the notochord of amphioxus. Nature 222, 87-88.
FLORKIN, M. (1944): L'Evolution Biochimique. Paris: Masson.
FLORKIN, M. (1966): A Molecular Approach to Phylogeny. New York: Elsevier.
FONTAINE, M., DAMAS, H., ROCHON-DUVIGNEAUD, A., PASTEELS, J. (1958): Formes actuelles: Super=ordre des petromyzonoidea et des myxinoidea. In: Traité de Zoologie (ed. P.-P. GRASSÉ), Vol. 13, pp. 13-172. Paris: Masson.
FOSTER, J.A., BRUENGER, E., GRAY, W.R., SANDBERG, L.B. (1973): Isolation and amino acid sequences of tropoelastin peptides. J. Biol. Chem. 248, 2876-2879.
FOSTER, J.A., GRAY, W.R., FRANZBLAU, C. (1973): Isolation and characterization of crosslinked peptides from elastin. Biochim. Biophys. Acta 3o3, 363-369.
FOUCART, M.F. (1966): Localisation du collagène dans le test d'un ousin (Echinoderme). Bull. Clin. Sci. Acad. Roy. Belg. Ser. 5, 52, 316-319.
FOWLER, L.J., BAILEY, A.J. (1972): Current concepts of the crosslinking in bone collagen. Clin. Orthop. Relat. Res. 85, 193-2o6.

FRANCOIS, C.J., GLIMCHER, M.J. (1966): The hexose content of the α-chains of cod-fish skin collagen. Biochim. Biophys. Acta 127, 556-559.
FRANCOIS, C.J., GLIMSCHER, M.J. (1967a): The separation of the α-chains of collagen by free-flow electrophoresis. Biochem. J. 1o2, 148-152.
FRANCOIS, C.J., GLIMCHER, M.J. (1967b): The isolation and amino acid composition of the α-chains of chicken-bone collagen. Biochim. Biophys. Acta 133, 91-96.
FRANCIS, G., JOHN, R., THOMAS, J. (1973): Biosynthetic pathway of desmosines in elastin. Biochem. J. 136, 45-55.
FRANCO-BROWDER, S., DeRYDT, J., DORFMAN, A. (1963): The identification of a sulfated mucopolysaccharide in chick embryos. Stages 11-23. Proc. Nat. Acad. Sci. USA 49, 643-647.
FRANEK, M.D., DUNSTONE, J.R. (1968): Properties of a protein-bound chondroitin sulfate of pig aorta. Biochim. Biophys. Acta 165, 555-557.
FRANSSON, L. (197o): Structure and metabolism of the proteoglycans of dermatan sulfate. In: Chemistry and Molecular Biology of the Intercellular Matrix (ed. E.A. BALAZS), Vol. 2, pp. 823-842. New York: Academic Press.
FRANSSON, L.-Å. (1968): Structure of dermatan sulfate. III. The hybrid structure of dermatan sulfate from umbilical chord. J. Biol. Chem. 243, 15o4-151o.
FRANSSON, L.-Å., MALMSTROM, A. (1971): Structure of pig skin dermatan sulfate. I. Distribution of D-glucuronic acid residues. Eur. J. Biochem. 18, 422-43o.
FRANZBLAU, C. (1971): Elastin. In: Comprehensive Biochemistry (eds. M. FLORKIN, E.H. STOTZ), Vol. 26C, pp. 659-712. Amsterdam: Elsevier.
FREEMAN, I.L., STEVEN, F.S., JACKSON, D.S. (1968): Isolation and amino acid composition of bovine corneal polymeric collagen. Biochim. Biophys. Acta 154, 252-254.
FREEMAN, M.A.R. (ed.) (1973): Adult Articular Cartilage. London: Pitman.
FREY-WYSSLING, A. (1957): Macromolecules in cell structure. Nature 179, 941-943.
FRIEDEN, E. (1967): Thyroid hormones and the biochemistry of amphibian metamorphosis. Recent Prog. Hormone Res. 23, 139-194.
FRIEDMAN, M.H., GREEN, K. (1971): Ion binding and Donnan equilibrium in rabbit corneal stroma. Am. J. Physiol. 221, 356-362.
FROMMHAGEN, L.H., FAHRENBACH, M.J., BROCKMAN, Jr., J.A., STOCKSTAD, E.L.R. (1953): Heparin-like anticoagulants from mollusca. Proc. Soc. Exp. Biol. Med. 82, 28o-283.
FRY, H., ROBERSTON, W. VAN B. (1967): Interlocked stresses in cartilage. Nature 215, 53-54.
FUJIMOTO, D. (1968): Isolation of collagens of high hydroxyproline, hydroxylysine and carbohydrate content from muscle layer of *Ascaris lumbricoides* and pig kidney. Biochim. Biophys. Acta 168, 537-543.
FUJIMOTO, D. (1972): Amino acid composition of skin collagen from deep-sea fish. J. Biochem. 71, 9o5-9o6.
FUJIMOTO, D., ADAMS, E. (1964): Intraspecies composition difference in collagen from cuticle and body of *Ascaris* and *Lumbricus*. Biochem. Biophys. Res. Commun. 17, 437-442.
FUJIMOTO, D., IIZUKA, K. (1972): Isolation of noncollagen protein associated with collagen from the muscle layer of *Ascaris lumbricoides*. J. Biochem. 71, 1o89-1o91.
FUJIMOTO, D., IKEUCHI, T., NOZAWA, S. (1969): Intermolecular disulfide cross-linkages in the collagen from the muscle layer of *Ascaris lumbricoides*. Biochim. Biophys. Acta 188, 295-3o1.
FUJIMOTO, D., PROCKOP, D.J. (1968): Denatured collagen from the cuticle of *Ascaris Lumbricoides* as a substrate for protocollagen proline hydroxylase. J. Biol. Chem. 243, 4138-4142.
FUJIMOTO, D., PROCKOP, D.J. (1969): Protocollagen proline hydroxylase from *Ascaris lumbricoides*. J. Biol. Chem. 244, 2o5-21o.
FURTHMAYER, H., TIMPL, R. (197o): Acidic structural proteins of connective tissue. II. Evidence for disulfide cross-links. Biochim. Biophys. Acta 22o, 396-398.
FURUHASHI, T. (1961a): Polysulfated mucopolysaccharides of elasmobranch cartilage. J. Biochem. (Japan) 5o, 546-547.
FURUHASHI, T. (1961b): Studies on sulfated mucopolysaccharides in cartilage of elasmobranchii. II. Isolation of polysulfated mucopolysaccharides. J. Japan Biochem. Soc. 33, 746-752.

FURUHASHI, T., UCHIDA, K. (1961): Studies on sulfated mucopolysaccharides of elasmobranch cartilage. I. Chondroitin sulfuric acid from shark cartilage. J. Japan Biochem. Soc. 33, 537-54o.

GADEYNE, Ch., FRANCOIS, C. (197o): Distribution et proprietés du collagène de quelque tissus conjonctifs dans la classe des echinoides. Arch. Int. Physiol. Biochem. 78, 427-433.

GALLOP, P.M., BLUMENFELD, O.O., SEIFTER, S. (1972): Structure and metabolism of connective tissue proteins. Ann. Rev. Biochem. 41, 617-672.

GARDAIS, A., PICARD, J., HERMELIN, B. (1973): Glycosaminoglycan (GAG) distribution in aortic wall from five species. Comp. Biochem. Physiol. 44B, 5o7-515.

GARRONE, R. (1969): Collagène, spongine et squellette minéral chez l'eponge *Haliclona rosea* (O.S.). J. Micro. 8, 581-598.

GARRONE, R. (1971): Fibrogenèse du collagène chez l'eponge *Chondrosea reniformis nardo* (Démosponge Tetractinellide). Ultrastructure et fonction des lophocytes. C.R. Acad. Sci. (Paris), Sér. D, 273, 1832-1835.

GARRONE, R., PAVANS DE CECCATHY, M. (197o): Membranes cellulaires et fibrogènese du collagène. Proc. 7th Congrès Internatl. Micro. Electronique, Grenoble, pp. 639-64o.

GARRONE, R., THINEY, Y., PAVANS DE CECCATHY, M. (1971): Electron microscopy of a mucopolysaccharide cell coat in sponges. Experientia 27, 1324-1326.

GELMAN, R.A., BLACKWELL, J. (In Press): Interactions between mucopolysaccharides and cationic polypeptides in aqueous solution: chondroitin-4-sulfate and dermatan sulfate. Biopolymers

GELMAN, R.A., GLASER, D.N., BLACKWELL, J. (In Press): Interaction between chondroitin-6-sulfate and poly-L-arginine in aqueous solution. Biopolymers.

GELMAN, R.A., RIPPON, W.B., BLACKWELL, J. (1972): Interactions between chondroitin sulfate C and poly-L-lysine: Preliminary report. Biochem. Biophys. Res. Commun. 48, 7o8-711.

GERARD, R.W. (1969): Hierarchy, entitation, and levels. In: Hierarchical Structures (eds. L.L. WHYTE, A.G. WILSON, D. WILSON), pp. 215-228. New York: Elsevier.

GIBSON, T., KENEDI, R.M. (197o): The structural components of the dermis and their mechanical characteristics. In: The Dermis (eds. W. MONTAGNA, J.P. BENTLEY, R.L. DOBSON), pp. 19-38. New York: Appleton-Century-Crofts.

GILLARD, G.C., WUSTEMAN, F.S. (197o): Acid glycosaminoglycans of elastic cartilage. Biochem. J. 118, 25P.

GLAGOV, S. (1972): Hemodynamic risk factors: Mechanical stress, mural architecture, medial nutrition and the vulnerability of arteries to atherosclerosis. In: The Pathogenesis of Atherosclerosis (eds. R.W. WISSLER, J.C. GEER), pp. 164-199. Baltimore: Williams and Wilkins.

GLENNER, G.G., PAGE, D., ISERSKY, C., HARADA, M., CUATRECASAS, P., EANES, E.D., DeLELLIS, R.A., BLADEN, H.A., KEISER, H.R. (1971): Murine amyloid fibril protein: Isolation, purification and characterization. J. Histochem. Cytochem. 19, 16-28.

GLIMCHER, M.J., KRANE, S.M. (1968): The organization and structure of bone, and the mechanism of calcification. In: Treatise on Collagen, Vol. 2, Biology of Collagen (ed. B.S. GOULD), pp. 67-251. New York: Academic Press.

GLIMCHER, M.J., SEYER, J., BRICKLEY, D.M. (1969): The solubilization of collagen and protein-polysaccharides from the developing cartilage of lathyritic chicks. Biochem. J. 115, 923-926.

GOETINCK, P.F., SEKELLICK, M.J. (1972): Observations on collagen synthesis, lattice formation and morphology of scaleless and normal embryonic skin. Develop. Biol. 28, 636-648.

GOH, A.T., LOWTHER, D.A. (1966): Effect of age on the composition of bovine nasal cartilage. Nature 21o, 127o-1271.

GOLDSTEIN, A., ADAMS, E. (197o): Glycylhydroxyprolyl sequences in earthworm cuticle collagen: Glycylhydroxyprolylserine. J. Biol. Chem. 245, 5478-5483.

GOLDSTEIN, L., FORSTER, R.P. (1971): Urea biosynthesis and excretion in freshwater and marine elasmobranchs. Comp. Biochem. Physiol. 39B, 415-421.

GOODRICH, E.S. (1958): Studies on the Structure and Development of Vertebrates. New York: Dover.

GOTTE, L., MAMMI, M., PEZZIN, G. (1972): Scanning electron microscope observations on elastin. Conn. Tissue Res. 1, 61-67.

GOTTE, L., SERAFINI-FRACASSINI, A. (1963): Electron microscope observations on the structure of elastin. J. Atherosclerosis Res. 3, 247-251.
GOTTE, L., SERAFINI-FRACASSINI, A., MORET, V. (1963): The chemical composition of the NaCl-soluble fraction from autoclaved elastin. J. Atheroscler. Res. 3, 244-247.
GOTTSCHALK, A. (1969): Biosynthesis of glycoproteins and its relationship to heterogeneity. Nature 222, 452-454.
GOTTSCHALK, A. (ed.) (1972): Glycoproteins. Their Composition, Structure and Functions, 2nd Ed. Amsterdam: Elsevier.
GOULD, B.S. (1968a): Collagen biosynthesis. In: Treatise on Collagen, Vol. 2A, Biology of Collagen (ed. B.S. GOULD), pp. 139-188. New York: Academic Press.
GOULD, B.S. (1968c): The role of certain vitamins in collagen formation. In: Treatise on Collagen (ed. B.S. GOULD, Vol. 2A, pp. 323-365. New York: Academic Press.
GRANT, G.T., McNAB, C., REES, D.A., SKERRETT, R.J. (1969): Seed mucilages as examples of polysaccharide denaturation. Chem. Commun. 8o5-8o6.
GRANT, G.T., MORRIS, E.R., REES, D.A., SMITH, P.J.C., THOM, D. (1973): Biological interactions between polysaccharides and divalent cations: the egg-box model. FEBS Letters 32, 195-198.
GRANT, M.E., KEFALIDES, N.A., PROCKOP, D.J. (1972): The biosynthesis of basement membrane collagen in embryonic chick lens. II. Synthesis of a precursor form by matrix-free cells and a time-dependent conversion to α chains in intact lens. J. Biol. Chem. 247, 3545-3551.
GRANT, M.E., PROCKOP, D.J. (1972): The biosynthesis of collagen. New Eng. J. Med. 286, 194-199, 242-249, 291-3oo.
GRANT, M.E., SCHOFIELD, J.D., KEFALIDES, N.A., PROCKOP, D.J. (1973): The biosynthesis of basement membrane collagen in embryonic chick lens. III. Intracellular formation of the triple helix and the formation of aggregates through disulfide bonds. J. Biol. Chem. 248, 7432-7437.
GRANT, M.E., STEVEN, F.S., JACKSON, D.S. (1971): Carbohydrate content of insoluble elastins prepared from adult bovine and calf ligamentum nuchae and tropoelastin isolated from copper-deficient porcine aorta. Biochem. J. 121, 197-2o2.
GRANT, R.A. (1966): Variations in the amino acid composition of aortic elastic from different species. Brit. J. Exp. Pathol. 47, 163-167.
GRAY, W.R., SANDBERG, L.B., FOSTER, J.A. (1973): Molecular model for elastin structure and function. Nature 246, 461-466.
GREGORY, J.D. (1973): Multiple aggregation factors in cartilage proteoglycan. Biochem. J. 133, 383-386.
GREEN, D.E., GOLDBERGER, R.F. (1967): Molecular Insights into the Living Process. New York: Academic Press.
GREEN, H., GOLDBERG, B. (1968): Differentiation for collagen synthesis in cultured cells. Symp. Int. Soc. Cell Biol. 7, 123-134.
GREEN, K., HASTINGS, B., FRIEDMAN, M.H. (1971): Sodium ion binding in isolated corneal stroma. Am. J. Physiol. 22o, 52o-525.
GREER, R.B., JANICKE, G.H., MANKIN, H.J. (1968): Protein-polysaccharide synthesis at three levels of the normal growth plate. Calcif. Tissue Res. 2, 157-164.
GREGOIRE, P.E., DICTUS-VERMEULEN, C., AMERYCKX, I.P. (1972): Sulfate de dermatane et sulfate d'heparitine du foie de boeuf. Biochim. Biophys. Acta 279, 1o2-117.
GREGORY, J.D., HASCALL, V.C. (1972): Aggregation properties of chondroitinase-treated proteoglycan of bovine nasal septum. Federation Proc. 31, 433 abs.
GREILING, H., HERBERTZ, T., STUHLSATZ, H.W. (1964): Chromatographische Untersuchungen über die Mucopolysaccharidverteilung im hyalinen Knorpel. Hoppe-Seylers Z. Physiol. Chem. 336, 149-162.
GREILING, H., STUHLSATZ, H.W. (1966): Struktur und Stoffwechsel von Glykosaminoglykan-Proteinen. I. Die Keratan-Peptide der Rinder-Cornea. Hoppe-Seylers Z. Physiol. Chem. 345, 236-248.
GREILING, H., STUHLSATZ, H.W. (1969): Glykosaminoglykan-peptide aus dem humanen Kniegelenkknorpel. Hoppe-Seylers Z. Physiol. Chem. 35o, 449-456.
GREILING, H., STUHLSATZ, H.W., KISTERS, R. (197o): Structure and metabolism of proteokeratan sulfate. In: Chemistry and Molecular Biology of the Intercellular Matrix (ed. E.A. BALAZS), pp. 873-877. New York: Academic Press.

GRIFFITH, R.W., PANG, P.K.T., SRIVASTAVA, A.K., PICKFORD, G.E. (1973): Serum composition of freshwater stingrays (Potamotrygonidae) adapted to fresh and dilute sea water. Biol. Bull. 144, 3o4-32o.

GRIFFITH, R.W., UMMINGER, B.L., GRANT, B.F., PANG, P.K.T., PICKFORD, G.E. (1974): Serum composition of the coelacanth *Latimeria chalumnae* Smith. J. Exp. Zool. 187, 87-1o2.

GROBSTEIN, C. (1969): Epithelio-mesenchymal interactions in relation to reparative processes. In: Repair and Regeneration (eds. J.E. DUNPHY, W. VAN WINKLE), pp. 57-69. New York: McGraw-Hill.

GROSS, J. (1956): The behavior of collagen units as a model in morphogenesis. J. Biophys. Biochem. Cytol. Suppl. 2, 261-274.

GROSS, J. (1963): Comparative biochemistry of collagen. In: Comparative Biochemistry (eds. M. FLORKIN, H.S. MASON), Vol. 5, pp. 3o7-346. New York: Academic Press.

GROSS, J. (1969): Some aspects of collagen metabolism. In: Aging of Connective and Skeletal Tissues (eds. A. ENGEL, T. LARSSON), pp. 33-44. Stockholm: Nordiska Bokhandelns Förlag.

GROSS, J., DUMSHA, B. (1958): Elastoidin: A two component member of the collagen class. Biochim. Biophys. Acta 28, 268-27o.

GROSS, J., DUMSHA, B., GLAZER, N. (1958): Comparative biochemistry of collagen. Some amino acids and carbohydrates. Biochim. Biophys. Acta 3o, 293-297.

GROSS, J., KIRK, D. (1958): The heat precipitation of collagen from neutral salt solutions: Some rate-regulating factors. J. Biol. Chem. 233, 355-36o.

GROSS, J.I., MATHEWS, M.B., DORFMAN, A. (196o): Sodium chondroitin sulfate-protein complexes of cartilage II metabolism. J. Biol. Chem. 235, 2889-2892.

GROSS, J., MATOLTSY, A.G., COHEN, C. (1955): Vitrosin: A member of the collagen class. J. Biophys. Biochem. Cytol. 1, 215-22o.

GROSS, J., SOKAL, Z., RONGVIE, M. (1956): Structural and chemical studies on the connective tissue of marine sponges. J. Histochem. Cytochem. 4, 227-246.

GUSTAVSON, K.H. (1956): The Chemistry and Reactivity of Collagen, p. 224. New York: Academic Press.

GUTMAN, H. (1969): Structure and function in living systems. In: Hierarchical Structures (eds. L.L. WHYTE, A.G. WILSON, D. WILSON). New York: Elsevier.

HABUCHI, H., YAMAGATA, T., IWATA, H., SUZUKI, S. (1973): The occurrence of a wide variety of dermatan sulfate-chondroitin sulfate copolymers in fibrous cartilage. J. Biol. Chem. 248, 6o19-6o28.

HABUCHI, O., YAMAGATA, T., SUZUKI, S. (1971): Biosynthesis of the acetylgalactosamine 4,6-disulfate unit of squid chondroitin sulfate by transsulfation from 3'-phosphoadenosine 5'-phosphosulfate. J. Biol. Chem. 246, 7357-7365.

HAGOPIAN, A., BOSMANN, H.B., EYLAR, E.H. (1968): Glycoprotein biosynthesis: The localization of polypeptidyl: N-acetylgalactosaminyl collagen: glucosyl, and glycoprotein: galactosyltransferases in HeLa cell membrane fractions. Arch. Biochem. Biophys. 128, 387-396.

HALL, D.A. (1969): Connective tissues. In: Aging Life Processes (ed. S. BAKERMAN), pp. 79-122. Springfield, Illinois: Thomas.

HALL, D.A., SAXL, H. (1961): Studies of human and tunicate cellulose and of their relation to reticulin. Proc. Roy. Soc. B155, 2o2-217.

HALLÉN, A. (1962): The collagen and ground substance of human intervertebral disc at different ages. Acta Chem. Scand. 16, 7o5-71o.

HALSTEAD, H.B. (1969): Calcified tissues in the earliest vertebrates. Calcif. Tissue Res. 3, 1o7-124.

HALSTEAD, L.B. (1968): The pattern of vertebrate evolution. San Francisco: Freeman.

HAMERMAN, D., ROJKIND, M., SANDSON, J. (1966): Protein bound to hyaluronate: Chemical and immunologiacl studies. Federation Proc. 25, 1o4o-1o45.

HAMERMAN, D., SANDSON, J. (197o): Immunological cross-reactions between streptococcal hyaluronic acid and proteoglycans of human connective tissue. In: Chemistry and Molecular Biology of the Intercellular Matrix (ed. E.A. BALAZS), Vol. 3, pp. 1537-1549. New York: Academic Press.

HANDLER-BERNICH, E., LOTKE, P., RUBENSTEIN, R. (1972): Membrane characteristics of human articular cartilage. Biochim. Biophys. Acta 266, 732-736.

HANDLEY, C.J., PHELPS, C.F. (1972): Biosynthesis in vitro of chondroitin sulfate in neonatal rat epiphyseal cartilage. Biochem. J. 126, 417-432.

HANNIG, K., NORDWIG, A. (1967): The amino acid sequences in collagen. In: Treatise on Collagen, Vol. 1, Chemistry of Collagen (ed. G.N. RAMACHANDRAN), pp. 73-1o1. New York: Academic Press.

HANSEN, F.C. (19o5): Untersuchungen über die Gruppe der Bindegewebssubstanz: Der Hyalin-Knorpel. I. Anat. Anz. 27, 537-82o.

HARDINGHAM, T.E., FITTON JACKSON, S., MUIR, H. (1972): Replacement of proteoglycans in embryonic chick cartilage in organ culture after treatment with testicular hyaluronidase. Biochem. J. 129, 1o1-112.

HARDINGHAM, T.E., MUIR, H. (1972): The specific interaction of hyaluronic acid with cartilage proteoglycans. Biochim. Biophys. Acta 279, 4o1-4o5.

HARDINGHAM, T.E., MUIR, H. (1973a): Binding of oligosaccharides of hyaluronic acid to proteoglycans. Biochem. J. 135, 9o5-9o8.

HARDINGHAM, T.E., MUIR, H. (1973b): Hyaluronic acid in cartilage. Biochem. Soc. Trans. 1, 282-284.

HARDINGHAM, T.E., PHELPS, C.F. (197o): The glycosaminoglycans of neonatal rat skin. Biochem. J. 117, 813-818.

HARDISTY, M.W., POTTER, I.C. (1971): The general biology of adult lampreys. In: The Biology of Lampreys (eds. M.W. HARDISTY, I.C. POTTER), Vol. 1, pp. 127-2o6. New York: Academic Press.

HARE, P.E., ABELSON, P.H. (1963): Amino acids in the proteins from aragonite and calcite in the shell of *Mytilus californianus*. Science 139, 216-217.

HARE, P.E., ABELSON, P.H. (1965): Amino acid composition of some calcified proteins. In: Annual Report of the Director Geophysical Laboratory, pp. 223-231. Washington, D.C.: Carnegie Institution.

HARKNESS, R.D. (1961): Biological functions of collagen. Biol. Rev. 36, 399-463.

HARKNESS, R.D. (1968): Mechanical properties of collagenous tissues. In: Treatise on collagen, Vol. 2A, Biology of Collagen (ed. B.S. GOULD), pp. 247-31o. New York: Academic Press.

HARKNESS, R.D. (1971): Mechanical properties of skin in relation to its biological function and its chemical components. In: Biophysical Properties of Skin (ed. H.R. ELDEN), pp. 394-436. New York: Interscience.

HARPER, E., SEIFTER, S., SHARRER, B. (1967): Electron microscopic and biochemical characterization of collagen in blatterian insects. J. Cell Biol. 33, 385-394.

HARTLEY, B.S., SHATTON, D.M. (1971): Pancreatic elastase. In: The Enzymes, 3rd Ed. (ed. P.D. BOYER), Vol. 3, pp. 323-373. New York: Academic Press.

HASCALL, V.C., HEINEGÅRD, D. (1974): Aggregation of cartilage proteoglycans: I. The role of hyaluronic acid. J. Biol. Chem. 249, 4232-4241.

HASCALL, V.C., RIOLO, R.L. (1972): Characteristics of the protein-keratan sulfate core and of keratan sulfate prepared from bovine nasal cartilage proteoglycan. J. Biol. Chem. 247, 4529-4538.

HASCALL, V.C., RIOLO, R.L., HAYWORD, Jr., J., REYNOLDS, C.C. (1972): Treatment of bovine nasal cartilage proteoglycan with chondroitinases from *Flavobacterium heparinum* and *Proteus vulgaris*. J. Biol. Chem. 247, 4521-4528.

HASCALL, V.C., SAJDERA, S.W. (197o): Physical properties and polydispersity of proteoglycan from bovine nasal cartilage. J. Biol. Chem. 245, 492o-493o.

HASCALL, V.C., SAJDERA, S.W. (1969): Protein-polysaccharide complex from bovine nasal cartilage. The function of glycoprotein in the formation of aggregates. J. Biol. Chem. 244, 2384-2396.

HASHIMOTO, A., LUDOWIEG, J.J. (1968): The protein polysaccharides from the nucleus pulposus of whale intervertebral discs. Effect of disulfide reducing reagents. Biochemistry 7, 2469-2475.

HASS, G.M. (1939): Elastic tissue. Arch. Pathol. 27, 334-365, 582-613.

HASSID, W.Z. (1969): Biosynthesis of oligosaccharides and polysaccharides in plants. Science 165, 137-144.

HASSID, W.Z. (1971): Biosynthesis of cellulose and related plant cell-wall polysaccharides. High Polym. 5, 679-693.

HASSON, A. (1962): Interaction of quaternary ammonium bases with a purified acid polysaccharide and other macromolecules from the electric organ of electric eel. Biochim. Biophys. Acta 56, 275-292.

HATA, R., NAGAI, Y. (1973): Distribution of acidic glycosaminoglycans in tadpole backskin. Biochim. Biophys. Acta 3o4, 4o8-412.

HAY, E.D., DODSON, J.W. (1973): Secretion of collagen by corneal epithelium I. Morphology of the collagenous products produced by isolated epithelia grown on frozen-killed lens. J. Cell Biol. 57, 19o-213.

HAY, E.D., REVEL, J.-P. (1969): Fine Structure of the Developing Avian Cornea. Basel: Karger.

HEDLEY, R.H., WAKEFIELD, J.St.J. (1967): A collagen-like sheath in the arenaceous foraminifer *Haliphysema* (Protozoa) J. Roy. Microscop. Soc. 87, 475-481.

HEIDRICH, H.G., WYNSTON, L.K. (1965): A report on the presence of a third α-component in calf and rat skin tropocollagen. Hoppe-Seylers Z. Physiol. Chem. 342, 166-169.

HEINEGÅRD, D. (1972): Hyaluronidase digestion and alkaline treatment of bovine tracheal cartilage proteoglycans. Isolation and characterization of different keratan sulfate proteins. Biochim. Biophys. Acta 285, 193-2o7.

HEINEGÅRD, D., GARDELL, S. (1967): Protein-polysaccharide complex (proteoglycan) from human nucleus pulposus. Biochim. Biophys. Acta 148, 164-171.

HEINTZ, A. (1963): Phylogenetic aspects of myxinoids. In: The Biology of Myxine (eds. A. BRODAL, R. FANGE), pp. 9-21. Oslo: Universitetsforlaget.

HELMREICH, E. (1969): Control of synthesis and breakdown of glycogen, starch and cellulose. In: Comprehensive Biochemistry (Carbohydrate Metabolism) (eds. M. FLORKIN, E.H. STOTZ), Vol. 17, pp. 17-91. New York: Elsevier.

HELTING, T., LINDAHL, U. (1971): Occurrence and biosynthesis of β-glucuronidic linkages in heparin. J. Biol. Chem. 246, 5442-5447.

HELTING, T., RODÉN, L. (1968): The carbohydrate-protein linkage region of chondroitin 6-sulfate. Biochim. Biophys. Acta 17o, 3o1-3o8.

HENKART, P., HUMPHREYS, S., HUMPHRIES, T. (1973): Characterization of sponge aggregation factor. A unique proteoglycan complex. Biochemistry 12, 3o45-3o5o.

HEPTINSTAL, S., ARCHIBALD, A.R., BADDILEY, J. (197o): Teichoic acids and membrane function in bacteria. Nature 225, 519-521.

HERBAI, G., LINDAHL, U. (197o): Regional differences in the incorporation rates of [^{3}H]acetate and [^{35}S]sulfate into chondroitin sulfate of mouse costal cartilages in vitro. Acta Physiol. Scand. 8o, 5o2-5o9.

HERRING, G.M. (1968a): The chemical structure of tendon cartilage, dentin and bone matrix. Clin. Orthopaed. 6o, 261-299.

HERRING, G.M. (1968b): Studies on the protein-bound chondroitin sulfate of bovine cortical bone. Biochem. J. 1o7, 41-49.

HERRING, G.M. (1969): Mucosubstances and ion-binding in bone. In: Bibliotheca Nutritio et Dieta (ed. J.C. SOMOGYI), No. 13, pp. 147-154. Basel: Karger.

HERRING, G.M. (1972): The organic matrix of bone. In: The Biochemistry and Physiology of Bone (ed. E.H. BOURNE), 2nd Ed., Vol. 1, pp. 127-189. New York: Academic Press.

HERRING, G.M., KENT, P.W. (1963): Some studies on mucosubstances of bovine cortical bone. Biochem. J. 89, 4o5-414.

HIGHBERGER, J.H., KANG, A.H., GROSS, J. (1971): Comparative studies on the amino acid sequence of the α2-CB2 peptides from chick and rat skin collagens. Biochemistry 1o, 61o-615.

HIGHTON, T.C., MYERS, D.B., RAYNS, D.E. (1968): The intercellular spaces of synovial tissue. New Zealand Med. J. 67, 315-325.

VON HIPPEL, P.H. (1967): Structure and stabilization of the collagen molecule in solution. In: Treatise on Collagen (ed. G.N. RAMACHANDRAN), Vol. 1, pp. 253-338. New York: Academic Press.

VON HIPPEL, P.H., SCHLEICH, T. (1969): The effects of neutral salts on the structure and conformational stability of macromolecules in solution. In: Structure and Stability of Biological Macromolecules (eds. S.N. TIMASHEFF, G.D. FASMAN), pp. 417-574. New York: Dekker.

VON HIPPEL, P.H., WONG, K. (1963): Collagen ⇌ gelatin phase transition. I. Further studies of the effects of solvent environment and polypeptide chain composition. Biochemistry 2, 1387-1398.

HIRANO, S. (1972): Molecular structure of heparin as examined by nuclear magnetic resonance, optical rotatory dispersion, and electron microscopy. Int. J. Biochem. 3, 677-683.

HIRANO, S., MEYER, K. (1971): Enzymatic degradation of corneal and cartilagenous keratosulfates. Biochem. Biophys. Res. Commun. 44, 1371-1375.
HJERTQUIST, S., LEMPERG, R. (1972): Identification and concentration of the glycosaminoglycans of human articular cartilage in relation to age and osteoarthritis. Calcif. Tissue Res. 1o, 223-237.
HJERTQUIST, S., VEJLENS, L. (1968): The glycosaminoglycans of dog compact bone and epiphyseal cartilage in the normal state and in experimental hyperparathyroidism. Calcif. Tissue Res. 2, 314-333.
HJERTQUIST, S., WASTESON, A. (1972): The molecular weight of chondroitin sulfate from human articular cartilage. Effect of age and osteoarthritis. Calcif. Tissue Res. 1o, 31-37.
HO, T.-Y. (1966): The isolation and amino acid composition of the bone collagen in Pleistocene mammals. Comp. Biochem. Physiol. 18, 353-358.
HO, T.-Y. (1967a): The amino acids of bone and dentine collagens in Pleistocene mammals. Biochim. Biophys. Acta 133, 568-573.
HO, T.-Y. (1967b): Relationship between amino acid contents of mammalian bone collagen and body temperature as a basis for estimation of body temperature of prehistoric mammals. Comp. Biochem. Physiol. 22, 113-119.
HODARA, M.A., VON FIGURA, K., FILIPOVIC, I., BUDDECKE, E. (1973): Chemistry of the arterial wall XVI. Topochemical studies on the glycosaminoglycan and lipid metabolism in bovine arterial tissue. Hoppe-Seylers Z. Physiol. Chem. 354, 445-452.
HODGE, A.J. (1967): Structure at the electron microscope level. In: Treatise on Collagen, Vol. 1, Chemistry of Collagen (ed. G.N. RAMACHANDRAN), pp. 185-2o5. New York: Academic Press.
HODGE, J.A., PETRUSKA, J.A. (1963): Recent studies with the electron microscope on ordered aggregates of the tropocollagen molecule. In: Aspects of Protein Structure (ed. G.N. RAMACHANDRAN), pp. 289-3oo. New York: Academic Press.
HODSON, S. (1971): Why the cornea swells. J. Theor. Biol. 33, 419-427.
HODSON, S., MEENAN, S. (1969): The distribution of acidic polysaccharides in corneal stroma. Experientia 25, 13o5.
HOEVE, C.A.J., FLORY, P.J. (1958): The elastic properties of elastin. J. Am. Chem. Soc. 8o, 6523-6526.
HOFFERT, J.R., FROMM, P.O. (1969): Glycosaminoglycans in the vitrous and aqueous humors of the teleost. Comp. Biochem. Physiol. 3o, 771-777.
HOFFMAN, P., LINKER, A., MEYER, K. (1957): The acid mucopolysaccharides of connective tissues. II. Further experiments on chondroitin sulfate B. Arch. Biochem. Biophys. 69, 435-44o.
HOFFMAN, P., MASHBURN, Jr., T.A. (197o): Collagen-proteoglycan interactions in bovine nasal cartilage. In: Chemistry and Molecular Biology of the Intercellular Matrix (ed. E.A. BALAZS), pp. 1179-1199. New York: Academic Press.
HOPWOOD, J.J. (1972): Cartilage Proteoglycans. Biosynthesis and Chemical Structure. Ph.D. Dissertation submitted to Monash University, Monash, Australia.
HORNER, A.A. (1971): Macromolecular heparin from rat skin. J. Biol. Chem. 246, 231-239.
HOTTA, K., HAMAYAKI, H., KUROKAWA, M. (197o): Isolation and properties of a new type of sialopolysaccharide-protein complex from the jelly coat of sea urchin eggs. J. Biol. Chem. 245, 5434-544o.
HOW, M.J., JONES, J.V.S., DOYLE, J. (1969): Comparative studies of Lorenzini jelly from two species of elasmobranchs. I. Preparation of glycopeptides. Carbohyd. Res. 11, 2o7-216.
HOW, M.J., JONES, J.V.S., STACEY, M. (197o): Comparative studies of Lorenzini jelly from two species of elasmobranch. II. Structural studies of glycopeptides. Carbohyd. Res. 12, 171-181.
HRANISAVLJEVIC', J., SIMPSON, D.L., DAVIDSON, E.A. (1972). Properties of cartilage proteoglycans. Biochemistry 11, 2983-299o.
HUFFER, W.E. (197o): Glycosaminoglycans in the cartilage of developing chick embryos. Calcif. Tissue Res. 6, 55-69.
HULMES, D.J.S., MILLER, A., PARRY, D.A.D., PIEZ, K.A., WOODHEAD-GALLOWAY, J. (1973): Analysis of the primary structure of collagen for the origins of molecular packing. J. Mol. Biol. 79, 137-148.

HUMPHRIES, Jr., A.A. (197o): Incorporation of sulfate-^{35}S into the oviducts and egg jelly of newt. Notophthalmus viridescencs. Exp. Cell Res. 59, 157-161.
HUNT, S. (197o): Polysaccharide-protein Complexes in Invertebrates. New York: Academic Press.
HUNT, S., GRANT, M.E., LIEBOVICH, S.J. (197o): Polymeric collagen isolated from squid *(Loligo peallii)* connective tissue. Experientia 26, 12o4-12o5.
HUNT, S., JEVONS, F.R. (1965): The polysaccharide-peptide complex in the hypobranchial mucin of *Buccinum undatum* L. Biochim. Biophys. Acta 1o1, 214-216.
HUNT, S., JEVONS, F.R. (1966): The hypobranchial mucin of the whelk *Buccinum undatum* L. The polysaccharide sulfate component. Biochem. J. 98, 522-529.
HYMAN, L.H. (194o): The Invertebrates, Vol. 1, p. 281, pp. 334-36o. New York: McGraw-Hill.
HYMAN, L.H. (1959): The Invertebrates, Vol. 5, Smaller coelomate groups. New York: McGraw-Hill.
IGARASHI, S., KANG, A.H., GROSS, J. (197o): Renaturation and ordering by electron microscopy of the cyanogen bromide pepetides from the α2 chain of chick skin collagen. Biochem. Biophys. Res. Commun. 38, 697-7o2.
IGARASHI, S., TRELSTAD, R.L., KANG, A.H. (1973): Physical chemical properties of chick cartilage collagen. Biochim. Biophys. Acta 295, 514-519.
INOUE, S., YOSIZAWA, Z. (1966): Purification and properties of sulfated sialopolysaccharides isolated from pig colonic mucosa. Arch. Biochem. Biophys. 117, 257-265.
ISAAC, D.H., ATKINS, E.D.T. (1973): Molecular conformations of chondroitin-4-sulphate. Nature New Biol. 244, 252-253.
ISEMURA, M., IKENAKA, T., MATSUSHIMA, Y. (1973): Comparative study of carbohydrate complexes. I. The strucutres of glycopeptides derived from cuttlefish skin collagen. J. Biochem. (Tokyo) 74(1), 11-21.
ISENBERG, H.D., DOUGLAS, S.D., LAVINE, L.S., WEISFELLNER, H. (1965): Laboratory studies with coccolithophorid calcification. Stud. Trop. Oceanogr. 155-177.
ISENBERG, H.D., LAVINE, L.S., MANDELL, C., WEISSFELLNER, H. (1965): Qualitative chemical composition of the calcifying organic matrix obtained from cell-free coccoliths. Nature 2o6, 1153-1154.
ISOBE, N., SENO, N. (1971): Chondroitin-protein complex from squid skin. Biochim. Biophys. Acta 252, 612-615.
IVERIUS, Per H. (1972): Interaction between human plasma lipoproteins and connective tissue glycosaminoglycans. J. Biol. Chem. 247, 26o7-2613.
IWATA, H. (1969): Determination and the fine structure of chondroitin sulfate isomers of human cartilage and pathological cartilage and tissues. Nippon Seikeigekagakkai Zasshi 43, 455-473.
IWATA, H., URIST, M.R. (1972): Protein polysaccharide of bone morphogenetic matrix. Clin. Orthop. 87, 257-274.
JACKSON, D.S., BENTLEY, J.P. (1968): Collagen-glycosaminoglycan interactions. In: Treatise on Collagen (ed. B.S. GOULD), Vol. 2A, pp. 189-214. New York: Academic Press.
JACKSON, D.S., CLEARY, E.G. (1967): The determination of collagen and elastin. Methods Biochem. Anal. 15, 25-76.
JANADO, M., DUNSTONE, J. (1972): A glycoprotein from bovine nasal cartilage J. Biochem. (Tokyo) 71, 635-643.
JANCZURA, E., PERKINS, H.R., ROGERS, H.J. (1961): Teichuronic acid: A mucopolysaccharide present in wall preparations from vegetative cells of *Bacillus subtilis*. Biochem. J. 8o, 82-93.
JANSSON, L., LINDAHL, U. (197o): Evidence for the existence of a multichain proteoglycan of heparan sulfate. Biochem. J. 117, 699-7o2.
JAQUES, L.B., KAVANAGH, L.W., MAZUREK, M., PERLIN, A.S. (1966): The structure of heparin. Proton magnetic resonance spectral observations. Biochem. Biophys. Res. Commun. 24, 447-451.
JARVIK, E. (1968): Aspects of vertebrate phylogeny. In: Current Problems of Lower Vertebrate Phylogeny (ed. T. ØRVIG), pp. 497-527. Stockholm: Almqvist and Wiksell.
JEANLOZ, R.W. (197o): Mucopolysaccharides of higher animals. In: The Carbohydrates (eds. W. PIGMAN, D. HORTON, A. HERP), Vol. 2B, pp. 589-625. New York: Academic Press.

JEANLOZ, R.W., BALAZS, E.A. (eds.) (1965): The Amino Sugars. Vols. 1 and 2 published, others in preparation, New York: Academic Press.
JENSEN, C.E., ZACHARIAE, F. (1958a): Studies on the mechanism of ovulation. Histochemical and physico-chemical investigations on genuine follicular fluids. Acta Endocrinol. 27, 343-355.
JENSEN, C.E., ZACHARIAE, F. (1958b): Studies on the mechanism of ovulation. Isolation and analysis of acid mucopolysaccharides in bovine follicular fluid. Acta Endocrinol. 27, 356-368.
JEUNIAUX, C. (1971): Chitinous structures. In: Comprehensive Biochemistry (eds. M. FLORKIN, E.H. STOTZ), Vol. 26C, pp. 595-632. New York: Elsevier.
JOHN, R., THOMAS, J. (1972): Chemical compositions of elastins isolated from aortas and pulmonary tissues of humans of different ages. Biochem. J. 127, 261-269.
JOPE, M. (1967a): The protein of brachiopod shell. I. Amino acid composition and implied protein taxonomy. Comp. Biochem. Physiol. 2o, 593-6oo.
JOPE, M. (1967b): The protein of brachiopod shell. II. Shell protein from fossil articulates; amino acid composition. Comp. Biochem. Physiol. 2o, 6o1-6o5.
JOPE, M. (1969a): Protein of brachiopod shell. III. Comparison with structural protein of soft tissue. Comp. Biochem. Physiol. 3o, 2o9-224.
JOPE, M. (1969b): Protein of brachiopod shell. IV. Shell protein from fossil inarticulates: Amino acid composition and disc electrophoressis of fossil articulate shell protein. Comp. Biochem. Physiol. 3o, 225-232.
JOPE, M. (1971): Constitutents of brachiopod shells. In: Comprehensive Biochemistry (eds. M. FLORKIN, E.H. STOTZ), Vol. 26C, pp. 749-784. Amsterdam: Elsevier.
JOSSE, J., HARRINGTON, W.F. (1964): Role of pyrrolidine residues in the structure of stabilization of collagen. J. Molec. Biol. 9, 269-287.
JUKES, T.H. (1966): Molecules and Evolution. New York: Columbia University Press.
JUNGE-HÜLSING, G., WAGNER, H. (1968): Altern und Bindegewebe. Z. Gerontol. 1, 345-359.
JÜRASEK, L., COLVIN, J.R., WHITAKER, D.R. (1967): Microbiological aspects of the formation and degradation of cellulosic fibres. Adv. Appl. Microbiol. 9, 131-17o.
KANG, A.H., GROSS, J. (197oa): Amino acid sequence of cyanogen bromide peptides from the amino-terminal region of chick skin collagen. Biochemistry 9, 796-8o3.
KANG, A.H., GROSS, J. (197ob): Relationship between the *intra* and *inter*molecular cross-links of collagen. Proc. Nat. Acad. Sci. USA 67, 13o7-1314.
KANG, A.H., IGARASHI, S., GROSS, J. (1969): Characterization of the cyanogen bromide peptides from the α2 chain of chick skin collagen. Biochemistry 8, 32oo-32o4.
KANG, A., PIEZ, K.A., GROSS, J. (1969a): Characterization of the α chains of chick skin collagen and the nature of the NH_2-terminal cross-link region. Biochem. 8, 3648-3655.
KANG, A.H., PIEZ, K.A., GROSS, J. (1969b): Characterization of the cyanogen bromide peptides from the α1 chain of chick skin collagen. Biochem. 8, 15o6-1514.
KAO, K.T., LESLIE, J.G., HITT, W.E., McGAVACK, T.H. (197o): Isolation and partial characterization of protein polysaccharides from rat costal cartilage. Proc. Soc. Exp. Biol. Med. 135, 781-784.
KAPLAN, D., MEYER, K. (1959): Ageing of human cartilage. Nature 183, 1267-1268.
KAPLAN, D., MEYER, K. (196o): Mucopolysaccharides of aorta at various ages. Proc. Soc. Exp. Biol. Med. 1o5, 78-81.
KAPLAN, N.O. (1965): Evolution of dehydrogneases. In: Evolving Genes and Proteins (eds. V. BRYSON, H.J. VOGEL), pp. 243-277. New York: Academic Press.
KATCHALSKY, A. (1964): Polyelectrolytes and their biological interactions. Biophys. J. 4, 9-41.
KATZ, E.P. (197o): NMR study of the slowly exchanging hydrogens of 64oÅ steerskin collagen. Biopolymers 9, 745-748.
KATZ, E.P., LI, S. (1972): The molecular packing of collagen in mineralized and non-mineralized tissues. Biochem. Biophys. Res. Commun. 46, 1368-1373.
KATZMAN, R.L. (1971): On arabinose as a constituent of hyaluronic acid from bovine brain. J. Neurochem. 18, 1187-119o.
KATZMAN, R.L., BHATTACHARYYA, A.K., JEANLOZ, R.W. (1969): Invertebrate connective tissue. I. The amino acid and carbohydrate composition of the collagen from *Thyone briareus*. Biochim. Biophys. Acta 184, 523-528.

KATZMAN, R.L., HALFORD, M.H., REINHOLD, V.N., JEANLOZ, R.W. (1972): Isolation and structure determination of glucosylgalactosylhydroxylysine from sponge and sea anemone collagen. Biochemistry 11, 1161-1167.

KATZMAN, R.L., JEANLOZ, R.W. (1969): Acid polysaccharides from invertebrate connective tissue: Phylogenetic aspects. Science 166, 758-759.

KATZMAN, R.L., JEANLOZ, R.W. (197oa): The carbohydrate chemistry of invertebrate connective tissue. In: Chemistry and Molecular Biology of the Intercellular Matrix (ed. E.A. BALAZS), Vol. 1, pp. 217-227. New York: Academic Press.

KATZMAN, R.L., JEANLOZ, R.W. (197ob): Invertebrate connective tissue. V. Isolation of a peptide containing fucose, mannose and hexosamine from *Metridium dianthus* collagen. Biochem. Biophys. Res. Commun. 4o, 628-635.

KATZMAN, R.L., JEANLOZ, R.W. (197oc): Evidence for a covalent linkage between heteropolysaccharide and collagen from the sea anemone *(Metridium dianthus)*. Federation Proc. 29, 599 abs.

KATZMAN, R.L., JEANLOZ, R.W. (197od): Are acidic polysaccharides involved in collagen fibril formation or stabilization? Biochim. Biophys. Acta 22o, 516-521.

KATZMAN, R.L., JEANLOZ, R.W. (1973): Thyonatan 4-sulfate [(1→2)-α-L-Fucopyranan 4-sulfate]. A sulfated polysaccharide from *Thyone briareus* connective tissue. J. Biol. Chem. 248, 5o-55.

KATZMAN, R.L., KANG, A.H. (1972): The presence of fucose, mannose and glucosamine-containing heteropolysaccharide in collagen from the sea anemone *Metridium dianthus*. J. Biol. Chem. 247, 5486-5489.

KATZMAN, R.L., KANG, A.H., BEACHEY, E.H. (1973): Collagen-induced platelet aggregation: Involvement of an active glycopeptide (α1-CB5). Science 181, 67o-672.

KATZMAN, R.L., LISOWSKA, E., JEANLOZ, R.W. (197o): Invertebrate connective tissue. Isolation of D-arabinose from sponge acidic polysaccharide. Biochem. J. 119, 17-19.

KAWAI, Y. (1968): Mucopolysaccharides from the connective tissues of the squid *Ommastrephes sloani pacificus*. Natural Science Report Ochanomizue Univ. 19, 23-45.

KAWAI, Y., SENO, N., ANNO, K. (1966): Chondroitin polysulfate of squid cartilage. J. Biochem. 6o, 317-321.

KAWALEC, M., KOWALSKI, H., KAWIAK, J. (1972): Transport of sulfate in cartilage. Acta Med. Pol. 13, 251-259.

KEEDGSTRA, K., TALMADGE, K.W., BAUER, W.D., ALBERSHEIM, P. (1973): The structure of plant cell walls. III. A model of the walls of suspension-cultured sycamore cells based on the interconnections of the macromolecular components. Plant Physiol. 51, 188-196.

KEELEY, F.W., LA BELLA, F.S. (1972a): Amino acid composition of elastin in the developing chick aorta. Conn. Tissue Res. 1, 113-12o.

KEELEY, F.W., LA BELLA, F.S. (1972b): Isolation of dityrosine from an alkali-soluble connective tissue protein. Biochim. Biophys. Acta 263, 52-69.

KEELEY, F.W., LA BELLA, F., QUEEN, G. (1969): Dityrosine in a non-hydroxyproline, alkali-soluble protein isolated from chick aorta and bovine ligament. Biochem. Res. Commun. 34, 156-161.

KEFALIDES, N.A. (197o): Comparative biochemistry of mammalian basement membranes. In: Chemistry and Molecular Biology of the Intercellular Matrix (ed. E.A. BALAZS), Vol. 1, pp. 535-573. New York: Academic Press.

KEFALIDES, N.A. (1971): Chemical properties of basement membranes. Int. Rev. Exp. Pathol. 1o, 1-39.

KEFALIDES, N.A. (1973): Structure and biosynthesis of basement membranes. Int. Rev. Conn. Tissue Res. 6, 63-1o4.

KEISER, H.D., SANDSON, J. (1973): Immunological studies of fragments of bovine nasal cartilage proteoglycan subunit. Presented at the meeting of the Society for Complex Carbohydrates, Chicago, Nov. 14-16, 1973.

KEISER, H.D., SHULMAN, H.J., SANDSON, J.I. (1972): Immunochemistry of cartilage proteoglycan. Immunodiffusion and gel-electrophoretic studies. Biochem. J. 126, 163-169.

KEMPSON, B.E., MUIR, H., SWANSON, S.A.V., FREEMAN, M.A.R. (197o): Correlations between stiffness and the chemical constituents of cartilage on the human femoral head. Biochim. Biophys. Acta 215, 7o-77.

KENT, P.W. (1964): Chitin and mucosubstances. In: Comparative Biochemistry (eds. M. FLORKIN, H.S. MASON), Vol. 7, pp. 93-136. New York: Academic Press.

KERN, H.A., BRASSIL, D. (1967): Fractionation of the mucoproteins of bovine corneal stroma. Arch. Biochem. Biophys. 118, 115-121.

KERSTEN, H., ZÜRN, L. (1959): Hydroxylysine in wool. Nature 184, 149o-1491.

KERWAR, S.S., KOHN, L.D., LAPIERE, C.M., WEISSBACH, H. (1972): The *in vitro* synthesis of procollagen on polysomes. Proc. Nat. Acad. Sci. USA 69, 2727-2731.

KIMATA, K., OKAYAMA, M., OOHIRA, A., SUZUKI, S. (1973): Cytodifferentiation and proteoglycan biosynthesis. Mol. Cell. Biochem. 1, 211-228.

KIMATA, K., OKAYAMA, M., SUZUKI, S., SUZUKI, I., HOSHINO, M. (1971): Nascent mucopolysaccharides attached to the golgi membrane of chondrocytes. Biochim. Biophys. Acta 237, 6o6-61o.

KIMURA, S. (1971a): Marine invertebrate collagens. III. Characterization of cuticle collagen in annelids. Nippon Suisan Gakkaishi 37, 419-431.

KIMURA, S. (1971b): Marine invertebrate collagens. IV. Structural differences in collagen from cuticles and intestines of *Neanthes diversicolor*. Nippon Suisan Gakkaishi 37, 432-44o.

KIMURA, S. (1972a): Determination of the glycosylated hydroxylysines in several invertebrate collagens. J. Biochem. 71, 367-37o.

KIMURA, S. (1972b): Marine invertebrate collagens. V. Neutral sugar compositions and glycosylated hydroxylysine contents of several collagens. Nippon Suisan Gakkaishi 38, 1153-1161.

KIMURA, S. (1973): Marine invertebrate collagens. VII. Galactose mono and oligosaccharides from Neanthes cuticle collagen. Nippon Suisan Gakkaishi 39, 311-316.

KIMURA, S., KOBAYASHI, N., KUBOTA, M. (1972): Marine invertebrate collagens. VI. Properties of collagen from a ascidian, *Halocynthia roretzi*. Nippon Suisan Gakkaishi 38, 1163-1169.

KIMURA, S., KUBOTA, M. (1966): Studies on elastoidin. I. Some chemical and physical properties of its components. J. Biochem. (Tokyo) 6o, 615-621.

KIMURA, S., KUBOTA, M. (1967): Elastoidin. II. Dialyzable peptides released by pepsin digestion of elastoidin. Nippon Suisan Gakkaishi 33, 43o-437.

KIMURA, S., KUBOTA, M. (1968a): Elastoidin. III. Protein components of soluble elastoidin. Nippon Suisan Gakkaishi 34, 535-54o.

KIMURA, S., KUBOTA, M. (1968b): Elastoidin. IV. Hydrothermal stability of elastoidin. Nippon Suisan Gakkaishi 34, 1o36-1o39.

KIMURA, S., KUBOTA, M. (1968c): Properties of collagen from the abalone. Nippon Suisan Gakkaishi 34, 925-929. English abstract: Chem. Abst. 72, 2o (197o), Abs. No. 51o22c.

KIMURA, S., NAGOAKA, Y., KUBOTA, M. (1969): Marine invertebrate collagens. I. Collagens from crustaceans and molluscans. Nippon Suisan Gakkaishi 35, 743-748.

KIRRANE, J.A., GLYNN, L.E. (1968): Immunology of Collagen. Int. Rev. Conn. Tissue Res. 4, 1-31.

KIVIRIKKO, K.I., PROCKOP, D.J. (1972): Partial purification and characterization of protocollagen lysine hydroxylase from chick embryos. Biochim. Biophys. Acta 258, 366-379.

KLEINE, T.O., KIRSIG, H.J., HILZ, H. (1971): Biosynthesis of chondroitin sulfate proteins. III. Metabolic and chemical heterogeneity in calf rib cartilage as revealed by cesium chloride gradient centrifugation and CPC (cetyl pyridinium chloride) cellulose chromatography. Hoppe-Seylers Z. Physiol. Chem. 352, 479-487.

KLEMPERER, P. (1955): The significance of the intermediate substances of the connective tissue in human disease. The Harvey Lectures, Series XIL, 1953/54, 1oo-123.

KOBAYASHI, T.K., PEDRINI, V. (1973): Proteoglycans-collagen interactions in human costal cartilage. Biochim. Biophys. Acta 3o3, 148-16o.

KONDO, K., SENO, N., ANNO, K. (1971): Mucopolysaccharides from chicken skin of three age groups. Biochim. Biophys. Acta 244, 513-522.

KONDO-OTSUKA, K., SENO, N., ANNO, K. (1972): Isolation of heparan sulfate from chicken skin. Biochim. Biophys. Acta 279, 26o-264.

KOSHER, R.A., LASH, J.W., MINOR, R.R. (1973): Environmental enhancement of *In vitro* chondrogenesis. IV. Stimulation of somite chondrogenesis by exogenous chondromucoprotein. Develop. Biol. 35, 21o-22o.
KOTUKU, T., YOSIZAWA, Z., YAMAUCHI, F. (1967): Comparison of heparin specimens isolated from bovine, porcine, and whale organs. Arch. Biochem. Biophys. 12o, 553-562.
KRAEMER, P.M. (1971a): Complex carbohydrates of animal cells: Biochemistry and physiology of the cell periphery. In: Biomembranes (ed. L.A. MANSON), Vol. 1, pp. 67-19o. New York: Plenum Press.
KRAEMER, P.M. (1971b): Heparan sulfates of cultured cells. I. Membrane-associated and cell-sap species in Chinese hamster cells. Biochemistry 1o, 1437-1445.
KRAEMER, P.M. (1971c): Heparan sulfates of cultrued cells. II. Acid-soluble and precipitable species of different cell lines. Biochemistry 1o, 1445-1451.
KRESSE, H., GROSSMAN, A. (197o): Vergleichende Untersuchung des Mucopolysaccharid- und Kollagengehaltes in verschiedenen topographischen Zonen der Niere von Ratte, Hund und Schwein. Z. Klin. Chem. klin. Biochem. 8, 42o-424.
KRESSE, H., HEIDEL, H., BUDDEKCE, E. (1971): Chemical and metabolic heterogeneity of a bovine aorta chondroitin sulfate-dermatan sulfate proteoglycan. Eur. J. Biochem. 22, 557-562.
KRÖZ, v.W., BUDDECKE, E. (1967): Chemische und makromolekulare Altersveränderungen von Polysaccharid-Proteinen aus menschlichem Rippenknorpel. Hoppe-Seylers Z. Physiol. Chem. 348, 665-674.
KUBOTA, M., KIMURA, S. (1967): Skin collagen of the great blue shark. Nippon Suisan Gakkaishi 33, 338-342.
KUBOTA, M., KIMURA, S., TAJIMA, H. (1967): Collagen in aquatic animals. III. Physicochemical properties of the pepsin-solubilized collagen of dolphin hide. Nippon Suisan Gakkaishi 33, 217-223.
KUBOTA, M., UCHIDA, N., KIMURA, S. (1969): Studies on neutral salt soluble collagen from whale nose cartilage. Nippon Suisan Gakkaishi 35, 1oo6-1o11.
KUBOTA, M., UCHIDA, N., KIMURA, S. (197oa): On the subunit of whale collagens. Nippon Suisan Gakkaishi 36, 1246-1249.
KUBOTA, M., UCHIDA, N., KIMURA, S. (197ob): Comparison of collagens from different tissues of sperm whale. Nippon Suissan Gakkaishi 36, 1242-1245.
KUETTNER, K.E., WEZEMAN, F.H., SIMMONS, D.J., LISK, P.Y., CROXEN, R.L., SOBLE, L.W., EISENSTEIN, R. (1972): Lysozyme in preosseous cartilage. V. Response of embryonic chick cartilage to antilysozyme antibodies in organ culture. Lab. Invest. 27, 324-33o.
KUHN, T.S. (197o): The Structure of Scientific Revolutions, 2nd Ed. Chicago: The University of Chicago Press.
KÜHN, K., GRASSMAN, W., HOFMANN, U. (1959): Über die Bildung der Kollagenfibrillen aus gelöstem Kollagen und die Funktion der kohlenhydrathaltigen Begleitkomponenten. Z. Naturforschung 14, 436-443.
KUHNKE, E. (1958): Neuere Ergebnisse zum Feinbau der Kollagenfibrille. Z. Rheumaforschung 17, 259-274.
KULONEN, E., PIKKARAINEN, J. (197o): Comparative studies on the chemistry and chain structure of collagen. In: Chemistry and Molecular Biology of the Intercellular Matrix (ed. E.A. BALAZS), Vol. 1, pp. 81-97. New York: Academic Press.
KURUP, P.A., LEELAMNA, S., VYAYAMMAL, P.L., VYAYELEKSHMI, N.R. (1969): Nature of cuticle of the scorpion *Heterometrus seaber*. Indian J. Exp. Biol. 7, 27-28.
KVIST, T.N., FINNEGAN, C.V. (197oa): The distribution of glycosaminoglycans in the axial region of the developing chick embryo. I. Histochemical evidence. J. Exp. Zool. 175, 221-24o.
KVIST, T.N., FINNEGAN, C.V. (197ob): The distribution of glycosaminoglycans in the axial region of the developing chick embryo. II. Biochemical analysis. J. Exp. Zool. 175, 241-258.
LA BELLA, F., KEELEY, F., VIVIAN, S., THORNHILL, D. (1967): Evidence for dityrosine in elastin. Biochem. Biophys. Res. Commun. 26, 748-753.
LAMPORT, D.T.A., KATONA, L., ROERIG, S. (1973): Galactosylserine in extensin. Biochem. J. 133, 125-132.
LAMPORT, D.T.A., MILLER, D.H. (1971): Hydroxyproline arabinosides in the plant kingdom. Plant Physiol. 48, 454-456.

LANGHAM, M.E., HART, R.W., COX, J. (1969): The interaction of collagen and mucopolysaccharides. In: The Cornea (ed. M.E. LANGHAM), pp. 157-184. Baltimore: Johns Hopkins Press.

LANGLEY, O.K. (197o): Interaction between peripheral nerve polyanions and Alcian Blue. J. Neurochem. 17, 1535-1541.

LASH, J.N., WHITEHOUSE, M.W. (196o): An unusual polysaccharide in chondroid tissue of the snail *Busycon*: polyglucose sulfate. Biochem. J. 74, 351-355.

LAURENT, T.C. (197o): Structure of hyaluronic acid. In: Chemistry and Molecular Biology of the Intercellular Matrix (ed. E.A. BALAZS), Vol. 2, 7o3-732. New York: Academic Press.

LAURENT, T.C., ANSETH, A. (1961): Studies on corneal polysaccharides. II. Characterization. Exp. Eye Res. 1, 99-1o5.

LAURENT, T.C., WASTESON, Å., ÖBRINK, B. (1969): Macromolecular properties of glycosaminoglycans (mucopolysaccharides) and proteoglycans. In: Ageing of Connective and Skeletal Tissue (eds. A. ENGEL, T. LARSSON), pp. 65-8o. Stockholm: Nordiska Bokhandelns Förlag.

LAVINE, L.S., ISENBERG, H.D. (1964): Comparative biology of calcification. J. Bone Joint Surg. 46-A, 1563-1576.

LAYMAN, D.L., SOKOLOFF, L., MILLER, E.J. (1972): Collagen synthesis by articular chondrocytes in monolayer culture. Exp. Cell Res. 73, 1o7-112.

LEAKE, C.D. (1969): Historical aspects of the concept of organizational levels of living material. In: Hierarchical Structures (eds. L.L. WHYTE, A.G. WILSON, D. WILSON), pp. 147-159. New York: Elsevier.

LEE, D.E. (1965): The Physiology of Nematodes. San Francisco: Freeman.

LEESON, C.R., THREADGOLD, L.T., SINCLAIR, N.R.St.C. (1961):Histochemical observations upon the development of the notochord in the chick. Acta Anat. 46, 91-97.

LEHMAN, J.P. (1958a): Super-ordre des chondrostéens. Formes fossiles. In: Traité de Zoologie (ed. P.-P. GRASSÉ), Vol. 13, pp. 213o-217o. Paris: Masson.

LEHMAN, J.P. (1966): Dinoi et Crossopterygii. In: Traité de Paleontologie (ed. J. PIVETEAU, Vol. IV, Part 3. Paris: Masson.

LEHNHOFF, H.M., KLINE, E.S., HURLEY, R. (1957): A hydroxyproline-rich, intracellular, collagen-like protein of *Hydra* nematocysts. Biochim. Biophys. Acta 26, 2o4-2o5.

LEHTONEN, A., KARKKAINEN, J., HAAHTI, E. (1966): Carbohydrate components in the epithelial mucin of hagfish, *Myxine glutinosa*. Acta Chem. Scand. 2o, 1456-1462.

LENAERS, A., ANSAY, M., NUSGENS, B.V., LAPIÈRE, C.M. (1971): Collagen made of extended α-chains, procollagen in genetically defective dermatosporaxic calves. Eur. J. Biochem. 23, 533-543.

LEONE, C.A. (ed.) (1964): Taxonomic Biochemistry and Serology. New York: Ronald Press.

LEVENE, P.A. (1925): Hexosamines and Mucoproteins. New York: Longmans, Green.

LEVINE, P.T., GLIMCHER, M.J., SEYER, J.M., HUDDLESTON, J.I., WEIN, J.W. (1966): Noncollagenous nature of the proteins of shark enamel. Science 154, 1192-1193.

LEVINE, P.T., SEYER, J., HUDDLESTON, J., GLIMCHER, M.J. (1967): The comparative biochemistry of the organic matrix proteins of developing enamel. Int. Arch. Oral Biol. 12, 4o7-41o.

LEVITT, D., DORFMAN, A. (1972): The irreversible inhibition of differentiation of limb-bud mesenchyme by bromodeoxyuridine. Proc. Nat. Acad. Sci. USA 69, 1253-1257.

LEVITT, D., DORFMAN, A. (1974): Concepts and mechanisms of cartilage differentiation. In: Current Topics in Developmental Biology (eds. A. MONROY, A.A. MOSCONA), Vol. 8, pp. 1o3-149. New York: Academic Press.

LEWIS, M.S., PIEZ, K.A. (1964): The characterization of collagen from the skin of the dogfish shark, *Squalus acanthias*. J. Biol. Chem. 239, 3336-334o.

LIBERTI, P.A., STIVALA, S.S. (1967): Physicochemical studies of fractionated bovine heparin. Arch. Biochem. Biophys. 119, 51o-518.

LINDAHL, U. (197oa): Attempted isolation of a heparin proteoglycan from bovine liver capsule. Biochem. J. 116, 27-34.

LINDAHL, U. (197ob): Structure of heparin, heparan sulfate and their proteoglycans. In: Chemistry and Molecular Biology of the Intercellular Matrix (ed. E.A. BALAZS), Vol. 2, pp. 943-96o. New York: Academic Press.

LINDAHL, U., AXELSSON, O. (1971): Identification of iduronic acid as the major sulfated uronic acid of heparin. J. Biol. Chem. 246, 74-82.

LINDAHL, U., BÄCKSTRÖM, G., MALMSTRÖM, A., FRANSSON, L. (1972): Biosynthesis of L-iduronic acid in heparin: Epimerization of D-glucuronic acid on the polymer level. Biochem. Biophys. Res. Commun. 46, 985-991.
LINDAHL, U., RODÉN, L. (1972): Carbohydrate-peptide linkages in proteoglycans of animal, plant and bacterial origin. In: Glycoproteins (ed. A. GOTTSCHALK), 2nd Ed., pp. 491-518. Amsterdam: Elsevier.
LINDNER, J. (1963): Über die Synthese der organischen Interzellularsubstanz in Knorpel und Knochen. Verh. Dtsch. Ges. Path. 47, 15o-157.
LINKER, A., HOVINGH, P. (1973): Heparitin sulfates. (Heparan sulfates). Carbohyd. Res. 29, 41-62.
LINSENMAYER, T.F., TOOLE, B.P., TRELSTAD, R.L. (1973): Temporal and spatial transitions in collagen types during chick limb development. Develop. Biol. 35, 232-239.
LINSENMAYER, T.F., TRELSTAD, R.L., GROSS, J. (1973): The collagen of chick embryonic notochord. Biochem. Biophys. Res. Commun. 53, 39-45.
LIPPMAN, M. (1968): Glycosaminoglycans and cell division. In: Epithelial-Mesenchymal Interactions (eds. R. FLEISCHMAJER, R.E. BILLINGHAM), pp. 2o8-229. Baltimore: Williams & Wilkins.
LIPPMAN, M., MATHEWS, M.B., CIFONELLI, J.A. (1971): Complete and reversible inhibition of cell division by mactin. Proc. Am. Soc. Cancer Res. 12, 38.
LIPSON, M.J., CERSKUS, R.A., SILBERT, J.E. (1971): Glycosaminoglycans and glycosaminoglycan-degrading enzyme of *Rana catesbiana* back skin during late stages of metamorphosis. Dev. Biol. 25, 198-2o8.
LIPSON, M.J., SILBERT, J.E. (1965): Mucopolysaccharides of tadpole fin and back skin. Biochim. Biophys. Acta 1o1, 279-284.
LIPSON, M.J., SILBERT, J.E. (1968): Glycosaminoglycans of adult frog back skin. Biochim. Biophys. Acta 158, 344-35o.
LLOYD, A.G., BLOOM, G.D., BALAZS, E.A. (1967): Evidence for the covalent association of heparin and protein in mast-cell granules. Biochem. J. 1o3, 76P-77P.
LLOYD, A.G., DOGSON, K.S., PRICE, R.G. (1963): Comparative studies on cartilage aminopolysaccharide sulfates. I. Polysaccharides from shark, skate, dogfish and fin whale. Biochim. Biophys. Acta 69, 496-5o4.
LOEWI, G. (1961): The acid mucopolysaccharides of human skin. Biochim. Biophys. Acta 52, 435-44o.
LOEWI, G., MEYER, K. (1958): The acid mucopolysaccharides of embryonic skin. Biochim. Biophys. Acta 27, 453-456.
LOEWI, G., MUIR, H. (1965): The antigenicity of chondromucoprotein. Immunology 9, 119-127.
LOVE, R., FROMMHAGEN, L.H. (1953): Histochemical studies on the clam *Mactra solidissima*. Proc. Soc. Exp. Biol. Med. 83, 838-844.
LOVELL, D., BRIGGS, J.C., SCHORAH, C.J. (1966): Chemical analysis of acid mucopolysaccharides of mixed salivary tumors. Brit. J. Cancer 2o, 463-468.
LOVELL, D., SCHORAH, C.J., CURRAN, R.C. (1966): Isolation and identification of a mucopolysaccharide from eggwhite. Nature 212, 86o-861.
LOVELL, D., TIGHE, J.R., CURRAN, R.C. (1967): The chemical determination of acid mucopolysaccharides and collagen of the normal human placenta. J. Obstet. Gynaec. Brit. Cwlth. 74, 568-571.
LOWTHER, D.A., BAXTER, E. (1966): Isolation of a chondroitin sulfate protein complex from bovine intervertebral disks. Nature 211, 595-597.
LOWTHER, D.A., NATARAJAN, M. (1972): The influence of glycoprotein on collagen fibril formation in the presence of chondroitin sulfate proteoglycan. Biochem. J. 127, 6o7-6o8.
LOWTHER, D.A., PRESTON, B.N., MEYER, F.A. (197o): Isolation and properties of chondroitin sulfates from bovine heart valves. Biochem. J. 118, 595-6o1.
LOWTHER, D.A., TOOLE, B.P., HERRINGTON, A.C. (197o): Interaction of proteoglycans with collagen. In: Chemistry and Molecular Biology of Intercellular Matrix (ed. E.A. BALAZS), Vol. 2, pp. 1135-1153. New York: Academic Press.
LUBOSCH, W. (1938): Die permanenten knorpeligen Skeletteile. In: Handbuch der vergleichenden Anatomie der Wirbeltiere (eds. L. BOLK, E. GOPPERT, E. KALLIUS, W. LUBOSCH), 5. Band, pp. 249-274. Berlin: Urban & Schwarzenberg.

LUCAS, F., RUDALL, K.M. (1968): Extracellular fibrous proteins: The silks. In: Comprehensive Biochemistry, Vol. 26B, Extracelluar and Supporting Structures (eds. M. FLORKIN, E.H. STOTZ), pp. 475-558. New York: Elsevier.
LUDOWIEG, J.J., SHARMA, M.L., HASHIMOTO, A., EMMONS, E., NISHIZAWA, I., OKAZAKI, K. (1972): The mammalian intervertebral disc. Composition of the nucleus pulposus of whales during development and aging. Conn. Tissue Res. 1, 79-91.
LUNT, M.R., KENT, P.W. (1961): Evidence for the occurrence of uridinediphospho-N-acetylglucosamine in crustacean tissues. Biochem. J. 78, 128-134.
LUTZ, P.L., ROBERTSON, J.D. (1971): Osmotic constituents of the coelacanth *Latimeria chalumnae* Smith. Biol. Bull. 141, 553-56o.
LYMAN, B., ROSENBERG, L., KARPATKIN, S. (1971): Biochemical and biophysical aspects of human platelet adhesion to collagen fibers. J. Clin. Invest. 5o, 1854-1863.
MacGREGOR, E.A., BOWNESS, J.M. (1971): Interaction of proteoglycans and chondroitin sulfates with calcium or phosphate ions. Can. J. Biochem. 49, 417-425.
MACKIE, W., PARKER, K.D., SMOLKO, E.E. (1971): Crystalline structures of poly-D-mannuronic acid and poly-L-guluronic acid. Polymer. Letters 9, 311-316.
MALMSTROM, A., FRANSSON, L.A. (1971): Structure of pig skin dermatan sulfate. 2. Demonstration of sulfated iduronic acid residues. Eur. J. Biochem. 18, 431-435.
MAMMI, M., GOTTE, L., PEZZIN, G. (1968): Evidence for order in the structure of α-elastin. Nature 22o, 371-373.
MANERY, J.F. (1966): Connective tissue electrolytes. Federation Proc. 25, 1799-18o3.
MANKIN, H.J., BARON, P.A. (1965): The effect of aging on protein synthesis in articular cartilage of rabbits. Lab. Invest. 14, 658-664.
MANNING, G.S. (1969): Limiting laws and counterion condensation in polyelectrolyte solutions. I. Colligative properties. J. Chem. Phys. 51, 924-938.
MANWELL, C. (1963): The blood proteins of cyclostomes. A study in phylogenetic and ontogenic biochemistry. In: The Biology of Myxine (eds. A. BRODAL, R. FÄNGER), pp. 372-455. Oslo: Universitets-Forloget.
MARGOLIASH, E. (1969): Homology: A definition. Science 163, 127.
MARGOLIS, R.U. (1967): Acid mucopolysaccharides and proteins of bovine whole brain, white matter, and myelin. Biochim. Biophys. Acta 141, 91-1o2.
MARGOLIS, R.U. (1969): Mucopolysaccharides. In: Handbook of Neurochemistry (ed. A. LAJTHA), Vol. 1, pp. 245-26o. New York: Plenum Press.
MARGOLIS, R.U., ATHERTON, D.M. (1972): The heparan sulfate of rat brain. Biochim. Biophys. Acta 273, 368-373.
MARGOLIS, R.K., MARGOLIS, R.U. (197o): Sulfated glycoproteins from rat brain glycoproteins. Biochem. 9, 4389-4396.
MARGOLIS, R.K., MARGOLIS, R.U. (1973): The turnover of hexosamine and sialic acid in glycoproteins and mucopolysaccharides of brain. Biochim. Biophys. Acta 3o4, 413-42o.
MARGOLIS, R.U., MARGOLIS, R.K., ATHERTON, D.M. (1972): Carbohydrate-peptide linkages in glycoproteins and mucopolysaccharides from brain. J. Neurochem. 19, 2317-2324.
MARGOLIS, R.U., MARGOLIS, R.K., SANTELLA, R., ATHERTON, D.M. (1972): The hyaluronidase of brain. J. Neurochem. 19, 2325-2332.
MARK, K. von der, BORNSTEIN, P. (1973): Characterization of the pro α1 chain of procollagen. J. Biol. Chem. 248, 2285-2289.
MARK, K. von der, REXROTH, F., WENDT, P., KÜHN, K. (197o): Isolation and characterization of cyanogen bromide-cleaved peptides from a collagenase digest of α1-chain of calf skin collagen. Eur. J. Biochem. 16, 143-151.
MAROUDAS, A. (1968): Physicochemical properties of cartilage in the light of ion exchange theory. Biophys. J. 8, 575-595.
MAROUDAS, A. (197o): Effect of fixed charge density on the distribution and diffusion coeffecients of solutes in cartilage. In: Chemistry and Molecular Biology of the Intercellular Matrix (ed. E.A. BALAZS), Vol. 3, pp. 1389-14o1. New York: Academic Press.
MAROUDAS, A. (1972): X-Ray microprobe analysis of articular cartilage. Conn. Tissue Res. 1, 153-163.

MAROUDAS, A. (1973): Physico-chemical properties of articular cartilage. In: Adult Articular Cartilage (ed. M.A.R. FREEMAN), pp. 131-17o. London: Pitman.
MAROUDAS, A., EVANS, H. (1972): A study of ionic equilibria in cartilage. Conn. Tissue Res. 1, 69-77.
MAROUDAS, A., MUIR, H., WINGHAM, J. (1969): The correlation of fixed negative charge with glycosaminoglycan content of human articular cartilage. Biochim. Biophys. Acta 177, 492-5oo.
MARSHALL, R.D. (1972): Glycoproteins. Ann. Rev. Biochem. 41, 673-7o2.
MARSHALL, R.D., NEUBERGER, A. (197o): Structure and metabolism of glycoproteins. Adv. Carbohyd. Chem. Biochem. 25, 4o7-478.
MASER, M.D., RICE, R.V. (1962): Biophysical and biochemical properties of earthworm-cuticle collagen. Biochim. Biophys. Acta 63, 255-265.
MASER, M.D., RICE, R.V. (1963): Soluble earthworm cuticle collagen: A possible dimer of tropocollagen. J. Cell Biol. 18, 569-577.
MASHBURN, T.A., HOFFMAN, P. (1967): The presence of collagen in protein-polysaccharide from shark cartilage. Biochem. Biophys. Res. Commun. 29, 686-691.
MASON, R.M. (197o): Protein-polysaccharide complexes from adult human trachial cartilage. Biochem. J. 119, 599-6o1.
MASON, R.M. (1971): Observations on the glycosaminoglycans of aging bronchial cartilage studied with Alcian Blue. Histochem. J. 3, 421-434.
MASON, R.M., WUSTEMAN, F.S. (197o): The glycosaminoglycans of human tracheobronchial cartilage. Biochem. J. 12o, 777-785.
MATHEWS, M.B. (1953): Chondroitinsulfuric acid - a linear polyelectrolyte. Arch. Biochem. Biophys. 43, 181-193.
MATHEWS, M.B. (1956): The molecular weight of sodium chondroitin sulfate by light scattering. Arch. Biochem. Biophys. 61, 367-377.
MATHEWS, M.B. (1958): Isomeric chondroitin sulfates. Nature 181, 421-422.
MATHEWS, M.B. (1959): Macromolecular properties of isomeric chondroitin sulfates. Biochim. Biophys. Acta 35, 9-17.
MATHEWS, M.B. (196o): Trivalent cation binding of acid mucopolysaccharides. Biochim. Biophys. Acta 37, 288-295.
MATHEWS, M.B. (1961): Acid strength of carboxylgroups in isomeric chondroitin sulfates. Biochim. Biophys. Acta 48, 4o2-4o3.
MATHEWS, M.B. (1962): Sodium chondroitin sulfate-protein complexes of cartilage. III. Preparations from shark. Biochim. Biophys. Acta 58, 92-1o1.
MATHEWS, M.B. (1964): Structural factors in cation binding to anionic polysaccharides of connective tissue. Arch. Biochem. Biophys. 1o4, 394-4o4.
MATHEWS, M.B. (1965a): Molecular evolution of connective tissue. A comparative study of acid mucopolysaccharide-protein complexes. In: Structure and Function of Connective and Skeletal Tissue (eds. S. FITTON JACKSON, R.D. HARKNESS, S.M. PARTRIDGE, G.R. TRISTAM), pp. 181-191. London: Butterworths.
MATHEWS, M.B. (1965b): The interaction of collagen and acid mucopolysaccharides. A model for connective tissue. Biochem. J. 96, 71o-716.
MATHEWS, M.B. (1967a): Macromolecular evolution of connective tissue. Biol. Rev. 42, 499-551.
MATHEWS, M.B. (1967b): Chondroitin sulfate and collagen in inherited skeletal defects of chickens. Nature 213, 1255-1256.
MATHEWS, M.B. (1967c): Biophysical aspects of acid mucopolysaccharides relevant to connective tissue structure and function. In: The Connective Tissue. Int. Acad. Pathol. Mono, No. 7 (ed. B.M. WAGNER), pp. 3o4-329. Baltimore: Williams & Wilkins.
MATHEWS, M.B. (1969): Molecular organization of connective tissue matrix. In: The Cornea (ed. M.E. LANGHAM). Baltimore: Johns Hopkins Press.
MATHEWS, M.B. (197oa): Binding of calcium by proteoglycan of chondroitin sulfate. In: Chemistry and Molecular Biology of the Intercellular Matrix (ed. E.A.BALAZS), pp. 1121-1123. New York: Academic Press.
MATHEWS, M.B. (197ob): The interactions of proteoglycans and collagen-model systems. In: Chemistry and Molecular Biology of the Intercellular Matrix (ed. E.A. BALAZS), pp. 1155-1169. New York: Academic Press.
MATHEWS, M.B. (1971): Comparative biochemistry of chondroitin sulfate-proteins of cartilage and notochord. Biochem. J. 125, 37-46.

MATHEWS, M.B. (1972): In: The Comparative Molecular Biology of Extracellular Matrices (ed. H.C. SLAVKIN), pp. 155-162. New York: Academic Press.
MATHEWS, M.B. (1973): Glycosaminoglycans in development and aging of vertebrate cartilage. In: Connective Tissue and Aging. Workshop Conference Hoechst, Internat. Congr. Ser. 264 (ed. H.G. VOGEL), pp. 151-159. Amsterdam: Excerpta Medica.
MATHEWS, M.B., CIFONELLI, J.A. (1965): Comparative biochemistry of keratosulfates. J. Biol. Chem. 24o, 414o-4145.
MATHEWS, M.B., DECKER, L. (1968a): The distribution of sulfate groups in chondroitin sulfate chains. Detection of "oversulfated" and "undersulfated" repeating periods. Biochim. Biophys. Acta 156, 419-421.
MATHEWS, M.B., DECKER, L. (1968b): The effect of acid mucopolysaccharides and acid mucopolysaccharide-proteins on fibril formation from collagen solutions. Biochem. J. 1o9, 517-526.
MATHEWS, M.B., DECKER, L. (1971): Determination of molecular weight of acid mucopolysaccharides by gel electrophoresis. Biochim. Biophys. Acta 244, 3o-34.
MATHEWS, M.B., DORFMAN, A. (1953): The molecular weight and viscosity of chondroitin sulfuric acid. Arch. Biochem.Biophys. 41-53.
MATHEWS, M.B., DUH, J., PERSON, P. (1962): Acid mucopolysaccharides of invertebrate cartilage. Nature 193, 378-379.
MATHEWS, M.B., GLAGOV, S. (1966a): Acid mucopolysaccharide patterns in ageing human cartilage. J. Clin. Invest. 45, 11o3-1111.
MATHEWS, M.B., GLAGOV, S. (1966b): Acid mucopolysaccharide patterns in development and aging of human and other vertebrate cartilage. In: Biochimie et Physiologie du Tissu Conjonctif (ed. P. COMTE), pp. 21-26. Lyon: Société Ormeco et Imprimerie du Sud-Est à Lyon.
MATHEWS, M.B., HINDS, L.deC. (1963): Acid mucopolysaccharides of frog cartilage in thyroxine-induced metamorphosis. Biochim. Biophys. Acta 74, 198-212.
MATHEWS, M.B., INOUYE, M. (1961): The determination of chondroitin sulfate C-type polysaccharides in mixtures with other acid mucopolysaccharides. Biochim. Biophys. Acta 53, 5o9-513.
MATHEWS, M.B., LOZAITYTE, I. (1958): Sodium chondroitin sulfate-protein complexes of cartilage. I. Molecular weight and shape. Arch. Biochem. Biophys. 74, 158-174.
MATHEWS, M.B., ROSEMAN, S., DORFMAN, A. (1951): Determination of the chondroitinase activity of bovine testicular preparations. J. Biol. Chem. 188, 328-334.
MATOLTSY, A.G. (1952): A study on the structural protein of the vitreous body (vitrosin). J. Gen. Physiol. 36, 29-37.
MATSUMURA, T. (1972): Relationship between amino-acid composition and differentiation of collagen. Int. J. Biochem. 3, 265-274.
MATTHEWS, B.F. (1952): Collagen/chondroitin sulfate ratio of human articular cartilage related to function. Brit. Med. J. 2, 1295.
MAURICE, D.M. (1969): The cornea and sclera. In: The Eye (ed. H. DAVSON), 2nd Ed., Vol. 1, pp. 489-6oo. New York: Academic Press.
MAURICE, D.M., RILEY, M.V. (197o): The Cornea. In: Biochemistry of the Eye (ed. C.N. CLIVEMORE, pp. 1-1o3. New York: Academic Press.
MAXIMOW, A.A. (1927): Bindegewebe und blutbildende Gewebe. In: Handbuch der mikroskopischen Anatomie des Menschen (ed. W.v.MÖLLENDORF), Band II/1, pp. 232-583. Berlin: Springer.
MAXIMOW, A.A., BLOOM, W. (1948): A Textbook of Histology, 5th Ed. Philadelphia: Saunders.
MAYR, E. (1964): From molecules to organic diversity. Federation Proc. 23, 1231-1235.
McBRIDE, O.W., HARRINGTON, W.F. (1967a): Helix-coil transition in collagen. Evidence for a single-stranded triple helix. Biochemistry 6, 1499-1514.
McBRIDE, O.W., HARRINGTON, W.F. (1967b): *Ascaris* cuticle collagen: On the disulfide cross-linkages and the molecular properties of the subunits. Biochemistry 6, 1484-1498.
McCABE, M. (1972): The diffusion coefficient of caffeine through agar gels containing a hyaluronic acid-protein complex. Biochem. J. 127, 249-253.
McCULLAGH, K.G., DEROUETTE, S., ROBERT, L. (1973): Elephant aortic tissue. II. Amino acid analysis, structural glycoproteins, and antigenicity. Exp. Mol. Pathol. 18, 2o2-213.

McGAVIN, S. (1965): The structure and properties of elastoidin. Biochim. Physiol. Tissue Conjonctif, Conf. Commun. Symp. Int. Lyon, 241-247 (publ. 1966).
McGAVIN, S., PYPER, A.S. (1964): An electron-microscope study of elastoidin. Biochim. Biophys. Acta 79, 600-605.
MECHANIC, G., GALLOP, P.M., TANZER, M.L. (1971): The nature of crosslinking in collagens from mineralized tissues. Biochem. Biophys. Res. Commun. 45, 644-653.
MEDAWAR, P.B. (1946): Old age and natural death. Modern Quart. 2, 30-49.
MEDOFF, J. (1967): Enzymic events during cartilage differentiation in the chick embryonic limb bud. Dev. Biol. 16, 118-143.
MEDVEDEV, Zh.A. (1966): Protein Biosynthesis and Problems of Heredity, Development and Ageing. New York: Plenum Press.
MEENAKSHI, V.R., SCHEER, B.T. (1959): Acid mucopolysaccharide of the crustacean cuticle. Science 130, 1189-1190.
MEENAKSHI, V.R., SCHEER, B.T. (1961): Metabolism of glucose in the crabs *Cancer magister* and *Hemipgrapsus nudus*. Comp. Biochem. Physiol. 3, 30-41.
MEIER, S., HAY, E.D. (1973): Synthesis of sulfated glycosaminoglycans by embryonic corneal epithelium. Develop. Biol. 35, 318-331.
MELLO, M.L.S., VIDAL, B.C. (1972): Acid mucopolysaccharides as components of the peritrophic membranes of *Melipona quadrifasciata*. Cienc. Bult. (Sao Paulo) 24, 261-267.
MELLO, M.L.S., VIDAL, B.C. (1973): Anisotropic properties of toluidine blue-stained collagen. Ann. D'Histochimie 18, 103-122.
MEYER, F.A., COMPER, W.D., PRESTON, B.D. (1971): Model connective tissue systems. A physical study of gelatin gels containing proteoglycans. Biopolymers 10, 1351-1364.
MEYER, K. (1969): Biochemistry and biology of mucopolysaccharides. Am. J. Med. 47, 664-672.
MEYER, K. (1970): Reflections on "mucopolysaccharides" and their protein complexes. In: Chemistry and Molecular Biology of the Intercellular Matrix (ed. E.A. BALAZS), Vol. 1, pp. 5-24. New York: Academic Press.
MEYER, K. (1972): Hyaluronidases. In: The Enzymes, 3rd Ed. (ed. P.D. BOYER), Vol. 3, pp. 307-320. New York: Academic Press.
MEYER, K., ANDERSON, B. (1965): The chemical specificity of the mucopolysaccharides of the cornea. Exp. Eye Res. 4, 346-348.
MEYER, K., BHAVANANDAN, V.P., YUNG, D., LEE, L.T., HOWE, C. (1967): The keratosulfate-like mucopolysaccharide of chick allantoic fluid. Proc. Nat. Acad. Sci. USA 58, 1655-1659.
MEYER, K., DAVIDSON, E., LINKER, A., HOFFMAN, P. (1956): The acid mucopolysaccharides of connective tissue. Biochim. Biophys. Acta 21, 506-518.
MIA, A.J., SETTERFIELD, G. (1966): Organization of cellulose in plant cell walls. Chem. Can. 18, 36-37, 63.
MICHAELI, D., MARTIN, G.R., KETTMAN, J.R., BENJAMINI, E., LEUNG, D.Y.K., BLATT, B.A. (1969): Localization of antigenic determinants in the polypeptide chains of collagen. Science 166, 1522-1524.
MICHEEL, F., MILLE, E. (1953): Über Aminosäureverbindungen der Alginsäure. Hoppe-Seylers Z. Physiol. Chem. 293, 121-126.
MIER, P.D., WOOD, M. (1969): The acid mucopolysaccharides of mammalian skin. Brit. J. Derm. 81, 528-533.
MILCH, R.A. (1965): Matrix properties of the aging arterial wall. Monogr. Surg. Sci. 2, 261-341.
MILLER, A., PARRY, D.A.D. (1973): Structure and packing of microfibrils in collagen. J. Mol. Biol. 75, 441-447.
MILLER, D.H., LAMPORT, D.T.A., MILLER, M. (1972): Hydroxproline heterooligosaccharides in Chlamydomonas. Science 176, 918-920.
MILLER, E.J. (1973): A review of biochemical studies on the genetically distinct collagens of the skeletal system. Clin. Orthop. Rel. Res. 92, 260-280.
MILLER, E.J., EPSTEIN, E.H., PIEZ, K.A. (1971): Identification of three genetically distinct collagens by cyanogen bromide cleavage of insoluble human skin and cartilage collagen. Biochem. Biophys. Res. Commun. 42, 1024-1029.
MILLER, E.J., LANE, J.M., PIEZ, K.A. (1969): Isolation and characterization of the peptides derived from the α1 chain of chick bone collagen after cyanogen bromide cleavage. Biochemistry 8, 30-39.

MILLER, E.J., LUNDE, L.G. (1973): Isolation and characterization of the cyanogen bromide peptides from the α1(11) chain of bovine and human cartilage collagen. Biochemistry 12, 3153-3159.

MILLER, E.J., MARTIN, G.R., PIEZ, K.A., POWERS, M.J. (1967): Characterization of chick bone collagen and compositional changes associated with maturation. J.Biol. Chem. 242, 5481-5489.

MILLER, E.J., MATUKAS, V. J. (1969): Chick cartilage collagen: A new type of α1 chain not present in bone or skin of the species. Proc. Nat. Acad. Sci. USA 64, 1264-1268.

MILLER, E.J., MATUKAS, V.J. (197o): A comparison of chick bone and cartilage collagens: Evidence for a unique type of α1 chain in cartilage collagen. In: Chemistry and Molecular Biology of the Intercellular Matrix (ed. E.A. BALAZS), Vol. 1, pp. 1o9-115. New York: Academic Press.

MILLER, II, M.F., WYCKOFF, R.W.G. (1968): Proteins in dinosaur bones. Proc. Nat. Acad. Sci. USA 6o, 176-178.

MILLER, R.L. (1971): Chromatographic separation of the enzymes required for hydroxylation of lysine and proline residues of protocollagen. Arch. Biochem. Biophys. 148, 339-342.

MILLOT, J. (1955): The coelacanth. Sci. Amer. 193, 34-39.

MILLOT, J., ANTHONY, J. (1958a): Crossopterygiens actuel. *Latimeria chalumnae.* In: Traité de Zoologie (ed. P.-P. GRASSÉ), Vol. 13, pp. 2533-2597. Paris: Masson.

MILLOT, J., ANTHONY, J. (1958b): Anatomie de *Latimeria chalumnae*. I. Squelette, Muscles et Formation de Sontien. Paris: Centre Nat. Recher. Sci.

MILLOT, J., ANTHONY, J. (1965): Anatomie de *Latimeria chalumnae*. II. Système Nerveux et Organes des Sens. Paris: Centre Nat. Recher. Sci.

MOCZAR, E., MOCZAR, M. (197o): Glycannes liés à la fraction insoluble de la cornée de roussette (*Scyliorhinus canicula* L.). Eur. J. Biochem. 13, 28-36.

MOCZAR, E., PAYRAU, P., ROBERT, L. (1969): Distribution of the carbohydrates in the soluble and insoluble fractions of the stroma of fish corneas. Comp. Biochem. Physiol. 3o, 73-82.

MOCZAR, M., MOCZAR, E. (1973): Phylogenetic evolution of the cornea. Macromolecular composition of the corneal stroma of the squid (*Sepia officinalis*). Comp. Biochem. Physiol. 45B, 213-223.

MOHANDRADHAKRISHNAN, V., RAMANATHAN, N. (1966): Studies on elastoidin. II. An electron microscope study. Leather Sci. 13, 137-144.

MONROY, A. (1965): Chemistry and Physiology of Fertilization. New York: Holt, Rhinehart and Winston.

MORGAN, P.H., JACOBS, H.G., SEGRIST, J.P., CUNNINGHAM, L.W. (197o): A comparative study of glycopeptides derived from selected vertebrate collagens. J. Biol. Chem. 245, 5o42-5o48.

MORRIS, R. (196o): General problems of osmoregulation with special reference to cyclostomes. Symp. Zool. Soc. Lon. 1, 1-16.

MOSS, M.L. (1963): The biology of acellular teleost bone. Ann. N.Y. Acad. Sci. 1o9, 337-35o.

MOSS, M.L. (1964): The phylogeny of mineralized tissues. Int. Rev. Gen. Exp. Zool. 1, 297-332.

MOSS, M.L. (1968a): Bone, dentine and enamel and the evolution of vertebrates. In: Biology of the Mouth (ed. P. PERSON), pp. 37-65. Washington: Am. Assoc. for the Advancement of Sci.

MOSS, M.L. (1968b): The origin of vertebrate calcified tissues. In: Current Problems of Lower Vertebrate Phylogeny (ed. T. ØRVIG), pp. 359-371.

MOTOHIRO, T. (196oa): Mucoprotein in marine products. I. Isolation of a mucoprotein from the meats of *Stichopus japonicus* and *Cucumaria japonica.* Nippon Suisan Gakkaishi 26, 1171-1174.

MOTOHIRO, T. (196ob): Mucoprotein in marine products. II. Isolation of a polyfucose sulfate from the mucoprotein of *Cucumaria japonicus*. Nippon Suisan Gakkaishi 26, 1175-1178.

MOTOHIRO, T. (196oc): Mucoprotein in marine products. III. Crystallization of the itin sulfate from the mucoprotein of *Cucumaria japonicus*. Nippon Suisan Gakkaishi 26, 1179-1182.

MOTTET, N.K. (1967): Activity of aminopeptidases (aminonaphthylamidases) during early limb bud differentiation in the chick embryo. J. Exp. Zool. 165, 279-291.
MOURANT, A.E. (1971): Transduction and skeletal evolution. Nature 231, 466-467.
MUIR, H. (1964): Chemistry and metabolism of connective tissue glycosaminoglycans (mucopolysaccharides). Int. Rev. Conn. Tissue Res. 2, 1o1-154.
MUIR, H. (1965): Glycosaminoglycans and glycoproteins in blood-vessel walls. In: The Amino Sugars (eds. E.A. BALAZS, R.W. JEANLOZ), Vol. 2A, pp. 311-336. New York: Academic Press.
MUIR, H., MUSTARD, J.F. (1968): Enhancement of platelet aggregation by glycosyminoglycans (mucopolysaccharides). Colloqu. Int. Cont. Nat. Rech.Sci. 169, 589-596.
MUIR, L., LEE, Y.C. (1969): Structures of the D-galactose oligosaccharides from earthworm cuticle collagen. J. Biol. Chem. 244, 2343-2349.
MUIR, L., LEE, Y.C. (197o): Glycopeptides from earthworm cuticle collagen. J. Biol. Chem. 245, 5o2-5o9.
MÜLLER, P., KÜHN, K. (1968): The differences between α1 and α3 chains of acid soluble calfskin collagen. FEBS Letters 1, 233-234.
MULLINGER, R.N., MANLEY, G. (1969): Glycosaminoglycans and atherosclerosis in animal aortas. J. Atherosclerosis Res. 9, 1o8-111.
MUTHIAH, P., KÜHN, K. (1973): Studies on proteoglycans obtained from bovine knee joint and ear cartilages: sequential extraction, fractionation, characterization and their effects on fibril formation. Biochim. Biophys. Acta 3o4, 12-19.
MURATA, K., HARADA, T., OKUBO, K. (1968): Enzymatic studies of chondroitin sulfates in human arterial tissue. J. Atherosclerosis Res. 8, 951-958.
MURATA, K., OGURA, T., OKUYAMA, T. (1972): Disaccharide subunits of chondroitin sulfates in human plasma. Biochem. Med. 6, 223-23o.
MURRAY, R.W. (1965): Receptor mechanisms in the ampullae of Lorenzini of elasmobranch fishes. Cold Spring Harbor Symp. Quant. Biol. 3o, 233-243.
MYERS, D.B., HIGHTON, T.C., RAYNS, D.G. (1969): Acid mucopolysaccharides closely associated with collagen fibrils in normal human synovium. J. Ultrastructure Res. 28, 2o3-213.
NAKAGAWA, H., SATAKE, K. (1971): Whale cartilage keratosulfate. I. Its fractionation and characterization. Bull. Jap. Soc. Sci. Fisheries 37, 919-926.
NAKAJIMA, T., VOLCANI, B.E. (1969): 3,4-dihydroxyproline: A new amino acid in diatom cell walls. Science 164, 14oo-14o1.
NEEDHAM, A.E. (1964): The Growth Process in Animals. New York: Van Nostrand.
NEEDHAM, J. (1968): Order and Life. Cambridge: M.I.T. Press (a reprint of 1936 Ed., New Haven: Yale University Press).
NÉMETH-CSÓKA, M. (196o): Untersuchungen über die kollagenen Fasern. I. Über die submikroskopische Struktur der *in vitro* präzipitierten kollagenen Fasern und die stabilisierende Rolle der saueren Mukopolysaccharide. Acta Histochem. 9, 282-294.
NÉMETH-CSÓKA, M. (1961): Untersuchungen über die kollagenen Fasern. II. Vergleichende chemische, polarisations- und elektronen-mikroskopische Untersuchungen über die Kollagen-Fasern mit Chondroitinschwefelsäure als Grundsubstanz. Acta Histochem. 12, 255-276.
NEURATH, H., WALSH, K.A., WINTER, W.P. (1967): Evolution of structure and function of proteases. Science 158, 1638-1645.
NEMOTO, T., YOSIZAWA, Z. (1969): Sulfated glycopeptides and glycosaminoglycan peptides isolated from intestinal mucosae of rabbit. Biochim. Biophys. Acta 192, 34-48.
NEVO, Z., DORFMAN, A. (1972): Stimulation of chondromucoprotein synthesis in chondrocytes by extracellular chondromucoprotein. Proc. Nat. Acad. Sci. USA 69, 2o69-2o72.
NEVO, Z., HORWITZ, A.L., DORFMAN, A. (1972): Synthesis of chondromucoprotein by chondrocytes in suspension culture. Dev. Biol. 28, 219-228.
NICKERSON, W.J., FALCONE, G., KESSLER, G. (1961): Protein-polysaccharide complexes of yeast cell walls. In: Macromolecular Complexes (ed. M.V. EDDS, Jr.), pp. 2o5-228. New York: Ronald Press.
NIEDERMEIER, W., GRAMLING, E.A. (1969): The pi complex. A review of hyaluronate-protein complexes. Ala. J. Med. Sci. 6, 239-246.

NIEDUSZYNSKI, I.A., ATKINS, E.D.T. (1973): Conformation of the mucopolysaccharides. X-ray fibre diffraction of heparin. Biochem. J. 135, 729-733.
NOBLE, G.K. (1931): Biology of the Amphibia. New York: McGraw-Hill.
NOLAN, C., MARGOLIASH, E. (1968): Comparative aspects of primary structures of proteins. Ann. Rev. Biochem. 37, 727-79o.
NOLD, J.G., KANG, A.H., GROSS, J. (197o): Collagen molecules: Distribution of alpha chains. Science 17o, 1o96-1o98.
NORDWIG, A. (1971): Collagenalytic enzymes. In: Advances in Enzymology (ed. F.F. NORD), Vol. 34, pp. 155-2o5. New York: Wiley.
NORDWIG, A., HAYDUK, V. (1969): Invertebrate collagens: Isolation, characterization and phylogenetic aspects. J. Molec. Biol. 44, 161-172.
NORDWIG, A., HAYDUK, V., GERISCH, G. (1969): Is collagen present in unicellular organisms? Hoppe-Seylers Z. Physiol. Chem. 35o, 245-248.
NORDWIG, A., NOWACK, H., HIEBER-ROGALL, E. (1973): Sea anemone collagen: further evidence for the existence of only one α-chain type. J. Mol. Evol. 2, 175-18o.
NORDWIG, A., PFAB, F.K. (1968): Independence of collagen hydroxylation on the conformational state of the precursor substrate. Biochim. Biophys. Acta 154, 6o3-6o5.
NORDWIG, A., PFAB, F.K. (1969): Specificities of protocollagen hydroxylases from different sources. Biochim. Biophys. Acta 181, 52-58.
NORMAN, J.R., GREENWOOD, P.H. (1963): A History of Fishes, 2nd Ed. New York: Hill and Wang.
NORTHCOTE, D.H. (1963): The biology and chemistry of the cell walls of higher plants, algae and fungi. Int. Rev. Cytol. 14, 223-265.
NOWIKOFF, M. (1912): Studien über das Knorpelgewebe von Wirbellosen. Z. Wiss. Zool. Abt. A, 1o3, 661-717.
ÖBRINK, B. (1971): Studies on the Interactions Between Collagen and Glycosaminoglycans. Dissertation: Department of Medical Chemistry, University of Uppsala, Uppsala.
ÖBRINK, B. (1972): Isolation and partial characterization of a dermatan sulfate proteoglycan from pig skin. Biochim. Biophys. Acta 264, 354-361.
ÖBRINK, B. (1973a): A study of the interactions between monomeric tropocollagen and glycosaminoglycans. Eur. J. Biochem. 33, 387-4oo.
ÖBRINK, B. (1973b): The influence of glycosaminoglycans on the formation of fibers from monomeric tropocollagen *in vitro*. Eur. J. Biochem. 34, 129-137.
ÖBRINK, B., WASTESON, Å. (1971): Nature of the interaction between chondroitin 4-sulfate and chondroitin sulfate-proteoglycan with collagen. Biochem. J. 121, 227-233.
O'CONNELL, J.J., LOW, F.N. (197o): A histochemical and fine structural study of early extracellular connective tissue in the chick embryo. Anat. Rec. 167, 425-438.
O'DELL, D.S. (1968): Immunology of collagen and related materials. In: Treatise on Collagen, Vol. 2A, Biology of Collagen (ed. B.S. GOULD), pp. 311-322. New York: Academic Press.
ÖGREN, S., LINDAHL, U. (1971): Degradation of heparin in mouse mastocytoma tissue. Biochem. J. 125, 1119-1129.
OGSTON, A.G. (1966a): On water binding. Federation Proc. 25, 986-989.
OGSTON, A.G. (1966b): When is pressure osmotic? Federation Proc. 25, 1112-1114.
OGSTON, A.G. (197o): The biological function of the glycosaminoglycans. In: Chemistry and Molecular Biology of the Intercellular Matrix (ed. E.A. BALAZS), Vol. 3, pp. 1231-124o. New York: Academic Press.
OGSTON, A.G., WELLS, J.D. (1972): The osmotic properties of sulfoethyl-sephadex. A model for cartilage. Biochem. J. 128, 685-69o.
OLAH, E.H., KOSTENSZKY, K.S. (1972): Effect of altered functional demand on the glycosaminoglycan content of the articular cartilage of dogs. Acta Biol. (Budapest) 23, 195-2oo.
OLSEN, B.R. (1965): Electron microscope studies on collagen. IV. Structure of vitrosin fibrils and interaction properties of vitrosin molecules. J. Ultrastructure Res. 13, 172-191.
OLSEN, B.R. (1967): Electron microscope studies on collagen. VI. The structure of segment-long-spacing aggregates consisting of molecules renatured from isolated α-fractions of codfish skin collagen. J. Ultrastructure Res. 19, 446-473.

OLSEN, B.R., ALPER, R., KEFALIDES, N.A. (1973): Structural characterization of a soluble fraction from lens-capsule basement membrane. Eur. J. Biochem. 38, 22o-228.

OLSSON, I. (1968): Biosynthesis of glycosaminoglycans (mucopolysaccharides) in leukemic myeloid cells. Biochim. Biophys. Acta 165, 324-334.

OLSSON, I. (1969a): Intracellular distribution and sites of synthesis of glycosaminoglycans (mucopolysaccharides) in human leukocytes. Exp. Cell Res. 54, 314-317.

OLSSON, I. (1969b): Intracellular transport of glycosaminoglycans (mucopolysaccharides) in human leucocytes. Exp. Cell Res. 54, 318-324.

OLSSON, I. (1969c): Isolation of human leukocyte granules, using colloidal silica-polysaccharide density gradients. Exp. Cell Res. 54, 325-33o.

OLSSON, I. (1969d): Chondroitin sulfate proteoglycan of human leukocytes. Biochim. Biophys. Acta 177, 241-249.

OLSSON, I. (1971): Mucopolysaccharides of rabbit bone marrow cells. Identification and properties. Exp. Cell Res. 67, 416-426.

OLSSON, I., GARDELL, S. (1967): Isolation and characterization of glycosaminoglycans from human leukocytes and platelets. Biochim. Biophys. Acta 141, 348-357.

OLSSON, I., GARDELL, S., THUNELL, S. (1968): Biosynthesis of glycosaminoglycans (mucopolysaccharides) in human leukocytes. Biochim. Biophys. Acta 165, 3o9-323.

OOSAWA, F. (1971): Polyelectrolytes. New York: Dekker.

OPARIN, A.L. (ed.) (1963): Evolutionary Biochemistry. London: Pergamon Press.

OREKHOVICH, V.N., CHIPIKITER, V.O., MAZOUROV, V.I., KOUNINA, V.O. (196o): Procollagènes. - Classification, mêtabolisme, action des proteinases. Bull. Soc. Chim. Biol. 42, 5o5-518.

ØRVIG, T. (1951): Histologic studies of placoderms and fossil elasmobranchs. 1. The endoskeleton with remarks on the hard tissue of the lower vertebrates in general. Arkiv. Zool. 2, 321-454.

ØRVIG, T. (1965): Paleohistological notes. 2. Certain comments on the phyletic significance of acellular bone tissue in lower vertebrates. Arkiv. Zool. 29, 551-556.

ØRVIG, T. (1967): Phylogeny of tooth tissues: Evolution of some calcified tissues in early vertebrates. In: Structural and Chemical Organization of Teeth (ed. A.E. MILES), pp. 45-11o. New York: Academic Press.

ØRVIG, T. (1968): The dermal skeleton; general considerations. In: Current Problems of Lower Vertebrate Phylogeny (ed. T. ØRVIG), pp. 373-397. Stockholm: Almquist & Wiksell.

OSANAI, K. (196o): On the cortical granules of the toad egg. Sci. Rep. Tohoku Univ., Ser. IV, 26, 69-75.

OSUOJI, C.I. (1971): Acid glycosaminoglycans of eggshell membranes. Biochim. Biophys. Acta 244, 481-483.

PALMOSKI, M.J., GOETINCK, P.F. (1972): Synthesis of proteochondroitin sulfate by normal, nanomelic, and 5-bromodeoxyuridine-treated chondrocytes in cell culture. Proc. Nat. Acad. Sci. USA 69, 3385-3388.

PAMER, T., GLASS, G.B.J., HOROWITZ, M. (1968): Purification and characterization of sulfated glycoproteins and hyaluronidase-resistant mucopolysaccharides from dog gastric mucosa. Biochemistry 7, 3821-3829.

PANCAKE, S.J., KARNOVSKY, M.L. (1971): The isolation and characterization of a mucopolysaccharide secreted by the snail, *Otella lactea*. J. Biol. Chem. 246, 253-262.

PANKOVICH, M., KORNGOLD, L. (1967): A comparison of the antigenic properties of nucleus pulposus and cartilage protein polysaccharide complexes. J. Immunol. 99, 431-437.

PARTRIDGE, S.M. (1962): Elastin. Adv. Prot. Chem. 17, 227-3o2.

PARTRIDGE, S.M. (1966): Biosynthesis of nature of elastin structures. Federation Proc. 25, 1o23-1o29.

PARTRIDGE, S.M. (1969): Elastin, biosynthesis and structure. Gerontologia 15, 85-1oo.

PASTERNACK, S.G., VEIS, A., BREEN, M. (1973): Solvent dependent changes in proteoglycan subunit conformation in aqueous GuHCl solutions. J. Biol. Chem. 249, 22o6-2211.

PATTEE, H. (1969): Physical conditions for primitive functional hierarchies. In: Hierarchical Structures (eds. L.L. WHYTE, A.G. WILSON, D. WILSON), pp. 161-177. New York: Elsevier.
PAUTARD, F.G.E. (1966): A biomolecular survey of calcification. In: Calcified Tissues (eds. H. FLEISCH, H.J.J. BLACKWOOD, M. OWEN), pp. 1o8-122. New York-Heidelberg-Berlin: Springer.
PAVANS DE CECCATHY, M., GARRONE, R. (1971): Fibrogenèse du collagène chez l'eponge *Chondrosea reniformis* Nardo (Démosponge Tétractinellide). Origine et évolution des lophocytes. C.R. Acad. Sci. (Paris) Sér. D, 273, 1957-1959.
PAWELEK, J.M. (1969): Effects of thyroxine and low oxygen tension on chondrogenic expression in cell cultrue. Dev. Biol. 19, 52-72.
PAZ, M.A., HENSON, E., BLUMENFELD, O.O., SEIFTER, S., GALLOP, P.M. (1971): Dehydromerodesmosine and merodesmosine in elastin. Biochem. Biophys. Res. Commun. 44, 1518-1523.
PAZ, M.A., SALAZAR, M.E., GAITE, E.E. (1967): The characterization of collagen from the swim bladder of the hake, *Merluccius hubsii*. Arch. Biochem. Biophys. 119, 485-49o.
PEARCE, R.H. (1965): Glycosaminoglycans and glycoproteins in skin. In: The Amino Sugars (eds. E.A. BALAZS, R.W. JEANLOZ), Vol. IIA, pp. 149-193. New York: Academic Press.
PEASE, D.C., BOUTEILLE, M. (1971): The tridimensional ultrastructure of native collagenous fibrils, cytochemical evidence for a carbohydrate matrix. J. Ultrastr. Res. 35, 339-358.
PEDERSEN, K.J. (1966): The organization of the connective tissue of *Discocelides longi* (Turbellaria, Polycladida). Z. Zellforsch. 71, 94-117.
PENNOCK, C.A. (1968): Association of acid mucopolysaccharides with isolated amyloid fibrils. Nature 217, 753-754.
PERCIVAL, E., McDOWELL, R.H. (1967): Chemistry and Enzymology of Marine Algae Polysaccharides. New York: Academic Press.
PERCIVAL, E.G.V., ROSS, A.G. (195o): Fucoidin. I. The isolation and purification of fucoidin from brown seaweeds. J. Chem. Soc. 717-72o.
PEREDA, J. (197o): Etude histochimique des mucopolysaccharides de l'oviduete et des gangues muqueuses des ovocytes de *Rana pipiens*: incorporation du $^{35}SO_4^{2-}$. Dev. Biol. 21, 318-33o.
PÉREZ-TAMAYO, R., ROJKIND, M. (eds.) (1973): Molecular Pathology of Connective Tissues. New York: Dekker.
PERLIN, A.S., CASU, B., SANDERSON, G.R., JOHNSON, L.F. (197o): MHz spectra of heparin, chondroitins and other mucopolysaccharides. Can. J. Chem. 48, 226o-2268.
PERSON, P. (1969): Cartilagenous dermal scales in cephalopods. Science 164, 14o4-14o5.
PERSON, P., MATHEWS, M.B. (1967): Endoskeletal cartilage in a marine polychaete, *Eudistylia polymorpha*. Biol. Bull. 132, 244-252.
PERSON, P., PHILPOTT, D.E. (1963): Invertebrate cartilages. Ann. N.Y. Acad. Sci. 1o9, 113-126.
PERSON, P., PHILPOTT, D.E. (1967): On the occurrence and the biologic significance of cartilage tissues in invertebrates. Clin. Orthopaedics and Related Research 53, 185-212.
PERSON, P., PHILPOTT, D.E. (1969a): The nature and significance of invertebrate cartilages. Biol. Rev. 44, 1-16.
PERSON, P., PHILPOTT, D.E. (1969b): The biology of cartilage. I. Invertebrate cartilages: *Limulus* gill cartilage. J. Morphol. 128, 67-94.
PERSON, P., PHILPOTT, D.E. (1969c): The nature and significance of invertebrate cartilages. Biol. Rev. 44, 1-16.
PESSAC, B., DEFENDI, V. (1972): Cell aggregation: Role of acid mucopolysaccharides. Science 175, 898-9oo.
PHILLIPS, J.H. (1956): Isolation of active nematocysts of *Metridium senile* and their chemical compostion. Nature 178, 932.
PHILPOTT, D.E., PERSON, P. (197o): The biology of cartilage. II. Invertebrate cartilages: squid head cartilage. J. Morphol. 131, 417-43o.
PICKEN, L. (196o): The organization of Cells. Oxford: Clarendon Press.

PICKFORD, G.E., GRANT, F.B. (1967): Serum osmolality in the coelacanth, *Latimeria chalumnae*: urea retention and ion regulation. Science 155, 568-57o.
PIETSCHMANN, V. (1962a): Acrania. In: Handbuch der Zoologie (ed. W. KÜKENTHAL, T. KRUMBACH), Vol. 6, pp. 1-112. Berlin: DeGruyter.
PIETSCHMANN, V. (1962b): Cyclostomata. In: Handbuch der Zoologie (ed. W. KÜKENTHAL, T. KRUMBACH), Vol. 6, 1st half, Part 1, pp. 127-547. Berlin: DeGruyter.
PIEZ, K.A. (1965): Characterization of a collagen from codfish skin containing three chromatographically different αchains. Biochemistry 4, 259o-2596.
PIEZ, K.A. (1967): Soluble collagen and the components resulting from its denaturation. In: Treatise on Collagen, Vol. 1, Chemistry of Collagen (ed. G.N. RAMACHANDRAN), pp. 2o7-252. New York: Academic Press.
PIEZ, K.A. (1968): Cross-linking of collagen and elastin. Ann. Rev. Biochem. 37, 547-57o.
PIEZ, K.A. (1969): Chemistry of collagen and elastin and biosynthesis of covalent cross-links. In: Aging of Connective and Skeletal Tissue (eds. A. ENGEL, T. LARSSON), pp. 15-32. Stockholm: Nordiska Bokhandelns Förlag.
PIEZ, K.A., BLADEN, H.A., BORNSTEIN, P., BUTLER, W.T., KANG, A.H., LANE, J.M., MILLER, E.J. (1968): Comparative studies on the chemistry of collagen utilizing cyanogen bromide cleavage. Brookhaven Symp. Biol. 21, 345-357.
PIEZ, K.A., EIGNER, E.A., LEWIS, M.S. (1963): The chromatographic separation and amino acid composition of the subunits of several collagens. Biochem. 2, 58-66.
PIEZ, K.A., GROSS, J. (1959): The amino acid composition and morphology of some invertebrate and vertebrate collagens. Biochim. Biophys. Acta 34, 24-39.
PIEZ, K.A., GROSS, J. (196o): The amino acid composition of some fish collagens: The relation between composition and structure. J. Biol. Chem. 235, 995-998.
PIKKARAINEN, J. (1968): The molecular structures of vertebrate skin collagens. Acta Physiol. Scand. Suppl. 3o9.
PIKKARAINEN, J., KULONEN, E. (1964): Electrophoretic fractionation of the various subunits from heat-denatured insoluble collagen. Acta Chem. Scand. 18, 2o11-2o12.
PIKKARAINEN, J., KULONEN, E. (1969): Comparative biochemistry of collagen. Nature 223, 839-841.
PIKKARAINEN, J., KULONEN, E. (1971): Structural studies on the basement membrane collagen. Scand. J. Clin. Lab. Invest. 27, Suppl. 116, 29 abs.
PIKKARAINEN, J., RANTANEN, J., VASTAMÄKI, M., LAMPIAHO, K., KARI, A., KULONEN, E. (1968): On collagens of invertebrates with special reference to *Mytilus edulis*. Europ. J. Biochem. 4, 555-56o.
PINNELL, S.R., MARTIN, G.R. (1968): The cross-linking of collagen and elastin: Enzymatic conversion of lysine in peptide linkage to α-aminoadipio-δ-semialdehyde (allysine) by an extract from bone. Proc. Nat. Acad. Sci. USA 61, 7o8-714.
PIRIE, A. (1969): The vitreous body. In: The Eye, 2nd Ed. (ed. H. DAVISON), Vol. 11, pp. 273-297. New York: Academic Press.
PITA, J.C., CUERVO, L.A., MODRUGA, J.E., MULLER, F.J., HOWELL, D.S. (197o): Evidence for a role of proteinpolysaccharides in regulation of mineral phase separation in calcifying cartilage. J. Clin. Invest. 49, 2188-2197.
PONTZ, B., MEIGEL, W., RAUTERBERG, J., KÜHN, K. (197o): Localization of two species specific antigenic determinants on the peptide chains of calfskin collagen. Eur. J. Biochem. 16, 5o-54.
POPENOE, E.A., ARONSON, R.B. (1972): Partial purification and properties of collagen lysine hydroxylase from chick embryos. Biochim. Biophys. Acta 258, 38o-386.
PRAME, G. (1967): Glycosaminoglycans of retrobulbar tissue of the crucian carp (*Carassius carassius* L.). Exp. Eye Res. 6, 171-178.
PRAME, G., GARDELL, S., ANTONOPOULOS, C.A. (1966): On the glycosaminoglycans of human retrobulbar tissues. Exp. Eye Res. 5, 79-85.
PRAUS, R., BRETTSCHNEIDER, I. (197o): Presence of a nonsulfated glycosaminoglycan in embryonic cornea. FEBS Letters 6, 221-224.
PRAUS, R., BRETTSCHNEIDER, I. (1971): Glycosaminoglycans in the developing chicken cornea. Ophthal. Res. 2, 367-373.
PRAUS, R., GOLDMAN, J.N. (197oa): Glycosaminoglycans in the nonswelling corneal stroma of dogfish dark. Invest. Ophthal. 9, 131-136.
PRAUS, R., GOLDMAN, J.N. (197ob): Microanalysis of corneal mucopolysaccharides of several species and the relationship to corneal swelling. Ophthalmol., Proc. Int. Congr. 21st, 2, 16o6-161o.

PRESTON, B. (1968): Physical characterization of dermatan sulfate-protein. Arch. Biochem. Biophys. 126, 974-976.
PRESTON, B.N., MEYER, F.A. (1971): Physical behavior of gelatin gels: A simple model for connective tissue. Biopolymers 1o, 35-46.
PRESTON, B.N., OBRINK, B., LAURENT, T.C. (1973): The rotational diffusion of albumin in solutions of connective-tissue polysaccharides. Eur. J. Biochem. 33, 4o1-4o6.
PRESTON, B.N., SNOWDEN, J.Mc K. (1972): Model connective tissue systems. The effect of proteoglycans on the diffusional behavior of small non-electrolytes and micro-ions. Biopolymers 11, 1627-1643.
PRESTON, B.N., SNOWDEN, J. Mc K., HOUGHTON, K.T. (1972): Model connective tissue systems. The effect of proteoglycans on the distribution of small non-electrolytes and micro-ions. Biopolymers 11, 1645-1659.
PRESTON, R.D. (1961): Cellulose-protein complexes in plant cell walls. In: Macromolecular Complexes (ed. M.V. EDDS, Jr.), pp. 229-253. New York: Ronald Press.
PRESTON, R.D., NICOLAI, E., REED, R., MILLARD, A. (1948): An electron microscope study of cellulose. Nature 162, 665-667.
PRIVALOV, P.L., SERDYNK, I.N., TIKTOPULO, E.I. (1971): Thermal conformational transformations of tropocollagen. II. Viscometric and light-scattering studies. Biopolymers 1o, 1777-1794.
PRIVALOV, P.L., TIKTOPULO, E.I. (197o): Thermal conformational transformation of tropocollagen. I. Calorimetric study. Biopolymers 9, 127-139.
PROCKOP, D.J. (197o): Intracellular biosynthesis of collagen and interactions of protocollagen hydroxylase with large peptides. In: Chemistry and Molecular Biology of the Intercellular Matrix (ed. E.A. BALAZS), Vol. 1, pp. 335-37o. New York: Academic Press.
PROCKOP, D.J., PETTENGILL, O., HOLTZER, H. (1964): Incorporation of sulfate and the synthesis of collagen by cultures of embryonic chondrocytes. Biochim. Biophys. Acta 83, 189-196.
PRODI, G. (1967): L'évolution biochimique du derme avec l'age. G. Gerontol. 15, 1299-1316.
PRODI, G., FERRERI, A.M., GRILLI, S., ROCCHI, P. (1971): Complex carbohydrates of the skin. Ital. J. Biochem. 2o, 16-26.
PUCCI-MINAFRA, I., CASANO, C., LA ROSA, C. (1972): Collagen synthesis and spicule formation in sea urchin embryos. Cell Differentiation 1, 157-165.
PUJOL, J.P., BOCQUET, J., TIFFON, Y. (197o): Analyse biochimique du byssus calcifé d'*Anomia ephippium* L. (Mollusque bivalve). Calcif. Tissue Res. 5, 317-326.
PUSZTAI, A., WATT, W.B. (1969): Fractionation and characterization of glycoproteins containing hydroxyproline from the leaves of *Vicia faba*. Eur. J. Biochem. 1o, 523-532.
PUYEAR, R.L. (1967): The glucuronic acid pathway in the blue crab, *Callinectes sapidus*. Rathbun: The enzymes of the UDP-glucose to glucuronic acid portion of the pathway. Comp. Biochem. Physiol. 2o, 499-5o8.
QUINTARELLI, G., DELLOVA, M.C. (1966): Age changes in the localization and distribution of glycosaminoglycans in human hyaline cartilage. Histochemie 7, 141-167.
RABINOWITZ, J.L., SHAPIRO, I.M. (1972): Lipids of bovine skin collagen. Arch. Oral. Biol. 17, 547-553.
RADHAKRISHNAMURTHY, B., BERENSON, G.S. (1973): Structures of glycopeptides from a glycoprotein from bovine aorta. J. Biol. Chem. 248, 2ooo-2oo6.
RADHAKRISHNAMURTHY, B., SRINIVASAN, S.R., DALFERES, Jr., E.R., BERENSON, G.S. (197o): Composition of glycopeptides from chondroitin sulfate-protein complex from squid skin. Comp. Biochem. Physiol. 36, 1o7-117.
RAFF, R.A., RAFF, E.C. (197o): Respiratory mechanisms and the metazoan fossil record. Nature 228, 1oo3-1oo5.
RAJULU, G.S., KRISHNAN, K.A. (1967): On the possible occurrence of hyaluronic acid in the cuticle of the earthworm *Megoscolex mauritii*. Indian J. Exp. Biol. 5, 55-56.
RAMACHANDRAN, G.N. (1967): Structure of collagen at the molecular level. In: Treatise on Collagen, Vol. 1, Chemistry of Collagen (ed. G.N. RAMACHANDRAN), pp. 1o3-183. New York: Academic Press.

RAMACHANDRAN, G.N., BANSAL, M., BHATNAGAR, R.S. (1973): A hypothesis on the role of hydroxyproline in stabilizing collagen structure. Biochim. Biophys. Acta 322, 166-171.
RAO, N.V., HARRINGTON, W.F. (1966): Relationship between the positioning of pyrrolidine residues and the stability of collagen. J. Molec. Biol. 21, 577-581.
RAO, V.S.R., FOSTER, J.F. (1963): Studies on the conformation of amylose in solution. Biopolymers 1, 527-544.
RAO, V.S.R., FOSTER, J.F. (1965): Studies of the conformation of carboxymethylamylose in aqueous solution. Biopolymers 3, 185-193.
RAUTERBERG, J., FIETZEK, P., REXROTH, F., BECKER, U., STARK, M., KÜHN, K. (1972): Amino acid sequence of the carboxyl terminal nonhelical cross-link region of the α1 chain of calf skin collagen. FEBS Letters 21, 75-79.
READ, K.R.H. (1967): Thermostability of proteins in poikilotherms. In: Molecular Mechanisms of Temperature Adaptation (ed. C.L. PROSSER), pp. 93-1o6. Publ. Am. Ass. Advan. Sci. No. 84, Washington.
READ, L.J. (1971): Chemical constituents of body fluids and urine of the holocephalan *Hydrolagus colliei*. Comp. Biochem. Physiol. 39A, 185-192.
REDDI, A.H., HUGGINS, C. (1972): Biochemical sequences in the transformation of normal fibroblasts in adolescent rats. Proc. Nat. Acad. Sci. USA 69, 16o1-16o5.
REES, D.A. (1967): The Shapes of Molecules. Edinburgh: Oliver and Boyd.
REES, D.A. (1972): Shapely polysaccharides. Biochem. J. 126, 257-273.
RIGBY, B.J. (1967a): Relation between the shrinkage of native collagen in acid solution and the melting temperature of the tropocollagen molecule. Biochim. Biophys. Acta 133, 272-277.
RIGBY, B.J. (1967b): Correlation between serine and thermal stability of collagen. Nature 214, 87-88.
RIGBY, B.J. (1968a): Thermal transitions in some invertebrate collagens and their relation to amino acid content and evironment temperature. In: Symposium on Fibrous Proteins (ed. W.G. CREWTHER), pp. 217-225. New York: Plenum Press.
RIGBY, B.J. (1968b): Amino-acid composition and thermal stability of the skin collagen of the Antarctic ice-fish. Nature 219, 166-167.
RIGBY, B.J. (1971): The thermal stability of collagen: Its significance in biology and physiology. Adv. Chem. Phys. 21, 537-555.
RIGBY, B.J., MASON, P. (1967): Thermal transitions in gastropod collagen and their correlation with environmental temperature. Aust. J. Biol. Sci. 2o, 265-271.
RINGERTZ, N.R. (1965): Glycosaminoglycan in mast cells and in mast-cell tumors. In: The Amino Sugars (eds. E.A. BALAZS, R.W. JEANLOZ), Vol. IIA, pp. 2o9-217. New York: Academic Press.
ROBERT, A.M., ROBERT, B., ROBERT, L. (197o): Chemical and physical properties of structural glycoproteins. In: Chemistry and Molecular Biology of the Intercellular Matrix (ed. E.A. BALAZS), pp. 237-242. New York: Academic Press.
ROBERT, L. (1966): Role des glycoprotéines de structure dans la morphogenèse de la trame conjonctive. Arch. Biochim. Cosmét. 83, 1-7.
ROBERT, L., DARRELL, R.W., ROBERT, B. (197o): Immunological properties of connective tissue glycoproteins. In: Chemistry and Molecular Biology of the Intercellular Matrix (ed. E.A. BALAZS), pp. 1591-1614. New York: Academic Press.
ROBERT, L., DISCHE, Z. (1963): Analysis of a sialofucoglucosaminogalactomannosidoglycan from corneal stroma. Biochem. Biophys. Res. Commun. 1o, 2o9-214.
ROBERT, L., ROBERT, B., MOCZAR, E., MOCZAR, M. (1972): Les glycoproteines de structure du tissu conjonctif. Path. Biol. 2o, 1oo1-1o12.
ROBERT, L., SCHILLINGER, G. (1967): Etude sur la composition chimique des cornées des poisson élasmobranches et téléostéens. Arch. Ophthal. 27, 813-818.
ROBERTS, S.J. (1956): Veterinary Obstetrics and Genital Diseases. Ann Arbor: Edwards.
ROBERTSON, J.D. (1957): The habitat of the early vertebrates. Biol. Rev. 32, 156-187.
ROBERTSON, J.D. (1963): Osmoregulation and ionic composition of cells and tissues. In: The Biology of Myxine (eds. A. BRODAL, R. FÄNGE), pp. 5o3-515. Oslo: Universitetsforlaget.
ROBINS, S.P., BAILEY, A.J. (1972): Age-related changes in collagen: The identification of reducible lysine-carbohydrate condensation products. Biochem. Biophys. Res. Commun. 48, 76-84.

ROBINSON, H.C., DORFMAN, A. (1969): The sulfation of chondroitin sulfate in embryonic chick cartilage epiphyses. J. Biol. Chem. 244, 348-352.
ROBINSON, D.S., HEANEY, R.K. (1973): Fractionation of soluble egg-shell mucopolysaccharides. Biochem. Soc. Trans. 1, 155-157.
ROBINSON, H.C., HOPWOOD, J.J. (1973): The alkaline cleavage and borohydride reduction of cartilage proteoglycan. Biochem. J. 133, 457-47o.
RODBARD, S. (197o): Negative feedback mechanisms in the architecture and function of the connective and cardiovascular tissues. Persp. Biol. Med. 14, 5o7-527.
RODÉN, L. (197o): Biosynthesis of acidic glycosaminoglycans (mucopolysaccharides). In: Metabolic Conjugation and Metabolic Hydrolysis (ed. W.H. FISHMAN), Vol. 3, pp. 345-442. New York: Academic Press.
RODÉN, L., BAKER, J.R., CIFONELLI, J.A., MATHEWS, M.B. (1972a): Isolation and characterization of connective tissue polysaccharides. In: Methods in Enzymology (ed. V. GINSBURG), Vol. 28B, pp. 73-14o. New York: Academic Press.
RODÉN, L., BAKER, J.R., SCHWARTZ, N.B., STOOLMILLER, A.C., YAMAGATA, S., YAMAGATA, T. (1972b): Some aspects of the structure and biosynthesis of connective tissue proteoglycans. In: Biochemistry of the Glycosidic Linkage (eds. R.A. PARAS, H. PONTIS), pp. 345-385. New York: Academic Press.
RODÉN, L., MATHEWS, M.B. (1968): Structure of hyaluronic acid-protein complex (HA-P). Federation Proc. 27, 529.
ROKOSOVÁ-ČMUCHALOVÁ, B., BENTLEY, J.P. (1968): Relation of collagen synthesis to chondroitin sulfate synthesis in cartilage. Biochem. Pharmacol., Suppl. 315-328.
ROMER, A.S. (1942): Cartilage, an embryonic adaptation. Am. Naturalist 76, 394-4o4.
ROMER, A.S. (1963): The ancient history of bone. Ann. N.Y. Acad. Sci. 1o9, 168-176.
ROMER, A.S. (1966): Vertebrate Paleontology, 3rd Ed. Chicago: University of Chicago Press.
ROMER, A.S. (1968): Notes and Comments on Vertebrate Paleontology. Chicago: University of Chicago Press.
ROMER, A.S. (197o): The Vertebrate Body, 4th Ed. Philadelphia: Saunders.
ROSENBERG, L., HELLMANN, W., KLEINSCHMIDT, A.K. (197o): Macromolecular models of proteinpolysaccharides from bovine nasal cartilage based on electron microscopic studies. J. Biol. Chem. 245, 4123-413o.
ROSENBERG, L., SCHUBERT, M., SANDSON, J. (1967): The protein-polysaccharide of bovine nucleus pulposus. J. Biol. Chem. 242, 4691-47o1.
ROSENBLOOM, J., HARSCH, M., JIMINEZ, S. (1973): Hydroxyproline content determines the denaturation temperature of chick tendon collagen. Arch. Biochem. Biophys. 158, 478-484.
ROSENBLOOM, J., PROCKOP, D.J. (1969): Biochemical aspects of collagen synthesis. In: Repair and Regeneration (eds. J.E. DUNPHY, W. VanWINKLE), pp. 117-135. New York: McGraw-Hill.
ROSS, R. (1968): The connective tissue fiber forming cell. In: Treatise on Collagen, Vol. 2A, Biology of Collagen (ed. B.S. GOULD), pp. 2-82. New York: Academic Press.
ROSS, R. (1973): The elastic fiber. J. Histochem. Cytochem. 21, 199-2o8.
ROSS, R., BORNSTEIN, P. (1969): The elastic fiber. I. The separation and partial characterization of its macromolecular components. J. Cell Biol. 4o, 366-381.
ROSS, R., BORNSTEIN, P. (197o): Studies of the components of the elastic fiber. In: Chemistry and Molecular Biology of the Intercellular Matrix (ed. E.A. BALAZS), pp. 641-655. New York: Academic Press.
ROSS, R., BORNSTEIN, P. (1971): Elastic fibers in the body. Sci. Am. 224, 44-52.
ROTHBARD, S., WATSON, R.F. (1965): Immunologic relations among various animal collagens. J. Exp. Med. 122, 441-454.
ROTHSCHILD, L. (1958): Fertilization in fish and lampreys. Biol. Rev. 33, 372-392.
RUBIN, A.L., DRAKE, M.P., DAVISON, P.F., PFAHL, D., SPEAKMAN, P.T., SCHMITT, F.O. (1965): Effects of pepsin treatment on the interaction of tropocollagen molecules. Biochemistry 4, 181-19o.

RUCKER, R.B., GOETTLICH-RIEMANN, TOM, K. (1973): Properties of chick tropoelastin. Biochim. Biophys. Acta 317, 193-2o1.

RUDALL, K.M. (1955): The distribution of collagen and chitin. In: Fibrous Proteins and Their Biological Significance (ed. M. BROWN, J.F. DANIELLI). Symp. Soc. Exper. Biol. 9, pp. 49-71. New York: Academic Press.

RUDALL, K.M. (1965): Protein-polysaccharide interactions in chitinous structures. In: Structure and Function of Connective and Skeletal Tissue (eds. S. FITTON JACKSON, R.D. HARKNESS, S.M. PARTRIDGE, G.R. TRISTRAM), pp. 191-196. London: Butterworths.

RUDALL, K.M. (1968): Comparative biology and biochemistry of collagen. In: Treatise on Collagen, Vol. 2A, Biology of Collagen (ed. B.S. GOULD), pp. 83-137. New York: Academic Press.

RUDALL, K.M., KENCHINGTON, W. (1971): Arthropod silks: The problem of fibrous proteins in animal tissues. Ann. Rev. Entomol. 16, 73-96.

RUDWICK, M.J.S. (197o): Living and Fossil Brachiopods. London: Hutchinson.

RUSKA, C., RUSKA, H. (1962): Die Culticula der Epidermis des Regenwurms (*Lumbricus terrestris* L.). Z. Zellforsch. Mikroskop. Anat. 53, 759-764.

SADAVA, D., CHRISPEELS, M.J. (1971a): Hydroxyproline biosynthesis in plant cells peptidyl proline hydroxylase from carrot discs. Biochim. Biophys. Acta 227, 278-287.

SADAVA, D., CHRISPEELS, M.J. (1971b): Intracellular site of proline hydroxylation in plant cells. Biochemistry 1o, 429o-4294.

SAHA, N.N., DAS, S., SAHA, A.K., BHATTACHARYA, S.D. (1964): Study on the structure of collagen. Proc. Nucl. Phys. Solid State Phys. Symp., Chandigarh, India, Part B, 6o1-611.

SAIGO, K., EGAMI, F. (197o): Purification and some properties of acid mucopolysaccharides of bovine brain. J. Neurochem. 17, 633-647.

SAITO, H., YAMAGATA, T., SUZUKI, S. (1968): Enzymatic methods for the determination of small amounts of isomeric chondroitin sulfates. J. Biol. Chem. 243, 1536-1542.

SAJDERA, S.W., HASCALL, V.C. (1969): Protein-polysaccharide complex from bovine nasal cartilage. A comparison of low and high shear extraction procedures. J. Biol. Chem. 244, 77-87.

SAKAKIBARA, S., INOUYE, K., SHUDO, K., KISHIDA, Y., KOBAYASHI, Y., PROCKOP, D.J. (1973): Synthesis of $(Pro\text{-}Hyp\text{-}Gly)_n$ of defined molecular weights. Evidence for the stabilization of collagen triple helix by hydroxyproline. Biochim. Biophys. Acta 3o3, 198-2o2.

SALITERNIK-GIVANT, S., BERMAN, E.R. (197o): Biochemical heterogeneity of the corneal glycosaminoglycans. Ophthal. Res. 1, 94-1o8.

SANDBERG, L.B., WEISSMAN, N., SMITH, D.W. (1969): The purification and partial characterization of a soluble elastin-like protein from copper-deficient porcine aorta. Biochemistry 8, 294o-2945.

SANDBERG, L.B., ZEIKUS, R.D., COLTRAIN, I.M. (1971): Tropoelastin purification from copper-deficient swine: A simplified method. Biochim. Biophys. Acta 236, 542-545.

SANDSON, J., DAMON, H., MATHEWS, M.B. (197o): Molecular localization of a cross-reactive antigenic determinant of cartilage proteoglycan. In: Chemistry and Molecular Biology of the Intercellular Matrix (ed. E.A. BALAZS), Vol. 3, pp. 1563-1567. New York: Academic Press.

SANDSON, J., ROSENBERG, L., WHITE, D. (1966): The antigenic determinants of the protein-polysaccharides of cartilage. J. Exp. Med. 123, 817-828.

SANNASI, A., HERMANN, H.R. (197o): Chitin in the Cephalochordata, *Branchiostoma floridae* (amphioxus). Experientia 26, 351-352.

SCHAEFFER, B. (1952): The Triassic coelacanth from *Diplurus* with observations on the evolution of the coelacanthini. Bull. Am. Museum Nat. Hist. 99, 25-78.

SCHAEFFER, B. (1961): Differential ossification in the fishes. Trans. N.Y. Acad. Sci. 23, 5o1-5o5.

SCHAEFFER, B. (1967): Comments on elasmobranch evolution. In: Sharks, Skates, and Rays (eds. P.W. GILBERT, R.F. MATHEWSON, D.R. RALL), pp. 3-35. Baltimore: Johns Hopkins Press.

SCHAEFFER, B. (1968): The origin and basic radiation of the Osteichthyes. In: Current Problems of Lower Vertebrate Phylogeny (ed. T. ØRVIG), pp. 2o7-222. Stockholm: Almqvist and Wiksell.
SCHAFFER, J. (193o): Die Stützgewebe. In: Handbuch der mikroskopischen Anatomie des Menschen (ed. W. v. MÖLLENDORF), 2. Band, 1. Teil, pp. 1-39o. Berlin: Springer.
SCHER, I., HAMERMAN, D. (1972): Isolation of human synovial-fluid hyaluronate by density-gradient ultracentrifugation and evaluation of its protein content. Biochem. J. 126, 1o73-1o8o.
SCHIFFMAN, E., MARTIN, G.R., MILLER, E.J. (197o): Matrices that calcify. In: Biological Calcification: Cellular and Molecular Aspects (ed. H. SCHRAER), pp. 27-67. New York: Appleton-Century-Crofts.
SCHILLER, S. (1963): Mucopolysaccharaides in relation to growth and thyroid hormones. J. Chrom. Dis. 16, 291-3o4.
SCHILLER, S. (1966): Connective and supporting tissues. Mucopolysaccharides of connective tissue. Ann. Rev. Physiol. 28, 137-158.
SCHILLER, S., DORFMAN, A. (196o): Effect of age on the heparin content of rat skin. Nature 185, 111-112.
SCHMID, K. (1972): Recent advances in the study of glycoproteins. Chimia 26, 4o5-414.
SCHMID, K., EMURA, J., KAUFMAN, H., ISEMURA, S., NIMBERG, R.B., ISHIGURO, M., IKENADA, T. (1972): Homologies between α1-acid glycoprotein and the immunoglobulins and homologies within the amino acid sequence of α1-acid glycoprotein. Federation Proc. 31, 465 abs.
SCHMIDT, M., DMOCHOWSKI, A. (1964): The acid mucopolysaccharides of normal and atherosclerotic human aorta. Acta Biochim. Polonica 11, 371-377.
SCHMIDT, M., DMOCHOWSKI, A., WIERZBOWSKA, B. (1966): Galactose and xylose as structural components of vertebrate chondroitin sulfate-protein complexes. Biochim. Biophys. Acta 117, 258-261.
SCHUBERT, M., HAMERMAN, D. (1968): A Primer on Connective Tissue Biochemistry. Philadelphia: Lea and Febiger.
SCHUBERT, M., HAMERMAN, D. (1965): Amino sugar-containing compounds in cartilage, tendon and intervertebral disc. In: The Amino Sugars (eds. E.A. BALAZS, R.W. JEANLOZ), Vol. IIA, pp. 257-279. New York: Academic Press.
SCHWARZ, K. (1973): A bound form of silicon in glycosaminoglycans and polyuronides. Proc. Nat. Acad. Sci. USA 7o, 16o8-1612.
SECHACHAR, B.R., BAGGA, S. (1963): Cytochemistry of the oocyte of *Lovis tardigradus* (Cebr.) and *Macaca mulatta mulatta* (Zimmermann). J. Morphol. 113, 119-137.
SEEGMILLER, R., FERGUSON, C.C., SHELDON, H. (1972): Studies on cartilage. VI. A genetically determined defect in tracheal cartilage. J. Ultrastr. Res. 38, 288-3o1.
SEGREST, J.P., CUNNINGHAM, L.W. (1971): Molecular basis for fibrillar aggregation of tropocollagen. Nature New Biol. 234, 26-28.
SEIFTER, S., GALLOP, P.M. (1966): The structure proteins. In: The Proteins, 2nd Ed. (ed. H. NEURATH), Vol. 4, pp. 155-458. New York: Academic Press.
SEIFTER, S., HARPER, E. (1971): The collagenases. In: The Enzymes, 3rd Ed. (ed. P.D. BOYER), Vol. 3, pp. 649-697. New York: Academic Press.
SENO, N., AKIYAMA, F., ANNO, K. (1972): A novel dermatan polysulfate from hagfish skin, containing trisulfated disaccharide residues. Biochim. Biophys. Acta 264, 229-233.
SENO, N., ANNO, K. (1961): The presence of hyaluronic acid in whale cartilage. Biochim. Biophys. Acta 49, 4o7-4o8.
SENO, N., ARIIZUMI, K., NAGASE, S., ANNO, K. (1972): Isolation and characterization of heparan sulfate from rat kidney. J. Biochem. 72, 479-481.
SENO, N., MEYER, K. (1963): Comparative biochemistry of skin. The mucopolysaccharides of shark skin. Biochim. Biophys. Acta 78, 258-264.
SENO, N., TODA, N. (197o): The carbohydrate-peptide linkage region of keratan sulfate from whale cartilage. Biochim. Biophys. Acta 215, 544-546.
SERAFINI-FRACASSINI, A. (1968): The protein-polysaccharide complex of bovine nasal cartilage. Studies on the molecular organization. Biochim. Biophys. Acta 17o, 289-3oo.

SERAFINI-FRACASSINI, A., SMITH, J.W. (1966): Observations on the morphology of the protein-polysaccharide complex of bovine nasal cartilage and its relationship to collagen. Proc. Roy. Soc. B, 165, 44o-449.
SERAFINI-FRACASSINI, A., STIMSON, W.H. (1971): Studies on low-molecular weight proteoglycans isolated from rabbit ear cartilage. FEBS Letters 15, 245-248.
SERAFINI-FRACASSINI, A., STIMSON, W.H., FLOREANI, L. (1971): The subunits of chemically treated proteoglycan isolated from bovine nasal cartilage. Biochem. J. 122, 1o1-1o6.
SERAFINI-FRACASSINI, A., WHITE, C.J.B., HUNTER, J.C. (1973): The characterization of heparin in bovine liver capsule. FEBS Letters 32, 116-118.
SHERR, C.J., TAUBMAN, M.B., GOLDBERG, B. (1973): Isolation of a disulfide-stabilized, three-chain polypeptide fragment unique to the precursor of human collagen. J. Biol. Chem. 248, 7o33-7o38.
SHOSTAK, S., PATEL, N.G., BURNETT, A.L. (1965): The role of mesoglea in mass cell movement in hydra. Dev. Biol. 12, 434-45o.
SHULMAN, H.J., MEYER, K. (1968): Cellular differentiation and the aging process in cartilaginous tissues. J. Exp. Med. 128, 1353-1362.
SHULMAN, H.J., MEYER, K. (197o): Protein-polysaccharide of chicken cartilage and chondrocyte cell cultures. Biochem. J. 12o, 689-697.
SIEGEL, S.M. (1962): The Plant Cell Wall. New York: Pergamon Press.
SIEGEL, S.M. (1968): Biochemistry of the plant cell wall. In: Comprehensive Biochemistry (eds. M. FLORKIN, E.H. STOTZ), Vol. 26A, pp. 1-51. New York: Elsevier.
SIEGEL, R.C., MARTIN, G.R. (197o): Collagen cross-linking. Enzymatic synthesis of lysine-derived aldehydes and the production of cross-linked components. J. Biol. Chem. 245, 1653-1658.
SIEGEL, R.C., PINNEL, S.R., MARTIN, G.R. (197o): Cross-linking of collagen and elastin. Properties of lysyl oxidase. Biochemistry 9, 4486-4492.
SILBERBERG, M.S., CIERESZKO, L.S., JACOBSON, R.A., SMITH, E.C. (1972): Evidence for a collagen-like protein within spicules of coelenterates. Comp. Biochem. Physiol. 43B, 323-332.
SILBERT, J.E. (197o): Biosynthesis of heparin. In: Chemistry and Molecular Biology of the Intercellular Matrix (ed. E.A. BALAZS), Vol. 2, pp. 991-1oo4. New York: Academic Press.
SIMARD, S.J., FRIEDMAN, S.M. (197o): The binding of some sodium substitutes to chondroitin sulfate. Experientia 26, 834-836.
SIMPSON, D.L., DAVIDSON, E.A. (1972): Isolation of the peptide core of costal cartilage chondroitin 4-sulfate proteoglycan. Biochemistry 11, 1856-1861.
SIMPSON, G.G. (1949): The Meaning of Evolution. New Haven: Yale University Press.
SIMPSON, G.G. (1961): Principles of Animal Taxonomy, Ch. 3. New York: Columbia University Press.
SIMPSON, G.G. (1962): The status of the study of organisms. Am. Scientist 5o, 36-45.
SIMPSON, G.G. (1964): Organisms and molecules in Evolution. Science 146, 1535-1538.
ŠIMŮNEK, Z., MUIR, H. (1972): Changes in the protein-polysaccharides of pig articular cartilage during prenatal life, development and old age. Biochem. J. 126, 515-523.
SINEX, F.M. (1968): The role of collagen in aging. In: Treatise on Collagen, Vol. 2B, Biology of Collagen (ed. B.S. GOULD), pp. 4o9-448. New York: Academic Press.
SINGH, M., CHANDRASEKARAN, E.V., CHERIAN, R., BACHHAWAT, B.K. (1969): Isolation and characterization of glycosaminoglycans in brain of different species. J. Neurochem. 16, 1157-1162.
SIREK, O.V., SCHILLER, S., DORFMAN, A. (1964): Acid mucopolysaccharides in aortic tissue of the dog. Biochim. Biophys. Acta 83, 148-151.
SLAVKIN, H.C. (1972): In: The Comparative Molecular Biology of Extracellular Matrices (ed. H.C. SLAVKIN), pp. XV-XVII, 2-3, 74-75. New York: Academic Press.
SLOMIANY, B.L., MEYER, K. (1972): Isolation and structural studies of sulfated glycoproteins of hog gastric mucosa. J. Biol. Chem. 247, 5o62-5o7o.
SMIDSRØD, O., HAUG, A. (1968): Dependence upon uronic acid composition of some ion-exchange properties of alginates. Acta Chem. Scand. 22, 1989-1997.
SMITH, B.D., BYERS, P.H., MARTIN, G.R. (1972): Production of procollagen by human fibroblasts in culture. Proc. Nat. Acad. Sci. USA 69, 326o-3262.

SMITH, D.W., BROWN, D.M., CARNES, W.H. (1972): Preparation and properties of salt-soluble elastin. J. Biol. Chem. 247, 2427-2432.
SMITH, H.M. (196o): Evolution of Chordate Structure. New York: Holt, Rinehart & Winston.
SMITH, H.W. (1931): The absorption and excretion of water and salts by the elasmobranch fishes. I. Fresh-water elasmobranchs. Am. J. Physiol. 98, 279-295.
SMITH, H.W. (1936): The retention and physiological role of urea in the elasmobranchs. Biol. Rev. 11, 49-82.
SMITH, H.W. (1961): From Fish to Philospher. Garden City, N.Y.: Doubleday.
SMITH, J.W. (1968): Molecular pattern in native collagen. Nature 219, 157-158.
SMITH, J.W. (197o): The disposition of protein-polysaccharide in the epiphyseal plate cartilage of the young rabbit. J. Cell Sci. 6, 843-864.
SMITH, J.W., FRAME, J. (1969): Observations on the collagen and protein-polysaccharide complex of rabbit corneal stroma. J. Cell Sci. 4, 421-436.
SMITH, J.W., PETERS, T.J., SERAFINI-FRACASSINI, A. (1967): Observations on the distribution of the protein-polysaccharide complex and collagen in bovine articular cartilage. J. Cell Sci. 2, 129-136.
SMITH, J.W., SERAFINI-FRACASSINI, A. (1967): The relationship of hyaluronate and collagen in the bovine vitreous body. J. Anat. 1o1, 99-112.
SMITH, J.W., SERAFINI-FRACASSINI, A. (1968): The distribution of the protein-polysaccharide complex in the nucleus pulposus in young rabbits. J. Cell Sci. 3, 33-4o.
SMITH, P.H. (1972): Autoradiographic evidence for the concurrent synthesis of collagen and chondroitin sulfates by chick sternal chondrocytes. Conn. Tissue Res. 1, 181-188.
SMITH, Q.T., LINDENBAUM, A. (1971): Composition and calcium binding of protein-polysaccharides of calf nasal septum and scapula. Calc. Tissue Res. 7, 29o-298.
SOKOLOFF, L. (1963): Elasticity of articular cartilage: Effect of ions and viscous solutions. Science 141, 1o55-1o57.
SOKOLOFF, L. (1969): The Biology of Degenerative Joint Disease. Chicago: University of Chicago Press.
SOLHEIM, K. (1965): Glycosaminoglycans, hydroxyproline, calcium and phosphorus in healing fractures. Acta Universitatis Lundensis. Sectis II, No. 28, pp. 3-22.
SOLHEIM, K. (1968a): The glycosaminoglycans of the human epiglottis. J. Oslo City Hosp. 18, 45-5o.
SOLHEIM, K. (1968b): Microradiography and the glycosaminoglycans in myositis ossificous. J. Oslo City Hosp. 18, 51-58.
SOUTEYRAND-BOULENGER, J.-D. (1971): Arthrologie. In: Traité de Zoologie (ed. P.-P. GRASSÉ), Vol. 16III, pp. 837-1157. Paris: Masson.
SPEAKMAN, P.T. (1971): Proposed mechanism for the biological assembly of collagen triple helix. Nature 229, 241-243.
SPIRO, R.G. (1965a): Amino sugars and macromolecules containing amino sugars in kidney. In: The Amino Sugars (eds. E.A. BALAZS, R.W. JEANLOZ), Vol. IIA, pp. 79-93. New York: Academic Press.
SPIRO, R.G. (1965b): Amino sugars and macromolecules containing amino sugars in liver. In: The Amino Sugars (eds. E.A. BALAZS, R.W. JEANLOZ), Vol. IIA, pp. 47-78. New York: Academic Press.
SPIRO, R.G. (1967): The structure of the disaccharide unit of the renal glomerular basement membrane. J. Biol. Chem. 242, 4813-4823.
SPIRO, R.G. (1969a): Glycoproteins: Their biochemistry, biology and role in human disease. New Eng. J. Med. 281, 991-1oo1, 1o43-1o56.
SPIRO, R.G. (1969b): Characterization and quantitative determination of the hydroxylysine-linked carbohydrate units of several collagens. J. Biol. Chem. 244, 6o2-612.
SPRIO, R.G. (197o): Glycoproteins. Ann. Rev. Biochem. 39, 599-638.
SPIRO, R.G. (1972): Basement membranes and collagens. In: Glycoproteins, 2nd Ed. (ed. A. GOTTSCHALK), pp. 964-999. Amsterdam: Elsevier.
SPIRO, R.G., BHOYROO, V.D. (1971): The carbohydrate of invertebrate collagens: A glucuronic acid-mannose disaccharide unit. Federation Proc. 3o, 1223 abs.
SPIRO, R.G., LUCAS, F., RUDALL, K.M. (1971): Glycosylation of hydroxylysine in collagens. Nature New Biol. 231, 54-55.

SRINIVAS, H., LEELA, R., RAO, P., RAMA, B. (1972): Nature of acid mucopolysaccharides of the rat and pig brain. Indian J. Biochem. Biophys. 9, 83-88.
SRINIVASAN, S.R., A. LOPEZ-S., RADHAKRISHNAMURTHY, B., BERENSON, G.S. (197o): Complexing of serum pre-β and β-lipoproteins and acid mucopolysaccharides. Atheroscler. 12, 321-334.
SRINIVASAN, S.R., RADHAKRISHNAMURTHY, B., DALFERES, Jr., E.R., BERENSON, G.S. (1969): Glycosaminoglycans from squid skin. Comp. Biochem. Physiol. 28, 169-176.
SRINIVASAN, S.R., RADHAKRISHNAMURTHY, B., PARGAONKAR, P.S., BERENSON, G.S. (1971): Glycoproteins from connective tissue of twins. Nature 229, 58-59.
STACEY, M., BARKER, S.A. (1962): Carbohydrates of Living Tissues. New York: Van Nostrand.
STAFFORD, H.A. (196o): A guide to the nomenclature and classification of organisms. In: Comparative Biochemistry (eds. M. FLORKIN, H.S. MASON), Vol. 1, pp. XVII-XXV. New York: Academic Press.
STARCHER, B.C., SACCOMANI, G., URRY, D.W. (1973): Coacervation and ion-binding studies on aortic elastin. Biochim. Biophys. Acta 31o, 481-486.
STARCHER, B.C., URRY, D.W. (1973): Elastin coacervate as a matrix for calcification. Biochem. Biophys. Res. Commun. 53, 21o-216.
STARK, M., KÜHN, K. (1968): The properties of molecular fragments obtained on treating calfskin collagen with collagenase from *Clostridium histolyticum*. Europ. J. Biochem. 6, 534-541.
STARK, M., LENAERS, A., LAPIÈRE, C., KÜHN, K. (1971): Electronoptical studies of procollagen from the skin of dermatosparaxic calves. FEBS Letters 18, 225-227.
STARK, M., MILLER, E.J., KÜHN, K. (1972): Comparative electronmicroscopic studies on the collagens extracted from cartilage, bone and skin. Eur. J. Biochem. 27, 192-196.
STARK, M., RAUTERBERG, J., KÜHN, K. (1971): Evidence for a non-helical region at the carboxyl terminus of the collagen molecule. FEBS Letters 13, 1o1-1o4.
STEFANOVICH, V., AKIYAMA, K. (197o): Comparative studies of aortic acid mucopolysaccharides in fifteeen species. Comp. Biochem. Physiol. 34, 125-13o.
STEMPIEN, M.F. (1966): The structure and location of sulfated and non-sulfated acid mucopolysaccharides as an aid to the taxonomy of Porifera: The class Demospongiae. Am. Zool. 6, 363.
STENSIÖ, E. (1958): Les cyclostomes fossile on ostracodermes. In: Traité de Zoologie (ed. P.-P. GRASSÉ), Vol. 13, pp. 173-425. Paris: Masson.
STENSIÖ, E. (1968): The cyclostomes with special reference to the diphyletic origin of the Petromyzontida and Myxinoidea. In: Current Problems of Lower Vertebrate Phylogeny, pp. 13-71. Stockholm: Alqmvist and Wiksell.
STERN, E.L., LINDAHL, B., RODÉN, L. (1971): The linkage of dermatan sulfate to protein. II. Monosaccharide sequence of the linkage region. J. Biol. Chem. 246, 57o7-5715.
STEVEN, F.S., JACKSON, D.S. (1968): Isolation and amino acid composition of insoluble elastin bovine foetal and adult aorta and ligamentum nuchae. Biochim. Biophys. Acta 168, 334-34o.
STEVEN, F.S., KNOTT, J., JACKSON, D.S., PODRAZKY, V. (1969): Collagen-protein-polysaccharide interactions in human intervertebral disc. Biochim. Biophys. Acta 188, 3o7-313.
STOCKWELL, R.A., SCOTT, J.E. (1965): Observations on the acid glycosaminoglycan (mucopolysaccharide) content of the matrix of ageing cartilage. Ann. Rheu. Dis. 24, 341-35o.
STOCKWELL, R.A., SCOTT, J.E. (1967): Distribution of acid glycosaminoglycans in human articular cartilage. Nature 215, 1376-1377.
STOLTZ, M., FURTHMAYER, H., TIMPL, R. (1973): Increased lysine hydroxylation in rat bone and tendon collagen and localization of the additional residues. Biochim. Biophys. Acta 31o, 461-468.
STOLTZ, M., TIMPL, R., FURTHMAYER, H., KÜHN, K. (1973): Structural and immunogenic properties of a major antigenic determinant in neutral salt-extracted rat-skin collagen. Eur. J. Biochem. 37, 287-294.
STOLTZ, M., TIMPL, R., KÜHN, K. (1972): Nonhelical regions in rat collagen α1-chain. FEBS Letters 26, 61-65.
STONE, A.L. (1967): Aggregation of cationic dyes on acid polysaccharides. II. Quantitative parameters of metachromasy. Biochim. Biophys. Acta 148, 193-2o6.

STONE, A.L. (1969): Conformations of hexose polysaccharides in solution. In: Structural Stability of Biological Macromolecules (eds. G. FASMAN, S. TIMASHEFF), Vol. 2, pp. 353-415. New York: Dekker.
STONE, A.L., BRADLEY, D.F. (1967): Aggregation of cationic dyes on acid polysaccharides. I. Spectrophotometric titration with acridine orange and other metachromatic dyes. Biochim. Biophys. Acta 148, 172-192.
STOOLMILLER, A.C., DORFMAN, A. (1969): The metabolism of glycosaminoglycans. In: Comprehensive Biochemistry (eds. M. FLORKIN, E.H. STOTZ), Vol. 17, pp. 241-275. New York: Elsevier.
STRAWICH, E., NIMNI, M.E. (1971): Properties of a collagen molecule containing three identical components extracted from bovine articular cartilage. Biochemistry 1o, 39o5-3911.
STUHLSATZ, H.W. (1973): The relation of glycosaminoglycans and glycoproteins to collagen and non-collagenous protein in pressure and non-pressure areas of the hip-joint cartilage of the femur. In: Connective Tissue and Ageing (ed. H.G. VOGEL), pp. 149-15o. Amsterdam: Excerpta Medica.
STUHLSATZ, H.W., KESTERS, R., WOLLMER, A., GREILING, H. (1971): Zur Struktur von Proteoglykanen aus der Cornea. Hoppe-Seylers Z. Physiol. Chem. 352, 289-3o3.
SUMMERS, G.W., CLAYCOMB, C.K. (1965): Amino-acid composition of collagen isolated from the oral cavity of the sperm whale. Nature 2o6, 1o46-1o47.
SUNDBLAD, L. (1965): Glycosaminoglycans and glycoproteins in synovial fluid. In: The Amino Sugars (eds. E.A. BALAZS, R.W. JEANLOZ), Vol. 2A, pp. 229-25o. New York: Academic Press.
SUSI, H., ARD, J.S., CARROLL, R.J. (1971): The infrared spectrum and water binding of collagen as a function of relative humidity. Biopolymers 1o, 1597-16o4.
SUZUKI, S. (196o): Isolation of novel disaccharides from chondroitin sulfates. J. Biol. Chem. 235, 358o-3588.
SUZUKI, S., KOJIMA, K., UTSUMI, K.R. (197o): Production of sulfated mucopolysaccharides by established cell lines of fibroblastic and non-fibroblastic origin. Biochim. Biophys. Acta 222, 24o-243.
SUZUKI, S., SAITO, H., YAMAGATA, T., ANNO, K., SENO, N., KAWAI, Y. (1968): Formation of three types of disulfated disaccharides from chondroitin sulfates by chondroitinase digestion. J. Biol. Chem. 243, 1543-155o.
ŠVEJCAR, J. (1968): The content of glycosaminoglycans (mucopolysaccharides) in mouse fetus and placenta. Biochim. Biophys. Acta 165, 84-88.
SVENNERHOLM, L. (1965): Glycolipids and other amino sugar-containing compounds in the nervous system. In: The Amino Sugars (eds. E.A. BALAZS, R.W. JEANLOZ), Vol. 2A, pp. 381-4oo. New York: Academic Press.
SWANN, D.A. (1968): Studies on hyaluronic acid. I. The preparation and properties of rooster comb hyaluronic acid. Biochim. Biophys. Acta 156, 17-3o.
SWANN, D.A. (197o): On the state of hyaluronic acid in a connective tissue matrix. In: Chemistry and Molecular Biology of the Intercellular Matrix (ed. E.A. BALAZS), Vol. 2, pp. 743-748. New York: Academic Press.
SWANN, D.A., CONSTABLE, I.J., HARPER, E. (1972): Vitreous structure. III. Composition of bovine vitreous collagen. Invest. Ophthalmol. 11, 735-738.
SYKES, B.C., PARTRIDGE, S.M. (1972): Isolation of a soluble elastin from lathyritic chicks. Biochem. J. 13o, 1171-1172.
SYLVEN, B. (1965): Amino sugar-containing compounds in tumors. In: The Amino Sugars (eds. E.A. BALAZS, R.W. JEANLOZ), Vol. IIA, pp. 195-2o4. New York: Academic Press.
SZIRMAI, J.A. (1965a): Glycosaminoglycans in umbilical chord. In: The Amino Sugars (eds. E.A. BALAZS, R.W. JEANLOZ), Vol. IIA, pp. 219-227. New York: Academic Press.
SZIRMAI, J.A. (1965b): Glycosaminoglycans in the connective tissue of the electric organ. In: The Amino Sugars (eds. E.A. BALAZS, R.W. JEANLOZ), Vol. IIA, pp. 251-256. New York: Academic Press.
SZIRMAI, J.A. (1969): Structure of cartilage. In: Aging of Connective and Skeletal Tissue (ed. A. ENGEL, T. LARSSON), pp. 163-184. Stockholm: Nordiska Bokhandelns Förlag.
SZIRMAI, J.A. (1971): Structure of the intervertebral body. In: Chemistry and Molecular Biology of the Intercellular Matrix (ed. E.A. BALAZS), Vol. 3, pp. 1279-13o8. New York: Academic Press.

SZIRMAI, J.A., van BOVEN-DE TYSSONSK, E., GARDELL, S. (1967): Microchemical analysis of glycosaminoglycans, collagen, total protein and water in histological layers of nasal septum cartilage. Biochim. Biophys. Acta 136, 331-35o.
TANAKA, K., NISHI, C., TAKAYA, M., UCHIYAMA, T. (1972): A hexosamine-containing polyfucose sulfate-protein complex from *Stichopus japonicus* Selenka. J. Biochem. 72, 1265-1267.
TANFORD, C. (1961): Physical Chemistry of Macromolecules. New York: Wiley.
TANZER, M.L. (1973): Cross-linking of collagen. Science 18o, 561-566.
TANZER, M.L., FAIRWEATHER, R., GALLOP, P.M. (1972): Collagen crosslinks: Isolation of reduced N^{ε}-hexosylhydroxylysine from borohydride-reduced calf skin insoluble collagen. Arch. Biochem. Biophys. 151, 137-141.
TANZER, M.L., HOUSLEY, T., BERUBE, L., FAIRWEATHER, R., FRANZBLAU, C., GALLOP, P.M. (1973): Structure of two histidine-containing cross-links from collagen. J. Biol. Chem. 248, 393-4o2.
TARLO, L.B.H. (1964): The origin of bone. In: Bone and Tooth Symposium (ed. H.J.J. BLACKWOOD), pp. 3-17. New York: Pergamon.
TANCREDI, G., PECORELLI, F. (1966): Content of sialic acid in relation to the function of cartilage. Exp. Med. Surg. 24, 156-159.
TAYLOR, A.C., KOLLROSS, J.J. (1946): Stages in the normal development of *Rana pipiens* larvae. Anat. Rec. 94, 7-23.
TAYLOR, R.L., SHIVELY, J.E., CONRAD, H.E., CIFONELLI, J.A. (1973): The uronic acid composition of heparins and heparan sulfates. Biochemistry 12, 3633-3637.
TENBAUM, E. (1955): Polarisationsoptische Beiträge zur Kenntnis der Gewebe von *Branchiostoma lanceolatum* (P). Z. Zellforschung 42, 149-192.
TESTER, A.L., KENDALL, J.I. (1968): Cupulae in shark neuromasts: Composition, origin, generation. Science 16o, 772-774.
THINEY, Y., GARRONE, R. (197o): Association des mucopolysaccharides et des fibres de collagène dans le tissu conjonctif des spongiaires. Proc. 7th Congrès Intern. Microscopie Électronique, Grenoble, Vol. 1, p. 589.
THOMAS, L. (1964): The effects of papain, vitamin A and cortisone on cartilage matrix *in vivo*. In: Connective Tissue: Intercellular Macromolecules. Am. Heart Assoc., pp. 2o7-213. Boston: Little, Brown.
THOMAS, L.J. (1954): The localization of heparin-like blood anticoagulant substances in the tissues of spisula solidissima. Biol. Bull. 1o6, 129-138.
THOMAS, Jr., L.J. (1951): A blood anticoagulant from surf clams. Biol. Bull. 1o1, 23o.
THOMPSON, E.W., PRESTON, R.D. (1967): Proteins in the cell walls of some green algae. Nature 213, 684-685.
THOMPSON, H.C., THOMPSON, M. (197o): Isolation and amino acid composition of the collagen of white shrimp *(Penaeus setiferus)*. II. Comp. Biochem. Physiol. 35, 471-477.
THOMPSON, Jr., H.C., THOMPSON, M.H. (1968): Isolation and amino acid composition of the collagen of white shrimp *(Penaeus setiferous)*. I. Comp. Biochem. Physiol. 27, 127-132.
THOMPSON, K.S. (1969): The biology of the lobe-finned fishes. Biol. Rev. 44, 91-154.
THORP, F.K., DORFMAN, A. (1963): The occurrence of intracellular chondroitin sulfate. J. Cell Biol. 18, 13-17.
THORP, F.K., DORFMAN, A. (1967): Differentiation of connective tissues. In: Current Topics in Developmental Biology (eds. A.A. MOSCONA, A. MONROY), Vol. 2, pp. 151-189. New York: Academic Press.
THORSON, T.B. (1967): Osmoregulation in fresh-water elasmobranchs. In: Sharks, Skates and Rays (eds. P.W. PERRY, R.F. MATHEWSON, D.P. RALL). Baltimore: Johns Hopkins Press.
THORSON, T.B. (197o): Freshwater stingrays, *Potamotrygon* spp.: Failure to concentrate urea when exposed to saline medium. Life Science 9, 893-9oo.
THORSON, T.B. (1971): Movement of Bull sharks, *Carcharinus leucas*, between Caribbean Sea and Lake Nicaragua demonstrated by tagging. Copeia, No. 2, 336-338.
THORSON, T.B., COWAN, C.M., WATSON, D.E. (1966a): The status of the freshwater shark of Lake Nicauagua. Copeia (3), 385-4o2.

THORSON, T.B., COWAN, C.M., WATSON, D.E. (1966b): Sharks and sawfish in the Lake Izabel-Rio Dulce system, Guatamala. Copeia (3), 62o-622.
THORSON, T.B., COWAN, C.M., WATSON, D.E. (1967): *Potamotrygon* spp.: elasmobranchs with low urea content. Science 158, 375-377.
THUNELL, S., ANTONOPOULOS, C.A., GARDELL, S. (1967): Analysis of aortic glycosaminoglycans from various animal species by CPC-cellulose column procedures. J. Atherosclerosis Res. 7, 283-294.
TIMPL, R., WOLFF, I., WEISER, M. (1969): Acidic structural proteins of connective tissue. I. Solubilization and preliminary chemical characterization. Biochim. Biophys. Acta 194, 112-12o.
TKOCZ, C., KÜHN, C. (1969): Formation of triple-helical collagen molecules from α1 or α2 polypeptide chains. Eur. J. Biochem. 7, 454-462.
TODA, N., SENO, N. (197o): Sialic acid in the keratan sulfate fraction from whale cartilage. Biochim. Biophys. Acta 2o8, 227-235.
TOLPIN, D.W., KLYCE, S.D., DOHLMAN, C.H. (1969): Swelling properties of dogfish cornea. Exp. Eye Res. 8, 429-437.
TOOLE, B.P. (1973): Hyaluronate inhibition of chondrogenesis: antagonism of thyroxine, growth hormone, and calcitonin. Science 18o, 3o2-3o3.
TOOLE, B.P., GROSS, J. (1971): The extracellular matrix of the regenerating newt limb: Synthesis and removal of hyaluronate prior to differentiation. Dev. Biol. 25, 57-77.
TOOLE, B.P., KANG, A.H., TRELSTAD, R.L., GROSS, J. (1972): Collagen heterogeneity within different growth regions of long bones of rachitic and non-rachitic chicks. Biochem. J. 127, 715-72o.
TOOLE, B.P., LOWTHER, D.A. (1966): The organization of hexosamine-containing compounds in bovine skin. Biochim. Biophys. Acta 121, 315-325.
TOOLE, B.P., LOWTHER, D.A. (1968): Dermatan sulfate-protein: Isolation from and interaction with collagen. Arch. Biochem. Biophys. 128, 567-578.
TOOLE, B.P., TRELSTAD, R.L. (1971): Hyaluronate production and removal during corneal development in the chick. Dev. Biol. 26, 28-35.
TOSATTI, M.P.M., GOTTE, L., MORET, V. (1971): Some features of the binding of calcium ions to elastin. Calcif. Tissue Res. 6, 329-334.
TOWE, K.M. (197o): Oxygen-Collagen priority and the early metazoan fossil record. Proc. Nat. Acad. Sci. USA 65, 781-788.
TOWE, K.M., URBANEK, A. (1972): Collagen-like structures in Ordovician graptolite periderm. Nature 237, 443-445.
TRACEY, M.U. (1968): The biochemistry of supporting materials in organisms. In: Advances in Comparative Physiology and Biochemistry (ed. O. LOWENSTEIN), Vol. 3, pp. 233-27o. New York: Academic Press.
TRAMS, E.G., LAUTER, C.J. (1964): Properties of electroplax protein. III. Biochim. Biophys. Acta 83, 296-3o4.
TRAUB, W., PIEZ, K.A. (1971): The chemistry and structure of collagen. Adv. Prot. Chem. 25, 243-352.
TRAUB, W., YONATH, A., SEGAL, D.M. (1969): On the molecular structure of collagen. Nature 221, 914-917.
TRAVIS, D.F., FRANCOIS, C.J., BONAR, L.C., GLIMCHER, M.J. (1969): Comparative studies of the organic matrices of invertebrate mineralized tissues. J. Ultrastr. Res. 18, 519-55o.
TRELSTAD, R.L. (1973): The developmental biology of vertebrate collagens. J. Histochem. Cytochem. 21, 521-528.
TRELSTAD, R.L., COULOMBRE, A.J. (1971): Morphogenesis of the collagenous stroma in the chick cornea. J. Cell Biol. 5o, 84o-858.
TRELSTAD, R.L., KANG, A.H., COHEN, A.M., HAY, E.D. (1973): Collagen synthesis *in vitro* by embryonic spinal chord epithelium. Science 179, 295-296.
TRELSTAD, R.L., KANG, A.H., GROSS, J. (1971): Characterization of a new type of collagen in the embryonic chick skin. Federation Proc. 3o, 1196 abs.
TRELSTAD, R.L., KANG, A.H., IGARASHI, S., GROSS, J. (197o): Isolation of two distinct collagens from chick cartilage. Biochemistry 9, 4993-4998.
TRELSTAD, R.L., KANG, A.H., TOOLE, B.P., GROSS, J. (1972): High resolution separation of native $[\alpha 1(I)]_2\alpha 2$ and $[\alpha 1(II)]_3$ and their component α chains. J. Biol. Chem. 247, 6469-6473.

TRENCH, R.K. (197o): Synthesis of a mucous cuticle by a zoanthid. Nature 227, 1155-1156.
TRIFFIT, J.T. (1968): Binding of calcium and strontium by alginates. Nature 217, 457-458.
TSIGANOS, C.P., HARDINGHAM, T.E., MUIR, H. (1971): Proteoglycans of cartilage: An assessment of their structure. Biochim. Biophys. Acta 229, 529-534.
UDENFRIEND, S. (197o): Biosynthesis of hydroxyproline in collagen. In: Chemistry and Molecular Biology of the Intercellular Matrix (ed. E.A. BALAZS), Vol. 1, pp. 371-384. New York: Academic Press.
UENO, R., KITAMIKADO, M. (1972): Enzymatic degradation of whale cartilage keratosulfate. V. Release of sialic acid-containing oligosaccharide. Bull. Jap. Soc. Sci. Fisheries 38, 5o3-5o9.
UITTO, J., JIMINEZ, S.A., DEHM, P., PROCKOP, D.J. (1972): Characterization of the precursor forms of the α1 and α2 chains of collagen from matrix-free tendon cells. Biochim. Biophys. Acta 278, 198-2o5.
URIST, M.R. (1961): Calcium and phosphorus in the blood and skeleton of the elasmobranchii. Endocrinol. 69, 778-8o1.
URIST, M.R. (1962a): The bone-body fluid continuim: Calcium and phosphorus in the skeleton and blood of extinct and living vertebrates. Persp. Biol. Med. 6, 75-115.
URIST, M.R. (1962b): Calcium and other ions in blood and skeleton of Nicaraguan fresh-water shark. Science 137, 984-986.
URIST, M.R. (1963): The regulation of calcium and other ions in the serums of hagfish and lampreys. Ann. N.Y. Acad. Sci. 1o9, 294-311.
URIST, M.R., IWATA, H., CECCOTTI, P.L., DORFMAN, R.L., BOYD, S.D., McDOWELL, R.M., CHIEN, C. (1973): Bone morphogenesis in implants of insoluble bone gelatin. Proc. Nat. Acad. Sci. USA 7o, 3511-3515.
URIST, M.R., VAN DE PUTTE, K.A. (1967): Comparative biochemistry of the blood of fishes. In: Sharks, Skates and Rays (eds. P.W. GILBERT, R.F. MATHEWSON, D.P. RALL), pp. 271-292. Baltimore: Johns Hopkins Press.
URIST, M.R., SPEER, D.P., IBSEN, K.J., STRATES, B.S. (1968): Calcium binding by chondroitin sulfate. Calcif. Tissue Res. 2, 253-261.
URRY, D.W. (1971): Neutral sites for calcium ion binding to elastin and collagen: A charge neutralization theory for calcification and its relationship to atherosclerosis. Proc. Nat. Acad. Sci. USA 68, 81o-814.
URRY, D.W. (1974): Studies on the conformation and interactions of elastin. In: Arterial Mesenchyme and Arteriosclerosis (ed. T.B. CLARKSON). New York: Plenum Publishing.
URRY, D.W., CUNNINGHAM, W.D., OHNISHI, T. (1973): A neutral polypeptide-calcium ion complex. Biochim. Biophys. Acta 292, 853-857.
URRY, D.W., CUNNINGHAM, W.D., OHNISHI, T. (1974): Studies on the conformation and interactions of elastin: proton magnetic resonance of the repeating pentapeptide. Biochemistry 13, 6o9-616.
URRY, D.W., KRIVACIC, J.R., HAIDER, J. (1971): Calcium ion effects - a notable change in elastin conformation by interacting at neutral sites. Biochem. Biophys. Res. Commun. 43, 6-11.
USHAKOV, B.P. (1964): Thermostability of cells and proteins of poikilotherms and its significance in speciation. Physiol. Rev. 44, 518-56o.
USHAKOV, B.P. (1967): Coupled evolutionary changes in protein stability. In: Molecular Mechanisms of Temperature Adaptation (ed. C.L. PROSSER), pp. 1o7-129. Publ. Am. Ass. Adv. Sci. No. 84, Washington.
VAN LIS, J.M.J., KRUISWIJK, T., MAGER, W.H., KALSBEEK, G.L. (1973): Glycosaminoglycans in human skin. Brit. J. Dermat. 88, 355-361.
VAN PILSUM, J.F., STEPHENS, G., TAYLOR, D. (1967): A survey of the occurrence of creatine and creatine-synthesizing enzymes and the phylogenetic significance thereof. Federation Proc. 26, 68o.
VAN PRAAG, D., STONE, A.L., RICHTER, A.J., FARBER, S.J. (1972): Composition of glycosaminoglycans (mucopolysaccharides) in rabbit kidney. II. Renal cortex. Biochim. Biophys. Acta 273, 149-156.
VARADI, D.P. (1972): Chemistry and fine structure of elastic fibers from normal adult skin. J. Invest. Dermatol. 59, 238-246.

VARGA, L., BALAZS, E.A. (1954): Studies on the structure of the vitreous body. II. Electrophoresis and sedimentation properties. Am. J. Ophthal. 38, 29-37.
VASSEUR, E. (1954): The Chemistry and Physiology of the Jelly Coat of the Sea Urchin Egg. Stockholm: Kihlstromstryck.
VEIS, A. (1967): Intact collagen. In: Treatise on Collagen. Vol. VI, Chemistry of Collagen (ed. G.N. RAMACHANDRAN), pp. 367-439. New York: Academic Press.
VEIS, A., ANESEY, J. (1965): Modes of intermolecular cross-linking in mature insoluble collagen. J. Biol. Chem. 24o, 3899-39o8.
VEIS, A., ANESEY, J., COHEN, J. (1958): Acid precursor gelatins, structure and significance in the collagen-gelatin transition. In: Recent Advances in Gelatin and Glue Research (ed. G. STAINSBY), pp. 155-164.
VEIS, A., ANESEY, J.R., GARVIN, J.E., DIMUZIO, M.T. (1972): High molecular weight collagen: A long-lived intermediate in the biogenesis of collagen fibrils. Biochem. Biophys. Res. Commun. 48, 14o4-1411.
VEIS, A., ANESEY, A., MUSSELL, S. (1967): A limiting microfibril model for the three-dimensional arrangement within collagen fibres. Nature 215, 931-934.
VEIS, A., ANESEY, J., YUAN, L., LEVY, S.J. (1973): Evidence for an amino-terminal extension in high-molecular-weight collagens from mature bovine skin. Proc. Nat. Acad. Sci. USA 7o, 1464-1467.
VELICAN, C., VELICAN, D. (1967): Structural factors of macromolecular stabilization. Rev. Roum. Med. Int. 4, 79-91.
VIALLI, Di M., BOLOGNANI, L., CROCE, G., BOLOGNANI FANTIN, A.M. (1969): Richerche istochimiche e biochimiche sugli esosaminglicani (MPS) del "veleno" delle parotoidi di Bufo vulgaris e viridis. Arch. Sci. Biol. 53, 1o8-125.
VIIDIK, A. (1969): The aging of collagen as reflected in its physical properties. In: Aging of Connective and Skeletal Tissue (eds. A. ENGEL, T. LARSSON), pp. 125-148. Stockholm: Nordiska Bokhandelns Förlag.
VIIDIK, A. (1973): Functional properties of collagenous tissues. Intern. Rev. Conn. Tissue Res. 6, 127-215.
VISWANATHA, T., IRREVERRE, F. (196o): Occurrence of hydroxylysine in trypsin. Biochim. Biophys. Acta 4o, 564-565.
VITTUR, F., PUGLIARELLO, M.C., DE BERNARD, B. (1972a): Some properties of proteoglycans of pre-osseous cartilage. Biochim. Biophys. Acta 257, 389-397.
VITTUR, F., PUGLIARELLO, M.C., DE BERNARD, B. (1972b): The calcium binding properties of a glycoprotein isolated from pre-osseus cartilage. Biochem. Biophys. Res. Commun. 48, 143-152.
VOLPIN, D., VEIS, A. (1971): Isolation and characterization of the cyanogen bromide peptides from the α1 and α2 chains of acid-soluble bovine skin collagen. Biochemistry 1o, 1751-1755.
VOLPIN, D., VEIS, A. (1973): The cyanogen bromide peptides from insoluble skin and dentin bovine collagens. Biochemistry 12, 1452-1464.
VUUST, J., PIEZ, K.A. (1972): A kinetic study of collagen biosynthesis. J. Biol. Chem. 247, 856-862.
WADA, K. (1964a): Mineralization of the calcified tissues in mollusks. III. Localization and distribution of ^{45}Ca in the mantle tissue and on the growing shell surface in several marine bivalves by radioautography. Nippon Suisan Gakkaishi 3o, 385-392.
WADA, K. (1964b): Mineralization of the calcified tissues in mollusks. IV. Selective fixation of ^{45}Ca into or onto the metachromatic matter in the processes of shell mineralization. Nippon Suisan Gakkaishi 3o, 393-399.
WADA, K. (1966): Mineralization of the calcified tissue in mollusks. XII. Specific patterns of nonmineralized layer conchiolin in the amino acid compositon. Nippon Suisan Gakkaishi 32, 3o4-311.
WADA, K., FURUHASHI, T. (197o): Mineralization of calcified tissue in mollusks. XVII. Acid polysaccharides in the shell of *Hydriopsis schlegeli*. Nippon Suisan Gakkaishi 36, 1122-1126.
WADA, K., FURUHASHI, T. (1971): Mineralization of the calcified tissue in mollusks. XVIII. Acid polysaccharide of mantle mucus of some marine bivalves. Nippon Suisan Gakkaishi 37, 13-17.
WADELL, M.C.L. (1971): Alteration of protein synthesis in cartilage by a synthetic double-stranded polyribonucleotide. J. Dent. Res. 5o, 1o72-1o79.

WAGH, P.V., ROBERTS, B.I. (1972): Purification and characterization of a glycoprotein from the intimal region of porcine aorta. Biochemistry 11, 4222-4227.
WALD, G. (196o): The significance of vertebrate metamorphosis. Circulation 21, 916-938.
WALKER, E.P. (1968): Mammals of the World, Vol. II, 2nd Ed. Baltimore: Johns Hopkins Press.
WALTMAN, B. (1966): Electrical properties and fine structure of the ampullary canals of Lorenzini. Acta Physiol. Scand. 66 (Suppl.), 264, 1-6o.
WALTON, K.W. (1973): The role of glycosaminoglycans and calcium in atherosclerosis. In: Connective Tissue and Ageing (ed. H. VOGEL), pp. 21-22. Amsterdam: Excerpta Medica.
WARDI, A.H., ALLEN, W.S., TURNER, D.L., STARY, Z. (1969): Hyaluronate-peptide linkage group. Biochim. Biophys. Acta 192, 151-154.
WARREN, G.H. (195o): The isolation of a mucopolysaccharide from *Aerobacter aerogenes*. Science 111, 473-474.
WASSERMAN, F. (1956): The intercellular components of connective tissue; origin, structure and interrelationships of fibres and ground substance. Ergebn. Anat. Entw. Gesch. 35, 24o-333.
WASSERMAN, F., BER, A., ALLALOUF, D. (1972): Acidic glycosaminoglycan composition of the gills of *Cyprinus carpio*. Comp. Biochem. Physiol. B 42, 669-677.
WASTESON, Å. (1971): Properties of fractionated chondroitin sulfate from ox nasal septa. Biochem. J. 122, 477-485.
WASTESON, Å., LINDAHL, U. (1971): The distribution of sulfate residues in the chondroitin sulfate chain. Biochem. J. 125, 9o3-9o8.
WASTESON, Å., LINDAHL, U., HALLÉN, A. (1972): Mode of degradation of the chondroitin sulfate proteoglycan in rat costal cartilage. Biochem. J. 13o, 729-738.
WATSON, J.D. (197o): Molecular Biology of the Gene, 2nd Ed. New York: Benjamin.
WATSON, M.R. (1958): The chemical composition of earthworm cuticle. Biochem. J. 68, 416-42o.
WATSON, M.R., SILVESTER, N.R. (1958): Studies of invertebrate collagen preparations. Biochem. J. 71, 578-584.
WATTS, D.C. (1968): Variation in enzyme structure and function: The guidelines of evolution. Adv. Comp. Physiol. Biochem. 3, 1-114.
WEIL, C., FREI, P., WEIGEL, W., JASINSKI, B. (1969): Le vieillissement des cartilages. Reone Rhum. Mal. ostéo-articulaire 36, 732-736.
WEINSTOCK, A. (1972): Elaboration of enamel and dentin matrix glycoproteins. In: The Biochemistry and Physiology of Bone (ed. G.H. BOURNE), 2nd Ed., Vol. 2, pp. 121-154. New York: Academic Press.
WEINSTOCK, A., KING, P.C., WUTHIER, R.E. (1967): The ion-binding characteristics of reconstituted collagen. Biochem. J. 1o2, 983-988.
WEINSTOCK, M. (1972): Collagen formation - observations on its intracellular packaging and transport. Z. Zellforsch. 129, 455-47o.
WEISS, P.A. (1949): The biological basis of adaptation. In: Adaptation (ed. J. ROMANO), pp. 1-22. New York: Cornell Univ. Press.
WEISS, P.A. (1956): The compounding of complex macromolecular and cellular units into tissue fabrics. Proc. Nat. Acad. Sci. USA 42, 819-83o.
WEISS, P.A. (1958): Cell Contact Int. Rev. Cytol. 7, 391-423.
WEISS, P.A. (1959): Cellular dynamics. Rev. Mod. Phys. 31, 11-2o.
WEISS, P.A. (196o): Molecular reorientation as unifying principle underlying cellular selectivity. Proc. Nat. Acad. Sci. USA 46, 993-1ooo.
WEISS, P.A. (1961): From cell to molecule. In: The Molecular Control of Cellular Activity (ed. J.M. ALLEN), pp. 1-72. New York: McGraw-Hill.
WEISS, P.A. (1962): Cells and their environment, including other cells. In: Biological Interactions in Normal and Neoplastic Growth (ed. M.J. BRENNAN), Ch. 1. Boston: Little, Brown.
WEISS, P.A. (1966): From cell dynamics to tissue architecture. In: Structure and Function of Connective and Skeletal Tissue. Proc. St. Andrews Scot. 1964, pp. 256-263. London: Butterworths.
WEISS, P.A. (1968): Aging: A corollary of development. In: Dynamics of Development: Experiments and Inferences (ed. P.A. WEISS), pp. 329-34o. New York: Academic Press.

WEIS-FOGH, T., ANDERSEN, S.O. (197o): New molecular model for the long-range elasticity of elastin. Nature 227, 718-721.

WELLAUER, P., WYLER, T., BUDDECKE, E. (1972): Electron microscopic and physicochemical studies on bovine nasal cartilage proteoglycan. Hoppe-Seylers Z. Physiol. Chem. 353, 1o43-1o52.

WELLS, J.D. (1973a): Thermodynamics of polyectrolyte solutions. An empirical extension of the Manning theory to finite salt concentration. Biopolymers 12, 223-227.

WELLS, J.D. (1973b): Salt activity and osmotic pressure in connective tissue. I. A study of solutions of dextran sulfate as a model system. Proc. Roy. Soc. 183 Series B, 399-419.

WELLS, P.J., SERAFINI-FRACASSINI (1973): Molecular organization of cartilage proteoglycan. Nature New Biol. 243, 266-268.

WENDT, P., VON DER MARK, K., REXRODT, F., KÜHN, K. (1972): The covalent structure of collagen. The amino-acid sequence of the 112-residues amino-terminal part of peptide α1-CB6 from calf-skin collagen. Eur. J. Biochem. 3o, 169-183.

WHEAT, R.W. (196o): Isolation of uridine diphosphate-glycosyl compounds from the slug. Science 132, 131o-1311.

WHEELWRIGHT, P. (ed.) (1966): The Presocratics. New York: Odyssey Press.

WILBUR, K.M., SIMKISS, K. (1968): Calcified shells. In: Comprehensive Biochemistry (eds. M. FLORKIN, E.H. STOTZ), Vol. 26A, pp. 229-295. Amsterdam: Elsevier.

WILBUR, K.M., WATABE, N. (1967): Mechanisms of calcium carbonate deposition in coccolithophorids and molluscs. Stud. Trop. Oceanogr. 133-154.

WILLIAMS, A.P. (196o): The chemical composition of snail gelatin. Biochem. J. 74, 3o4-3o7.

WILLIAMSON, F.B. (1972): Double-helical conformations of sulfated polysaccharides and their potential biological function. Biochem. J. 129, 28P.

WILLMER, E.N. (1965): The origins of collagen. In: Structure and Function of Connective and Skeletal Tissue. Proc. St. Andrews, Scot. 1964, pp. 196-2o6. London: Butterworths.

WILLMER, E.N. (197o): Cytology and Evolution, 2nd Ed. New York: Academic Press.

WILSON, A.C., KAPLAN, N.O., LEVINE, L., PESCE, A., REICHLIN, M., ALLISON, W.S. (1964): Evolution of lactic dehydrogenases. Federation Proc. Federation Am. Soc. Exp. Biol. 23, 1258-1266.

WILSON, A.C., KITTO, G.B., KAPLAN, N.O. (1967): Enzymatic identification of fish products. Science 157, 82-83.

WINTER, W.P., WALSH, K.A., NEURATH, H. (1968): Homology as applied to proteins. Science 162, 1433.

WINZLER, R.J. (1965): Glycoproteins and glycosaminoglycans in plasma and in some other body fluids. In: The Amino Sugars (eds. E.A. BALAZS, R.W. JEANLOZ), Vol. 2A, pp. 337-352. New York: Academic Press.

WOESSNER, J.F. (1968): Biological mechanisms of collagen resorptions. In: Treatise on Collagen (ed. B.S. GOULD), Vol. 2B, pp. 253-33o. New York: Academic Press.

WOLFF, I., FUCHSWANS, W., WEISER, M., FURTHMAYER, H., TIMPL, R. (1971): Acidic structural proteins of connective tissue. Characterization of their heterogenous nature. Eur. J. Biochem. 2o, 426-431.

WOLMAN, M. (1957): Histochemical study of changes occurring during the degeneration of myelin. J. Neurochem. 1, 37o-376.

WOODGER, J.H. (1929): Biological Principles. London: Kegan Paul.

WOODIN, A.M., BORUCHOFF, S.A. (1955): Particle interaction in solutions derived from ox vitreous humor. J. Biophys. Biochem. Cytol. 1, 489-5oo.

WOODWARD, C., DAVIDSON, E.A. (1968): Structure-function relationhsips of protein polysaccharide complexes: Specific ion-binding properties. Proc. Nat. Acad. Sci. USA 6o, 2o1-2o5.

WOODWARD, C.B., HARANISAVLJEVIĆ, J., DAVIDSON, E.A. (1972): Physical properties of cartilage proteoglycans. Biochemistry 11, 1168-1176.

WORTH, H.G.J. (1967): The chemistry and biochemistry of pectic substances. Chem. Rev. 67, 465-473.

WUSTEMAN, F.S. (1972): Glycosaminoglycans of bovine lung parenchyma and pleura. Experientia 28, 887-888.

WYCKOFF, R.W.G. (1972): The Biochemistry of Animal Fossils. Bristol, England: Scientechnica and Baltimore: Williams and Wilkins.
YAMAMOTO, T. (1961): Physiology of fertilization in fish eggs. Int. Rev. Cytol. 12, 361-4o5.
YONEMASU, K., STROUD, R.M., NIEDERMEIER, W., BUTLER, W.T. (1971): Chemical studies on Clq; a modulator of immunoglobulin biology. Biochem. Biophys. Res. Commun. 43, 1388-1394.
YOSHIDA, H., EGAMI, F. (1965): Formation of 3'-phosphoadenosine-5'-phosphosulfate by mucous gland extracts of *Charonica lampos* and transfer of sulfate to charonin-sulfuric acid. J. Biochem. (Tokyo) 57, 215-22o.
YOSIZAWA, Z. (1964): A new Type of heparin "ω-heparin" isolated from whale organs. Biochem. Biophys. Res. Commun. 16, 336-341.
YOSIZAWA, Z. (1967): Enzymatic degradation of ω-heparin (whale heparin). Biochim. Biophys. Acta 141, 6oo-6o4.
YOUNG, J.Z. (1962): The Life of the Vertebrates, 2nd Ed. New York: Oxford University Press.
YUAN, L., VEIS, A. (1973): The self-assembly of collagen molecules. Biopolymers 12, 1437-1444.
ZUCKERKANDL, E. (1968): Hemoglobins, Haeckel's "biogenetic law", and molecular aspects of development. In: Structural Chemistry and Molecular Biology (eds. A. RICH, N. DAVIDSON), pp. 256-274. San Francisco: Freeman.
ZUCKERKANDL, E., PAULING, L. (1965a): Molecules as documents of evolutionary history. J. Theor. Biol. 8, 357-366.
ZUCKERKANDL, E., PAULING, L. (1965b): Evolutionary divergence and convergence in proteins. In: Evolving Genes and Proteins (eds. V. BRYSON, H.J. VOGEL), pp. 97-166. New York: Academic Press.

ADDENDUM

ARNOTT, S., GUSS, J.M., WINTER, W.T. (In press): Glycosaminoglycan conformations. In: Extracellular Matrix Influence on Gene Expression, eds. H.C. SLAVKIN and R.C. GREULICH. New York: Academic Press.
BAIRATI, A. (1972): Collagen: An analysis of phylogenetic aspects. Boll. Zool. 39, 205-248.
BALLOU, C.E., RASCHKE, W.C. (1974): Polymorphism of the somatic antigen of yeast. Science 184, 127-134.
BASSETT, C.A.L. (1971): The effect of force on skeletal tissues. In: Physiological Basis of Rehabilitation Medicine, eds. J.A. DOWNEY and R.C. DARLING, pp. 283-316. Philadelphia: W. B. Saunders Co.
BORNSTEIN, P. (1974): The biosynthesis of collagen. Ann. Rev. 43, 567-603.
BUTLER, W.T., MILLER, E.J., FINCH, Jr., J.E., INAGAMI, T. (1974): Homologous regions of collagen α1(I) and α1(II) chains: Apparent clustering of variable and invariant amino acid residues. Biochem. Biophys. Res. Commun. 57, 190-195.
BUTLER, W.T., UNDERWOOD, S.P., FINCH, Jr., J.E. (1974): Chemical studies on the cyanogen bromide peptides of rat skin collagen. Amino acid sequence of α1-CB3. Biochem. 13, 2946-2953.
CARLISLE, E.M. (1974): Silicon as an essential element. Fed. Proc. 33, 1758-1766.
CHAPMAN, J.A. (1974): The staining pattern of collagen fibrils. 1. An analysis of electron micrographs. Conn. Tissue Res. 2, 137-150.
CHOI, H.U., MEYER, K. (1974): The structure of a sulfated glycoprotein of chick allantoic fluid. J. Biol. Chem. 249, 932-939.
CHUNG, E., KEELE, E.M., MILLER, E.J. (1974): Isolation and characterization of the cyanogen bromide peptides from the α1(III) chain of human collagen. Biochem. 13, 3459-3464.
DAVISON, P.F. (1973): Homeostasis in extracellular tissues: Insights from studies on collagen. In: CRC Critical Reviews in Biochemistry, ed. G.D. FASMAN, pp. 201-245. Cleveland: CRC Press.

DOYLE, B.B., HULMES, D.J.S., MILLER, A., PARRY, D.A.D., PIEZ, K.A., WOODHEAD-GALLOWAY, J. (1974): A D-periodic narrow filament in collagen. Proc. Roy. Soc. Lon. B. 186, 67-74.

EHRLICH, J., STIVALA, S.S. (1973): Chemistry and pharmacology of heparin. J. Pharm. Sci. 62, 517-544.

EISENSTEIN, R., LARSSON, S., SORGENTE, N., KUETTNER, K.E. (1973): Collagen-proteoglycan relationships in epiphyseal cartilage. Amer. J. Pathol. 73, 443-452.

EPSTEIN, Jr., E.H. (1974): $[\alpha 1(III)]_3$ human skin collagen. J. Biol. Chem. 249, 3225-3231.

FIETZEK, P.P., FURTHMAYER, H., KÜHN, K. (1974): Comparative sequence studies on α2-CB2 from calf, human, rabbit and pig-skin collagen. Europ. J. Biochem. 47, 257-261.

GARRONE, R., VACELET, J. PAVANDE CECCATTY, M., JUNQUA, S., ROBERT, L., HUC, A. (1973): Une formation collagène particulière: les filaments des éponges cornées *Ircinia*. Etude ultrastructurale, physicochimique et biochimique. J. Micro. 17, 241-260.

GELMAN, R.A., BLACKWELL, J. (1974): Interactions between mucopolysaccharides and cationic polypeptides in aqueous solution. Hyaluronic acid, heparitin sulfate and keratan sulfate. Biopolymers 13, 139-156.

GELMAN, R.A., BLACKWELL, J., MATHEWS, M.B. (1974): Interactions of an intact proteoglycan and its fragments with basic homopolypeptides in dilute aqueous solution. Biochem. J. 141, 445-454.

GOTTE, L., GIRO, M.G., VOLPIN, D., HORNE, R.W. (1974): The ultrastructural organization of elastin. J. Ultrastr. Res. 46, 23-33.

GOUDSMIT, E.M. (1972): Carbohydrates and carbohydrate metabolism in Mollusca. In: Chemical Zoology, ed. M. FLORKIN, Vol. 7, pp. 219-243. New York: Academic Press.

GRAY, W.R., SANDBERG, L.B., FOSTER, J.A. (1973): Molecular model for elastin structure and function. Nature 246, 461-466.

GRÉGOIRE, C. (1972): Structure of the Molluscan shell. In: Chemical Zoology, ed. M. FLORKIN, Vol. 7, pp. 45-102. New York: Academic Press.

HARDINGHAM, T.E., MUIR, H. (1974): Hyaluronic acid in cartilage and proteoglycan aggregation. Biochem. J. 139, 565-581.

HASCALL, V.C., HEINEGÅRD, D. (1974): Aggregation of cartilage proteoglycans. II. Oligosaccharide competitors of the proteoglycan-hyaluronic acid interaction. J. Biol. Chem. 249, 4242-4249.

HAY, E.D. (1973): Origin and role of collagen in the embryo. Amer. Zool. 13, 1085-1107.

HEINEGÅRD, D., HASCALL, V.C. (1974): Aggregation of cartilage proteoglycans. III. Characteristics of the proteins isolated from trypsin digests of aggregates. J. Biol. Chem. 249, 4250-4256.

HÖÖK, M., LINDAHL, U., IVERIUS, P. (1974): Distribution of sulfate and iduronic acid residues in heparin and heparan sulfate. Biochem. J. 137, 33-43.

HÖÖK, M., LINDAHL, U., BÄCKSTRÖM, G., MALMSTRÖM, A., FRANSSON, L. (1974): Biosynthesis of heparin. III. Formation of iduronic acid residues. J. Biol. Chem. 249, 3908-3915.

HOEVE, C.A.J., FLORY, P.J. (1974): The elastic properties of elastin. Biopolymers 13, 677-686.

HOPWOOD, J.J., ROBINSON, H.C. (1974a): The structure and composition of cartilage keratan sulfate. Biochem. J. 141, 517-526.

HOPWOOD, J.J., ROBINSON, H.C. (1974b): The alkali-labile linkage between keratan sulfate and protein. Biochem. J. 141, 57-69.

KANAYA, S., FUJIMOTO, D. (1973): Fragments of polyproline-like structure from a structural protein, cuticulin. J. Mol. Biol. 81, 415-418.

KIMATA, K., OKAYAMA, M., OOHIRA, A., SUZUKI, S. (1974): Heterogeneity of proteochondroitin sulfates produced by chondrocytes at different stages of cytodifferentiation. J. Biol. Chem. 249, 1646-1653.

KNIGHT, D.P., HUNT, S. (1974): Fibril structure of collagen in egg capsule of dogfish. Nature 249, 379-380.

KRAEMER, P.M., SMITH, D.A. (1974): High molecular-weight heparan sulfate from the cell surface. Biochem. Biophys. Res. Commun. 56, 423-430.

KURODA, J., SAITO, S., SENO, N., NAGASE, S., ANNO, K. (1974): Isolation and chemical characterization of mucopolysaccharides from rat tumors. Cancer Res. 34, 308-312.

LANGNESS, U., UDENFRIEND, S. (1974): Collagen biosynthesis in nonfibroblastic cell lines. Proc. Natl. Acad. Sci. USA 71, 50-51.
LASH, J.W., SAXÉN, L., KOSHER, R.A. (1974): Human chondrogenesis: Glycosaminoglycan content of embryonic human cartilage. J. Exp. Zool. 189, 127-131.
LEUNG, D.Y.M., GLAGOV, S., CLARK, J.M., MATHEWS, M.B. (In Press): Mechanical influences on the biosynthesis of extracellular molecules. In: Extracellular Matrix Influences on Gene Expression, eds. H.C. SLAVKIN and R.C. GREULICH. New York: Academic Press.
MADGWICK, J., HAUG, A., LARSEN, B. (1973): Pollymannuronic acid 5-epimerase from the marine alga *Pelvetia canaliculata*. Acta Chem. Scand. 27, 3592-3594.
MAILMAN, M.L., KIRKPATRICK, D.S., DRESDEN, M.H. (1974): A partial characterization of collagens from *Notophthalmus viridescens* and *Rana catesbiana*. Comp. Biochem. Physiol. 47B, 617-630.
MEIER, S., HAY, E.D. (1974): Control of corneal differentiation by extracellular materials. Collagen as a promoter and stabilizer of epithelial stroma production. Dev. Biol. 38, 249-270.
MILLER, E.J., MATHEWS, M.B. (1974): Characterization of notochord collagen as a cartilage-type collagen. Biochem. Biophys. Res. Commun. 60, 424-430.
MURATA, K., OGURA, T., OKUYAMA, T. (1974): The acidic glycosaminoglycans in leukocytes: An application of enzymatic methods. Conn. Tissue Res. 2, 101-109.
NOWACK, H., NORDWIG, A. (1974): Sea-anemone collagen: Isolation and characterization of the cyanogen-bromide peptides. Eur. J. Biochem. 45, 333-342.
PASTERNACK, S.G., VEIS, A., BREEN, M. (1974): Solvent-dependent changes in proteoglycan subunit conformation in aqueous guanidinium hydrochloride solutions. J. Biol. Chem. 249, 2206-2211.
PICARD, J., PAUL-GARDAIS, A., VEDEL, M. (1973): Glycoproteines sulfates des membranes de l'oeuf de poule et de l'oviducte. Isolement et characterisation de glycopeptides sulfates. Biochem. Biophys. Acta 320, 427-441.
PIEZ, K.A. (In Press): The regulation of fibrillogenesis. In: Extracellular Matrix Influences on Gene Expression, eds. H.C. SLAVKIN and R.C. GREULICH. New York: Academic Press.
RADHAKRISHNAMURTHY, B., BERENSON, G.S. (1973): Structures of glycopeptides from a glycoprotein from bovine aorta. J. Biol. Chem. 248, 2000-2006.
RAUTERBERG, J. (1973): The C-terminal non-helical portion of the collagen molecule. Clin. Orthop. Rel. Res. 97, 196-212.
REID, K.B.M. (1974): A collagen-like amino acid sequence in a polypeptide chain of human CIq (a subcomponent of the first component of complement). Biochem. J. 141, 189-203.
RUDALL, K.M., KENCHINGTON, W. (1973): The chitin system. Biol. Rev. 49, 597-636.
SCHOFIELD, J.D., PROCKOP, D.J. (1973): Procollagen - a precursor form of collagen. Clin. Orthop. Rel. Res. 97, 175-195.
SCHWARZ, K. (1974): Recent dietary trace element research, exemplified by tin, fluorine and silicon. Fed. Proc. 33, 1748-1757.
SENO, N., YAMASHIRO, S., ANNO, K. (1974): Isolation and characterization of a new disaccharide disulfate: 2-acetamido-2-deoxy-3-O-(2 or 3-O-sulfo-β-D-glucopyranosyluronic acid)-4-O-sulfo-D-galactose. Biochim. Biophys. Acta 343, 423-426.
SEYER, J.M., VINSON, W.C. (1974): Synthesis of Type I and Type II collagen by embryonic chick cartilage. Biochem. Biophys. Res. Commun. 58, 272-279.
SHAPIRO, S.S., SHERMAN, M.I. (1974): Sulfated mucopolysaccharides of midgestation embryonic and extraembryonic tissues of the mouse. Arch. Biochem. Biophys. 162, 272-280.
SRIVASTAVA, V.M.L., MALEMUD, C.J., SOKOLOFF, L. (1974): Chondroid expression by lapine articular chondrocytes in spinner culture following monolayer growth. Conn. Tissue Res. 2, 127-136.
TRELSTAD, R.L., KANG, A.H. (1974): Collagen heterogeneity in the avian eye: lens, vitreous body, cornea and sclera. Exp. Eye Res. 18, 395-406.
TRENCH, R.K. (1973): Further studies on the mucopolysaccharide secreted by the pedal gland of the marine slug *Tridachia crispata* (Opisthobranchia, Sacoglossa). Bull. Mar. Sci. 23, 299-312.

Subject Index

Molecular Biology, Biochemistry and Biophysics

Volumes published:

Vol. 1 J. H. VAN'T HOFF: Imagination in Science. Translated into English, with Notes and a General Introduction by G. F. SPRINGER. With 1 portrait. VI, 18 pages. 1967. DM 6,60; US $ 2.70

Vol. 2 K. FREUDENBERG and A. C. NEISH: Constitution and Biosynthesis of Lignin. With 10 figures. IX, 129 pages. 1968. DM 28,—; US $ 11.40

Vol. 3 T. ROBINSON: The Biochemistry of Alkaloids. With 37 figures. X, 149 pages. 1968. DM 39,—; US $ 14.60

Vol. 4 A. S. SPIRIN and L. P. GAVRILOVA: The Ribosome. Out of print. Second edition in preparation

Vol. 5 B. JIRGENSONS: Optical Activity of Proteins and Other Macromolecules. 2nd Revised and Enlarged Edition. With 71 figures. IX, 199 pages. 1973. DM 59,80; US $ 24.60

Vol. 6 F. EGAMI and K. NAKAMURA: Microbial Ribonucleases. With 5 figures. IX. 90 pages. 1969. DM 28,—; US $ 11.40

Vol. 8 Protein Sequence Determination. A Sourcebook of Methods and Techniques. Edited by SAUL B. NEEDLEMAN. Out of print. New edition in preparation

Vol. 9 R. GRUBB: The Genetic Markers of Human Immunoglobulins. With 8 figures. XII, 152 pages. 1970. DM 42,—; US $ 17.30

Vol. 10 R. J. LUKENS: Chemistry of Fungicidal Action. With 8 figures. XIII, 136 pages. 1971. DM 42,—; US $ 14.90

Vol. 11 P. REEVES: The Bacteriocins. With 9 figures. XI, 142 pages. 1972. DM 48,—; US $ 19.60

Vol. 12 T. ANDO, M. YAMASAKI and K. SUZUKI: Protamines: Isolation, Characterization, Structure and Function. With 24 figures. IX, 114 pages. 1973. DM 48,—; US $ 19.60

Vol. 13 P. JOLLÈS and A. PARAF: Chemical and Biological Basis of Adjuvants. With 24 figures. VIII, 153 pages. 1973. DM 48,—; US $ 19.60

Vol. 14 Micromethods in Molecular Biology. Edited by V. NEUHOFF. With 275 figures. XV, 428 pages. 1973. DM 98,—; US $ 40.20

Vol. 15 M. WEISSBLUTH: Hemoglobin. Cooperativity and Electronic Properties. With 50 figures. VIII, 175 pages. 1974. DM 72,—; US $ 29.60

Vol. 16 S. SHULMAN: Tissue Specificity and Autoimmunity. With 32 figures. XI, 196 pages. 1974. DM 60,—; US $ 24.60

Vol. 17 YA. A. VINNIKOV: Sensory Reception. Cytology, Molecular Mechanisms and Evolution. With 124 figures (173 illustrations). IX, 392 pages. 1974. DM 118,—; US $ 48.40

Vol. 18 H. KERSTEN and W. KERSTEN: Inhibitors of Nucleic Acid Synthesis. Biophysical and Biochemical Aspects. With 73 figures. IX, 184 pages. 1974. DM 68,—; US $ 27.20

Vol. 19 M. B. MATHEWS: Connective Tissue. Macromolecular Structure and Evolution. With 31 figures. XII 318, pages. 1975. DM 78,—; US $ 32.00

Vol. 20 M. A. LAUFFER: Entropy-Driven Processes in Biology: Polymerization of Tobacco Mosaic Virus Protein and Similar Reactions. With 90 figures. X, 264 pages. 1975. DM 73,—; US $ 30.00

Volume in press

Vol. 21 R. C. BURNS and R. W. F. HARDY: Nitrogen Fixation in Bacteria and Higher Plants. With 27 figures. Approx 224 pages. 1975

GPSR Compliance

The European Union's (EU) General Product Safety Regulation (GPSR) is a set of rules that requires consumer products to be safe and our obligations to ensure this.

If you have any concerns about our products, you can contact us on ProductSafety@springernature.com

In case Publisher is established outside the EU, the EU authorized representative is:

Springer Nature Customer Service Center GmbH
Europaplatz 3
69115 Heidelberg, Germany

Batch number: 10640288

Printed by Printforce, the Netherlands